Molecular Theory of Solutions

Dedicated to Ruby and Kaye

Molecular Theory of Solutions

Arieh Ben-Naim
Department of Physical Chemistry
The Hebrew University, Jerusalem

Great Clarendon Street, Oxford OX2 6DP

Oxford University Press is a department of the University of Oxford.
It furthers the University's objective of excellence in research, scholarship,
and education by publishing worldwide in

Oxford New York

Auckland Cape Town Dar es Salaam Hong Kong Karachi
Kuala Lumpur Madrid Melbourne Mexico City Nairobi
New Delhi Shanghai Taipei Toronto

With offices in

Argentina Austria Brazil Chile Czech Republic France Greece
Guatemala Hungary Italy Japan Poland Portugal Singapore
South Korea Switzerland Thailand Turkey Ukraine Vietnam

Oxford is a registered trade mark of Oxford University Press
in the UK and in certain other countries

Published in the United States
by Oxford University Press Inc., New York

First published 2006

British Library Cataloguing in Publication Data
Data available

Library of Congress Cataloging in Publication Data

Ben-Naim, Arieh, 1934–
Molecular theory of solutions / Arieh Ben-Naim.
p. cm.
Includes bibliographical references and index.
ISBN-13: 978–0–19–929969–0 (acid-free paper)
ISBN-10: 0–19–929969–2 (acid-free paper)
ISBN-13: 978–0–19–929970–6 (pbk. : acid-free paper)
ISBN-10: 0–19–929970–6 (pbk. : acid-free paper)
1. Solution (Chemistry). 2. Molecular Theory. I. Title.
QD541.B458 2006
541′.34—dc22 2006015317

Typeset by Newgen Imaging Systems (P) Ltd., Chennai, India
Printed in Great Britain
on acid-free paper by
Biddles Ltd, www.biddles.co.uk

ISBN 0–19–929969–2 978–0–19–929969–0 (Hbk.)
ISBN 0–19–929970–6 978–0–19–929970–6 (Pbk.)

10 9 8 7 6 5 4 3 2 1

Preface

The aim of a molecular theory of solutions is to explain and to predict the behavior of solutions, based on the input information of the molecular properties of the individual molecules constituting the solution. Since Prigogine's book (published in 1957) with the same title, aiming towards that target, there has been considerable success in achieving that goal for mixtures of gases and solids, but not much progress has been made in the case of liquid mixtures. This is unfortunate since liquid mixtures are everywhere. In almost all industries and all biological sciences, we encounter liquid mixtures. There exists an urgent need to understand these systems and to be able to predict their behavior from the molecular point of view.

The main difficulty in developing a molecular theory of liquid mixtures, as compared to gas or solid mixtures, is the same as the difficulty which exists in the theory of pure liquids, compared with theories of pure gases and solids. Curiously enough, though various lattice theories of the liquid state have failed to provide a fair description of the liquid state, they did succeed in characterizing liquid mixtures. The reason is that in studying mixtures, we are interested in the excess or the mixing properties – whence the problematic characteristics of the liquid state of the pure components partially cancel out. In other words, the characteristics of the mixing functions, i.e., the difference between the thermodynamics of the mixture, and the pure components are nearly the same for solids and liquid mixtures. Much of what has been done on the lattice theories of mixture was pioneered by Guggenheim (1932, 1952). This work was well documented by both Guggenheim (1952) and by Prigogine (1957), as well as by many others.

Another difficulty in developing a molecular theory of liquid mixtures is the relatively poor knowledge of the intermolecular interactions between molecules of different species. While the intermolecular forces between simple spherical particles are well-understood, the intermolecular forces between molecules of different kinds are usually constructed by the so-called combination rules, the most well-known being the Lorentz and the Berthelot rules.

In view of the aforementioned urgency, it was necessary to settle on an intermediate level of a theory[†]. Instead of the classical aim of a molecular theory

[†] By intermediate level of theory, I do not mean empirical theories which are used mainly by chemical engineers.

of solutions, which we can write symbolically as

I: Molecular Information $\rightarrow$ Thermodynamic Information

An indirect route has been developed mainly by Kirkwood, which involves molecular distribution functions (MDF) as an intermediate step. The molecular distribution function approach to liquids and liquid mixtures, founded in the early 1930s, gradually replaced the various lattice theories of liquids. Today, lattice theories have almost disappeared from the scene of the study of liquids and liquid mixtures[†]. This new route can be symbolically written as

II: Molecular Information + MDF $\rightarrow$ Thermodynamic Information

Clearly, route II does not remove the difficulty. Calculation of the molecular distribution functions from molecular properties is not less demanding than calculation of the thermodynamic quantities themselves.

Nevertheless, *assuming* that the molecular distribution functions are given, then we have a well-established theory that provides thermodynamic information from a combination of molecular information and MDFs. The latter are presumed to be derived either from experiments, from simulations, or from some approximate theories. The main protagonists in this route are the pair correlation functions; once these are known, a host of thermodynamic quantities can be calculated. Thus, the less ambitious goal of a molecular theory of solutions has been for a long time route II, rather route I.

Between the times of Prigogine's book up to the present, several books have been published, most notably Rowlinson's, which have summarized both the experimental and the theoretical developments.

During the 1950s and the 1960s, two important theories of the liquid state were developed, initially for simple liquids and later applied to mixtures. These are the scaled-particle theory, and integral equation methods for the pair correlation function. These theories were described in many reviews and books. In this book, we shall only briefly discuss these theories in a few appendices. Except for these two theoretical approaches there has been no new *molecular* theory that was specifically designed and developed for mixtures and solutions. This leads to the natural question "why a need for a new book with the same title as Prigogine's?"

To understand the reason for writing a new book with the same title, I will first modify route II. The modification is admittedly, semantic. Nevertheless, it provides a better view of the arguments I am planning to present below.

[†] Perhaps liquid water is an exception. The reason is that water, in the liquid state, retains much of the structure of the ice. Therefore, many theories of water and aqueous solution have used some kind of lattice models to describe the properties of these liquids.

We first rewrite route II as

III: Microscopic Properties $\longrightarrow$ Thermodynamic Properties

Routes II and III are identical in the sense that they use the same theoretical tools to achieve our goals. There is however one important conceptual difference. Clearly, molecular properties are microscopic properties. Additionally, all that has been learned about MDF has shown that in the liquid phase, and not too close to the critical point, *molecular distribution functions* have a *local* character in the sense that they depend upon and provide information on *local* behavior around a given molecule in the mixture. By *local*, we mean a few *molecular diameters*, many orders of magnitude smaller than the macroscopic, or global, dimensions of the thermodynamic system under consideration. We therefore rewrite, once again, route II in different words, but meaning the same as III, namely

IV: Local Properties $\longrightarrow$ Global Properties

Even with this modification, the question we posed above is still left unanswered: Why a new book on molecular theory of solutions? After all, even along route IV, there has been no theoretical progress.

Here is my answer to this question.

Two important and profound developments have occurred since Prigogine's book, not along route I, neither along II or III, but on the *reverse* of route IV. The one-sided arrows as indicated in I, II, and III use the tools of statistical thermodynamics to bridge between the molecular or *local* properties and thermodynamic properties. This bridge has been erected and has been perfected for many decades. It has almost always been used to cross in a one-way direction from the *local* to the *global.*

The new development uses the same tool – the same bridge – but in reversed direction; to go backwards from the *global* to the *local* properties. Due to its fundamental importance, we rewrite IV again, but with the reversed directed arrow:

$-$IV: Global Properties $\longrightarrow$ Local Properties

It is along this route that important developments have been achieved specifically for solutions, providing the proper justification for a new book with the *same* title. Perhaps a more precise title would be the *Local Theory of Solutions.* However, since the tools used in this theory are identical to the tools used in Prigogine's book, we find it fitting to use the same title for the present book. Thus, the tools are basically unchanged; only the manner in which they are applied were changed.

There are basically two main developments in the molecular theory of solutions in the sense of route −IV: one based on the inversion of the Kirkwood–Buff (KB) theory; the second is the introduction of a new measure to study solvation properties. Both of these use measurable macroscopic, or *global* quantities to probe into the microscopic, or the *local* properties of the system. The types of properties probed by these tools are local densities, local composition, local change of order, or structure (of water and aqueous solutions) and many more. These form the core of properties discussed in this book. Both use exact and rigorous tools of statistical mechanics to define and to calculate local properties that are not directly accessible to measurements, from measurable macroscopic quantities.

The first development consists of the inversion of the Kirkwood–Buff theory. The Kirkwood–Buff theory has been with us since 1951. It was dormant for more than 20 years. Though it is exact, elegant and very general, it could only be applied when all the pair correlation functions are available. Since, for mixtures, the latter are not easily available, the theory stayed idle for a long time. It is interesting to note that both Prigogine (1957) and Hill (1956) mentioned the KB theory but not any of its applications. In fact, Hill (1956), in discussing the Kirkwood–Buff theory, writes that it is "necessarily equivalent to the McMillan–Mayer (1945) theory, since both are formally exact." I disagree with the implication of that statement. Of course, any two exact theories must be, in principle, *formally* equivalent. But they are not necessarily equivalent in their range and scope of applicability and in their interpretative power. I believe that in all aspects, the Kirkwood–Buff theory is immensely superior to the McMillan–Mayer theory, as I hope to convince the reader of this book. It is somewhat puzzling to note that many authors, including Rowlinson, completely ignored the Kirkwood–Buff theory.

One of the first applications of the Kirkwood–Buff theory, even before its inversion, was to provide a convincing explanation of one of the most mysterious and intellectually challenging phenomenon of aqueous solutions of inert gases – the molecular origin of the large and negative entropy and enthalpy of solvation of inert gases in water. This was discussed by Ben-Naim (1974, 1992). But the most important and useful application of the KB theory began only after the publication of its inversion. A search in the literature shows that the "KB theory" was used as part of the title of articles on the average, only *once* a year until 1980. This has escalated to about 20–25 a year since 1980, and it is still increasing.

Ever since the publication of the inversion of the KB theory, there had been an upsurge of papers using this new tool. It was widely accepted and appreciated and used by many researchers as an efficient tool to study *local* properties of mixtures and solutions.

The traditional characterization and study of the properties of liquid mixtures by means of the *global* excess thermodynamic functions has been gradually and steadily replaced by the study of the *local* properties. The latter provides richer and more detailed information on the immediate environment of each molecule in the mixture.

The second development, not less important and dramatic, was in the theory of solvation. Solvation has been defined and studied for many years. In fact, there was not only one but at least three different quantities that were used to study solvation. The problem with the traditional quantities of solvation was that it was not clear what these quantities really measure. All of the three involve a process of transferring a solute from one hypothetical state in one phase, to another hypothetical state in a second phase. Since these hypothetical states have no clear-cut interpretation on a molecular level, it was not clear what the free energy change associated with such transfer processes really means. Thus, within the framework of thermodynamics, there was a state of stagnation, where three quantities were used as tools for the study of solvation. No one was able to decide which the preferred one is, or which is really the right tool to measure solvation thermodynamics.

As it turned out, there was no right one. In fact, thermodynamics could not provide the means to decide on this question. Astonishingly, in spite of their vagueness, and in spite of the inability to determine their relative merits, some authors vigorously and aggressively promoted the usage of one or the other tools without having any solid theoretical support. Some of these authors have also vehemently resisted the introduction of the new tool.

The traditional quantities of solvation were applicable only in the realm of very dilute solutions, where Henry's law is obeyed. It had been found later that some of these are actually inadequate measures of solvation[†]. The new measure that was introduced in the early 1970s replaced vague and hazy measures by a new tool, sharply focusing into the local realm of molecular dimensions. The new quantity, defined in statistical mechanical terms, is a sharp, powerful, and very general tool to probe local properties of not only solutes in dilute solutions, but of any molecule in any environment.

The new measure has not only sharpened the tools for probing the surroundings around a single molecule, but it could also be applied to a vastly larger range of systems: not only a single A in pure B, or a single B in pure A,

[†] In fact using different measures led to very different values of the solvation Gibbs energy. In one famous example the difference in the Gibbs energy of solvation of a small solute in H_2O and D_2O even had different *signs*, in the different measures.

but the "double infinite" range of all compositions of A and B, including the solvation of A in pure A, and B in pure B, which traditional tools never touched and could not be applied to.

Specifically for liquid water, the solvation of water in pure water paved the way to answer questions such as "What *is* the structure of water" and "How much is this structure changed when a solute is added?" The details and the scope of application of the new measure were described in the monograph by Ben-Naim (1987).

While the inversion of the KB theory was welcomed, accepted, and applied enthusiastically by many researchers in the field of solution chemistry, and almost universally recognized as a powerful tool for studying and understanding liquid mixtures on a molecular level, unfortunately the same was far from true for the new measure of solvation. There are several reasons for that.

First, *solvation* was a well-established field of research for many years. Just as there were not one, but at least three different measures, or mutants, there were also different physical chemists claiming preference for one or another of its varieties. These people staunchly supported one or the other of the traditional measures and adamantly resisted the introduction of the new measure. In the early 1970s, I sent a short note where I suggested the use of a new measure of solvation. It was violently rejected, ridiculing my *chutzpa* in usurping old and well-established concepts. Only in 1978 did I have the courage, the conviction – *and yes, the chutzpa* – to publish a full paper entitled "Standard Thermodynamics of Transfer; Uses and Misuses." This was also met with hostility and some virulent criticism both by personal letters as well as published letters to the editor and comments. The struggle ensued for several years. It was clear that I was "going against the stream" of the traditional concepts. It elicited the rage of some authors who were patronizing one of the traditional tools. One scientist scornfully wrote: "You tend to wreck the structure of solution chemistry . . . you usurp the symbol which has always been used for other purposes . . . why don't you limit yourself to showing that one thermodynamic coefficient has a simple molecular interpretation?" These statements reveal utter misunderstanding of the merits of the new measure (referred to as the "thermodynamic coefficient", probably because it is related to the Ostwald absorption coefficient). Indeed, as will be clear in chapter 7, there are some subtle points that have evaded even the trained eyes of practitioners in the field of solvation chemistry.

Not all resisted the introduction of the new tool. I wish to acknowledge the very firm support and encouragement I got from Walter Kauzmann and John Edsal. They were the first to appreciate and grasp the advantage of a new tool and encouraged me to continue with its development. Today, I am proud, satisfied, and gratified to see so many researchers using and understanding the new tool. It now looks as if this controversial issue has "signed off."

The struggle for survival of the different mutants was lengthy, but as in biology, eventually, the fittest survives, whereas all the others fade out.

The second reason is more subtle and perhaps stems from misunderstanding. Since the new measure for the solvation Gibbs energy *looks similar* to one of the existing measures, people initially viewed it merely as one more traditional measure, even referring to it as Ben-Naim's standard state. As will be discussed in chapter 7, one of the advantages (not the major one) of the new measure is that it *does not* involve any *standard state* in the sense used in the traditional approach to the study of solvation.

There is one more development which I feel is appropriate to mention here. It deals with the concepts of "entropy of mixing" and "free energy of mixing." It was shown in 1987 that what is referred to as "entropy of mixing" has nothing to do with the mixing process. In fact, mixing of ideal gases, in itself, has no effect on any thermodynamic quantity. What is referred to as "entropy of mixing" is nothing more than the familiar entropy of expansion. Therefore, mixing of ideal gases is *not*, in general, an irreversible process. Also, a new concept of *assimilation* was introduced and it was shown that the *deassimilation* process is inherently an irreversible process, contrary to the universal claims that the *mixing* process is inherently an irreversible process. Since this topic does not fall into the claimed scope of this book, it is relegated to two appendices.

Thus, the main scope of this book is to cover the two topics: the Kirkwood–Buff theory and its inversion; and solvation theory. These theories were designed and developed for mixtures and solutions. I shall also describe briefly the two important theories: the integral equation approach; and the scaled particle theory. These were primarily developed for studying pure simple liquids, and later were also generalized and applied for mixtures.

Of course, many topics are deliberately omitted (such as solutions of electrolytes, polymers, etc.). After all, one must make some choice of which topics to include, and the choices made in this book were made according to my familiarity and my assessment of the relative range of applicability and their interpretive power. Also omitted from the book are lattice theories. These have been fully covered by Guggenheim (1952, 1967), Prigogine (1957), and Barker (1963).

The book is organized into eight chapters and some appendices. The first three include more or less standard material on molecular distribution functions and their relation to thermodynamic quantities. Chapter 4 is devoted to the Kirkwood–Buff theory of solutions and its inversion which I consider as the main pillar of the theories of mixtures and solutions. Chapters 5 and 6 discuss various ideal solutions and various deviations from ideal solutions; all of these are derived and examined using the Kirkwood–Buff theory. I hope that this simple and elegant way of characterizing various ideal solutions will remove much of the confusion that exists in this field. Chapter 7 is devoted to solvation. We briefly introduce the new concept of solvation and compare it with the traditional concepts. We also review some applications of the concept of solvation. Chapter 8 combines the concept of solvation with the inversion of the Kirkwood–Buff theory. Local composition and preferential solvation are defined and it is shown how these can be obtained from the inversion of the KB theory. In this culminating chapter, I have also presented some specific examples to illustrate the new way of analysis of the properties of mixtures on a *local* level. Instead of the global properties conveyed by the excess function, a host of new information may be obtained from local properties such as solvation, local composition, and preferential solvation. Examples are given throughout the book only as illustrations – no attempt has been made to review the extensive data available in the literature. Some of these have been recently summarized by Marcus (2002).

The book was written while I was a visiting professor at the University of Burgos, Spain. I would like to express my indebtedness to Dr. Jose Maria Leal Villalba for his hospitality during my stay in Burgos.

I would also like to acknowledge the help extended to me by Andres Santos in the numerical solution of the Percus–Yevick equations and to Gideon Czapski for his help in the literature research. I acknowledge with thanks receiving a lot of data from Enrico Matteoli, Ramon Rubio, Eli Ruckenstein, and others. I am also grateful to Enrico Matteoli, Robert Mazo, Joaquim Mendes, Mihaly Mezei, Nico van der Vegt and Juan White for reading all or parts of the book and offering important comments. And finally, I want to express my thanks and appreciation to my life-partner Ruby. This book could never have been written without the peaceful and relaxing atmosphere she had created by her mere presence. She also did an excellent job in typing and correcting the many versions of the manuscript.

Arieh Ben-Naim
January 2006

Table of Contents

List of Abbreviations

BE	Binding energy
CN	Coordination number
DI	Dilute ideal
FG	Functional group
GMDF	Generalized molecular distribution function
GPF	Grand partition function
HB	Hydrogen bond
HS	Hard sphere
IG	Ideal gas
KB	Kirkwood–Buff
KBI	Kirkwood–Buff integral
LCST	Lower critical solution temperature
LJ	Lennard-Jones
lhs	Left-hand side
MDF	Molecular distribution function
MM	McMillan–Mayer
PMF	Potential of mean force
PS	Preferential solvation
PY	Percus–Yevick
QCDF	Quasi-component distribution function
rhs	Right-hand side
SI	Symmetrical ideal
SPT	Scaled particle theory
UCST	Upper critical solution temperature
VP	Voronoi polyhedron

ONE

Introduction

In this chapter, we first present some of the notation that we shall use throughout the book. Then we summarize the most important relationship between the various partition functions and thermodynamic functions. We shall also present some fundamental results for an ideal-gas system and small deviations from ideal gases. These are classical results which can be found in any textbook on statistical thermodynamics. Therefore, we shall be very brief. Some suggested references on thermodynamics and statistical mechanics are given at the end of the chapter.

1.1 Notation regarding the microscopic description of the system

To describe the configuration of a rigid molecule we need, in the most general case, six coordinates, three for the location of some "center," chosen in the molecule, e.g., the center of mass, and three orientational angles. For spherical particles, the configuration is completely specified by the vector $\mathbf{R}_i = (x_i, y_i, z_i)$ where x_i, y_i, and z_i are the Cartesian coordinates of the center of the ith particles. On the other hand, for a non-spherical molecule such as water, it is convenient to choose the center of the oxygen atom as the center of the molecule. In addition, we need three angles to describe the orientation of the molecule in space. For more complicated cases we shall also need to specify the angles of internal rotation of the molecule (assuming that bond lengths and bond angles are fixed at room temperatures). An infinitesimal element of volume is denoted by

$$d\mathbf{R} = dx\,dy\,dz. \tag{1.1}$$

This represents the volume of a small cube defined by the edges dx, dy, and dz. See Figure 1.1. Some texts use the notation d^3R for the element of volume to

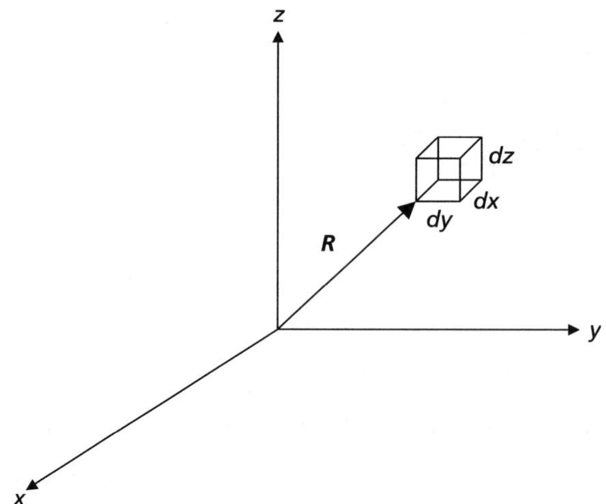

Figure 1.1 An infinitesimal element of volume $d\boldsymbol{R} = dxdydz$ at the point $\boldsymbol{R}$.

distinguish it from the vector, denoted by $d\boldsymbol{R}$. In this book, $d\boldsymbol{R}$ will always signify an element of volume.

The element of volume $d\boldsymbol{R}$ is understood to be located at the point $\boldsymbol{R}$. In some cases, it will be convenient to choose an element of volume other than a cubic one. For instance, an infinitesimal spherical shell of radius R and width dR has the volume†

$$d\boldsymbol{R} = 4\pi R^2 dR. \tag{1.2}$$

For a rigid nonspherical molecule, we use $\boldsymbol{R}_i$ to designate the location of the center of the ith molecule and $\boldsymbol{\Omega}_i$ the orientation of the whole molecule. As an example, consider a water molecule as being a rigid body. Let $\boldsymbol{\mu}$ be the vector originating from the center of the oxygen atom and bisecting the H–O–H angle. Two angles, say ϕ and θ, are required to fix the orientation of this vector. In addition, a third angle ψ is needed to describe the angle of rotation of the entire molecule about the axis $\boldsymbol{\mu}$.

In general, integration over the variable $\boldsymbol{R}_i$ means integration over the whole volume of the system, i.e.,

$$\int_V d\boldsymbol{R}_i = \int_0^L dx_i \int_0^L dy_i \int_0^L dz_i = L^3 = V \tag{1.3}$$

where for simplicity we have assumed that the region of integration is a cube of length L. The integration over $\boldsymbol{\Omega}_i$ will be understood to be over all possible orientations of the molecule. Using for instance, the set of Euler angles, we have

† Note that R is a scalar; $\boldsymbol{R}$ is a vector, and $d\boldsymbol{R}$ is an element of volume.

$$\int d\boldsymbol{\Omega}_i = \int_0^{2\pi} d\phi_i \int_0^{\pi} \sin\theta_i \, d\theta_i \int_0^{2\pi} d\psi_i = 8\pi^2. \tag{1.4}$$

Note that for a linear molecule, we have one degree of freedom less, therefore

$$\int d\boldsymbol{\Omega}_i = \int_0^{2\pi} d\phi_i \int_0^{\pi} \sin\theta_i \, d\theta_i = 4\pi. \tag{1.5}$$

The configuration of a rigid nonlinear molecule is thus specified by a six-dimensional vector, including both the location and the orientation of the molecule, namely,

$$\boldsymbol{X}_i = (\boldsymbol{R}_i, \boldsymbol{\Omega}_i) = (x_i, y_i, z_i, \phi_i, \theta_i, \psi_i). \tag{1.6}$$

The configuration of a system of N rigid molecules is denoted by

$$\boldsymbol{X}^N = \boldsymbol{X}_1, \boldsymbol{X}_2, \ldots, \boldsymbol{X}_N. \tag{1.7}$$

The infinitesimal element of the configuration of a single molecule is denoted by

$$d\boldsymbol{X}_i = d\boldsymbol{R}_i \, d\boldsymbol{\Omega}_i, \tag{1.8}$$

and, for N molecules,

$$d\boldsymbol{X}^N = d\boldsymbol{X}_1 d\boldsymbol{X}_2, \ldots, d\boldsymbol{X}_N. \tag{1.9}$$

1.2 The fundamental relations between statistical thermodynamics and thermodynamics

The fundamental equations of statistical thermodynamics are presented in the following subsections according to the set of independent variables employed in the characterization of a macroscopic system.

E, V, N ensemble

We consider first an isolated system having a fixed internal energy E, volume V, and number of particles N. Let $W(E, V, N)$ be the number of quantum mechanical states of the system characterized by the variables E, V, N. That is the number of eigenstates of the Hamiltonian of the system having the eigenvalue E. We assume for simplicity that we have a finite number of such eigenstates. The first relationship is between the entropy S of the system and the number of states, $W(E, V, N)$. This is the famous Boltzmann formula[†]

$$S(E, V, N) = k \ln W(E, V, N) \tag{1.10}$$

[†] This formula in the form $S = k \log W$ is engraved on Boltzmann's tombstone.

where $k = 1.38 \times 10^{-23}\ \mathrm{J\,K^{-1}}$ is the Boltzmann constant.

The fundamental thermodynamic relationship for the variation of the entropy in a system described by the independent variables E, V, N is

$$TdS = dE + PdV - \mu dN \tag{1.11}$$

from which one can obtain the temperature T, the pressure P, and the chemical potential μ as partial derivatives of S. Other thermodynamic quantities can be obtained from the standard thermodynamic relationships. For a summary of some thermodynamic relationships see Appendix A.

In practice, there are very few systems for which W is known. Therefore equation (1.10), though the cornerstone of the theory, is seldom used in applications. Besides, an isolated system is not an interesting system to study. No experiments can be done on an isolated system.

Next we introduce the fundamental distribution function of this system. Suppose that we have a very large collection of systems, all of which are identical, in the sense that their thermodynamic characterization is the same, i.e., all have the same values of E, V, N. This is sometimes referred to as a microcanonical ensemble. In such a system, one of the fundamental postulates of statistical thermodynamics is the assertion that the probability of a specific state i is given by

$$P_i = \frac{1}{W}. \tag{1.12}$$

This is equivalent to the assertion that all states of an E, V, N system have equal probabilities. Since $\sum P_i = 1$, it follows that each of the P_i is equal to W^{-1}.

T, V, N ensemble

The most useful connection between thermodynamics and statistical thermodynamics is that established for a system at a given temperature T, volume V, and the number of particles N. The corresponding ensemble is referred to as the isothermal ensemble or the canonical ensemble. To obtain the T, V, N ensemble from the E, V, N ensemble, we replace the boundaries between the isolated systems by diathermal (i.e., heat-conducing) boundaries. The latter permits the flow of heat between the systems in the ensemble. The volume and the number of particles are still maintained constant.

We know from thermodynamics that any two systems at thermal equilibrium (i.e., when heat can be exchanged through their boundaries) have the same temperature. Thus, the fixed value of the internal energy E is replaced by a fixed value of the temperature T. The internal energies of the system can now fluctuate.

The probability of finding a system in the ensemble having internal energy E is given

$$\Pr(E) = \frac{W(E, V, N) \exp(-\beta E)}{Q} \tag{1.13}$$

where $\beta = (kT)^{-1}$ and Q is a normalization constant. Note that the probability of finding a *specific* state having energy E is $\exp(-\beta E)/Q$. Since there are W such states, the probability of finding a state having energy E is given by (1.13). The normalization condition is

$$\sum_E \Pr(E) = 1, \tag{1.14}$$

the summation being over all the possible energies E. From (1.13) and (1.14), we have

$$Q(T, V, N) = \sum_E W(E, V, N) \exp(-\beta E) \tag{1.15}$$

which is the partition function for the canonical ensemble.

The fundamental connection between $Q(T, V, N)$, as defined in (1.15), and thermodynamics is given by

$$A(T, V, N) = -kT \ln Q(T, V, N) \tag{1.16}$$

where A is the Helmholtz energy of the system at T, V, N. Once the partition function $Q\,(T, V, N)$ is known, then relation (1.16) may be used to obtain the Helmholtz energy.† This relation is fundamental in the sense that all the thermodynamic information on the system can be extracted from it by the application of standard thermodynamic relations, i.e., from

$$dA = -SdT - PdV + \mu dN. \tag{1.17}$$

For a multicomponent system, the last term on the right-hand side (rhs) of (1.17) should be interpreted as a scalar product $\boldsymbol{\mu} \cdot d\boldsymbol{N} = \sum_{i=1}^{c} \mu_i\, dN_i$. From (1.17) we can get the following thermodynamic quantities:

$$S = -\left(\frac{\partial A}{\partial T}\right)_{V,N} = k \ln Q + kT\left(\frac{\partial \ln Q}{\partial T}\right)_{V,N} \tag{1.18}$$

$$P = -\left(\frac{\partial A}{\partial V}\right)_{T,N} = kT\left(\frac{\partial \ln Q}{\partial V}\right)_{T,N} \tag{1.19}$$

$$\mu = \left(\frac{\partial A}{\partial N}\right)_{T,V} = -kT\left(\frac{\partial \ln Q}{\partial N}\right)_{T,V}. \tag{1.20}$$

† We use the terms Helmholtz and Gibbs energies for what has previously been referred to as Helmholtz and Gibbs free energies, respectively.

Other quantities can be readily obtained by standard thermodynamic relationships.

T, P, N ensemble

In the passage from the *E, V, N* to the *T, V, N* ensemble, we have removed the constraint of a constant energy by allowing the exchange of thermal energy between the systems. As a result, the constant energy has been replaced by a constant temperature. In a similar fashion, we can remove the constraint of a constant volume by replacing the rigid boundaries between the systems by flexible boundaries. In the new ensemble, referred to as the isothermal–isobaric ensemble, the volume of each system may fluctuate. We know from thermodynamics that when two systems are allowed to reach mechanical equilibrium, they will have the same pressure. The volume of each system can attain any value. The probability distribution of the volume in such a system is

$$\Pr(V) = \frac{Q(T, V, N)\ \exp(-\beta PV)}{\Delta(T, P, N)} \tag{1.21}$$

where P is the pressure of the system at equilibrium. The normalization constant $\Delta(T, P, N)$ is defined by

$$\begin{aligned}\Delta(T, P, N) &= \sum_V Q(T, V, N) \exp(-\beta PV)\\ &= \sum_V \sum_E W(E, V, N)\ \exp(-\beta E - \beta PV).\end{aligned} \tag{1.22}$$

$\Delta(T, P, N)$ is called the isothermal–isobaric partition function or simply the *T, P, N* partition function. Note that in (1.22) we have summed over all possible volumes, treating the volume as a discrete variable. In actual applications to classical systems, this sum should be interpreted as an integral over all possible volumes, namely

$$\Delta(T, P, N) = c \int_0^\infty dV\, Q(T, V, N)\ \exp(-\beta PV) \tag{1.23}$$

where c has the dimension of V^{-1}, to render the rhs of (1.23) dimensionless. The partition function $\Delta(T, P, N)$, though less convenient in theoretical work than $Q\ (T, V, N)$, is sometimes very useful, especially when connection with experimental quantities measured at constant T and P is required.

The fundamental connection between $\Delta(T, P, N)$ and thermodynamics is

$$G(T, P, N) = -kT\ \ln \Delta(T, P, N) \tag{1.24}$$

where G is the Gibbs energy of the system.

The relation (1.24) is the fundamental equation for the T, P, N ensemble. Once we have the function $\Delta(T, P, N)$, all thermodynamic quantities may be obtained by standard relations, i.e.,

$$dG = -SdT + VdP + \mu dN. \tag{1.25}$$

Hence

$$S = -\left(\frac{\partial G}{\partial T}\right)_{P,N} = k \ln \Delta + kT\left(\frac{\partial \ln \Delta}{\partial T}\right)_{P,N} \tag{1.26}$$

$$V = \left(\frac{\partial G}{\partial P}\right)_{T,N} = -kT\left(\frac{\partial \ln \Delta}{\partial P}\right)_{T,N} \tag{1.27}$$

$$\mu = \left(\frac{\partial G}{\partial N}\right)_{T,P} = -kT\left(\frac{\partial \ln \Delta}{\partial N}\right)_{T,P}. \tag{1.28}$$

Other thermodynamic quantities may be obtained by standard thermodynamic relationships.

T, V, μ ensemble

An important partition function can be derived by starting from $Q\,(T, V, N)$ and replacing the constant variable N by μ. To do that, we start with the canonical ensemble and replace the impermeable boundaries by permeable boundaries. The new ensemble is referred to as the *grand* ensemble or the T, V, μ ensemble. Note that the volume of each system is still constant. However, by removing the constraint on constant N, we permit fluctuations in the number of particles. We know from thermodynamics that a pair of systems between which there exists a free exchange of particles at equilibrium with respect to material flow is characterized by a constant chemical potential μ. The variable N can now attain any value with the probability distribution

$$\Pr(N) = \frac{Q(T, V, N)\,\exp(\beta\mu N)}{\Xi(T, V, \mu)} \tag{1.29}$$

where $\Xi(T, V, \mu)$, the normalization constant, is defined by

$$\Xi(T, V, \mu) = \sum_{N=0}^{\infty} Q(T, V, N)\exp(\beta\mu N) \tag{1.30}$$

where the summation in (1.30) is over all possible values of N. The new partition function $\Xi(T, V, \mu)$ is referred to as the grand partition function, the open-system partition function, or simply the T, V, μ partition function.

In equation (1.30), we have defined the T, V, μ partition function for a one-component system. In a straightforward manner we may generalize the definition for a multicomponent system. Let $\mathbf{N} = N_1, \ldots, N_c$ be the vector representing the composition of the system, where N_i is the number of molecules of species i. The corresponding vector $\boldsymbol{\mu} = \mu_1, \ldots, \mu_c$ includes the chemical potential of each of the species. For an open system with respect to all components we have the generalization of (1.30)

$$\Xi(T, V, \boldsymbol{\mu}) = \sum_{N_1} \cdots \sum_{N_c} Q(T, V, \mathbf{N}) \exp[\beta \boldsymbol{\mu} \cdot \mathbf{N}] \tag{1.31}$$

where $\boldsymbol{\mu} \cdot \mathbf{N} = \sum_i \mu_i N_i$ is the scalar product of the two vectors $\boldsymbol{\mu}$ and $\mathbf{N}$.

An important case is a system open with respect to some of the species but closed to the others. For instance, in a two-component system of A and B we can define two partial grand partition functions as follows:

$$\Xi(T, V, N_A, \mu_B) = \sum_{N_B} Q(T, V, N_A, N_B) \exp(\beta \mu_B N_B) \tag{1.32}$$

$$\Xi(T, V, N_B, \mu_A) = \sum_{N_A} Q(T, V, N_A, N_B) \exp(\beta \mu_A N_A). \tag{1.33}$$

Equation (1.32) corresponds to a system closed with respect to A, but open with respect to B. Equation (1.33) corresponds to a system closed to B, but open to A.

The fundamental connection between the partition function defined in (1.30) and thermodynamics is

$$P(T, V, \mu) V = kT \ \ln \Xi(T, V, \mu) \tag{1.34}$$

where $P(T, V, \mu)$ is the pressure of a system characterized by the independent variables T, V, μ.

The fundamental relation (1.34) may be used to obtain all relevant thermodynamic quantities. Thus, using the general differential of PV we obtain

$$d(PV) = S\, dT + P\, dV + N\, d\mu \tag{1.35}$$

$$S = \left(\frac{\partial (PV)}{\partial T} \right)_{V,\mu} = k \ \ln \Xi + kT \left(\frac{\partial \ln \Xi}{\partial T} \right)_{V,\mu} \tag{1.36}$$

$$P = \left(\frac{\partial (PV)}{\partial V} \right)_{T,\mu} = kT \left(\frac{\partial \ln \Xi}{\partial V} \right)_{T,\mu} = kT \frac{\ln \Xi}{V} \tag{1.37}$$

$$N = \left(\frac{\partial(PV)}{\partial\mu}\right)_{T,V} = kT\left(\frac{\partial \ln \Xi}{\partial\mu}\right)_{T,V}. \tag{1.38}$$

Other quantities, such as the Gibbs energy or the internal energy of the system, may be obtained from the standard relations

$$G = \mu N \tag{1.39}$$

$$E = G + TS - PV. \tag{1.40}$$

1.3 **Fluctuations and stability**

One of the characteristic features of statistical mechanics is the treatment of fluctuations, whereas in thermodynamics we treat variables such as *E*, *V*, or *N* as having sharp values. Statistical mechanics acknowledge the fact that these quantities can fluctuate. The theory also prescribes a way of calculating the average fluctuation about the equilibrium values.

In the *T*, *V*, *N* ensemble, the average energy of the system is defined by

$$\langle E\rangle = \sum_E E\,\Pr(E) = \frac{\sum_E EW(E,V,N)\exp(-\beta E)}{Q(T,V,N)}. \tag{1.41}$$

Using the definition of $Q(T, V, N)$ in (1.15), we find that

$$\langle E\rangle = kT^2\left(\frac{\partial \ln Q(T,V,N)}{\partial T}\right)_{V,N}. \tag{1.42}$$

Note that the average energy of the system, denoted here by $\langle E\rangle$, is the same as the internal energy denoted, in thermodynamics, by *U*. In this book, we shall reserve the letter *U* for potential energy and use $\langle E\rangle$ for the total (potential and kinetic) energy. Sometimes when the meaning of *E* as an average is clear, we can use *E* instead of $\langle E\rangle$.

An important average quantity in the *T*, *V*, *N* ensemble is the average fluctuation in the internal energy, defined by

$$\sigma_E^2 = \langle (E - \langle E\rangle)^2\rangle. \tag{1.43}$$

Using the probability distribution (1.13), we can express σ_E^2 in terms of the constant-volume heat capacity, i.e.,

$$\begin{aligned}\langle (E - \langle E \rangle)^2 \rangle &= \sum_E (E - \langle E \rangle)^2 \Pr(E) \\ &= \sum_E [E^2 \Pr(E) - 2E\langle E \rangle \Pr(E) + \langle E \rangle^2 \Pr(E)] \\ &= \langle E^2 \rangle - \langle E \rangle^2. \end{aligned} \tag{1.44}$$

On the other hand, by differentiation of $\langle E \rangle$ in (1.41) with respect to T, we obtain the heat capacity at constant volume,

$$C_V = \left(\frac{\partial \langle E \rangle}{\partial T}\right)_{V,N} = \frac{\langle E^2 \rangle - \langle E \rangle^2}{kT^2}. \tag{1.45}$$

Thus the heat capacity C_V is also a measure of the fluctuation in the energy of the T, V, N system.

Similar relationships hold for the enthalpy in the T, P, N ensemble. Thus, using (1.22), we obtain

$$\langle H \rangle = kT^2 \left(\frac{\partial \ln \Delta}{\partial T}\right)_{P,N} = \langle E \rangle + P\langle V \rangle. \tag{1.46}$$

Here $\langle\ \rangle$ denotes averages in the T, P, N ensemble, using the probability distribution function

$$\Pr(E, V) = \frac{W(E, V, N) \exp(-\beta E - \beta P V)}{\Delta(T, P, N)}. \tag{1.47}$$

The constant-pressure heat capacity is obtained from (1.46) and from the definition of Δ. The result is

$$C_P = \left(\frac{\partial \langle H \rangle}{\partial T}\right)_{P,N} = \frac{\langle H^2 \rangle - \langle H \rangle^2}{kT^2} \tag{1.48}$$

where the average quantities in (1.48) are taken with the probability distribution (1.47).

In the T, P, N ensemble there exists fluctuations in the volume of the system, defined by

$$\langle (V - \langle V \rangle)^2 \rangle = \langle V^2 \rangle - \langle V \rangle^2 = kT\langle V \rangle \kappa_T \tag{1.49}$$

where the isothermal compressibility is defined by

$$\kappa_T = -\frac{1}{\langle V \rangle}\left(\frac{\partial \langle V \rangle}{\partial P}\right)_{T,N}. \tag{1.50}$$

Another quantity of interest in the T, P, N ensemble is the cross-fluctuations of volume and enthalpy. This is related to the thermal expansivity, α_P, by

$$\langle (V - \langle V \rangle)(H - \langle H \rangle) \rangle = \langle VH \rangle - \langle V \rangle \langle H \rangle = kT^2 \langle V \rangle \alpha_P \tag{1.51}$$

where

$$\alpha_P = \frac{1}{\langle V \rangle} \left(\frac{\partial \langle V \rangle}{\partial T} \right)_{P,N}. \tag{1.52}$$

Of foremost importance in the T, V, μ ensemble is the fluctuation in the number of particles, which, for a one-component system, is given by

$$\langle (N - \langle N \rangle)^2 \rangle = \langle N^2 \rangle - \langle N \rangle^2 = kT \left(\frac{\partial \langle N \rangle}{\partial \mu} \right)_{T,V} = kTV \left(\frac{\partial \rho}{\partial \mu} \right)_T. \tag{1.53}$$

In (1.53), all average quantities are taken with the probability distribution $\Pr(N)$ given in (1.29). The fluctuations in the number of particles in the T, V, μ ensemble can be expressed in terms of the isothermal compressibility, as follows.

From the Gibbs–Duhem relation

$$-S\,dT + V\,dP = N\,d\mu \tag{1.54}$$

we obtain

$$\left(\frac{\partial P}{\partial \mu} \right)_T = \frac{N}{V} = \rho. \tag{1.55}$$

Using the chain rule of differentiation, we have

$$\left(\frac{\partial \rho}{\partial \mu} \right)_T = \left(\frac{\partial \rho}{\partial P} \right)_T \left(\frac{\partial P}{\partial \mu} \right)_T = \kappa_T \rho^2. \tag{1.56}$$

Combining (1.53) and (1.56), we obtain the final result

$$\langle N^2 \rangle - \langle N \rangle^2 = kTV\rho^2 \kappa_T. \tag{1.57}$$

Further relations involving cross-fluctuations in the number of particles in a multicomponent system are discussed in chapter 4. Note that in (1.54)–(1.56) we used the thermodynamic notation for V, N, etc. In applying these relations in the T, V, μ ensemble, the density ρ in (1.57) should be understood as

$$\rho = \frac{\langle N \rangle}{V} \tag{1.58}$$

where the average is taken in the T, V, μ ensemble.

Note that (1.57) can be written as

$$\frac{\langle N^2\rangle - \langle N\rangle^2}{\langle N\rangle^2} = \frac{kT\kappa_T}{V}.$$

This should be compared with equation (1.49). Thus, the relative fluctuations in the volume in the T, P, N ensemble have the same values as the relative fluctuations in the number of particles in the T, V, μ ensemble, provided that $\langle V\rangle$ in the former is equal to V in the latter.

We have seen that C_V, C_p, κ_T, and $(\partial\mu/\partial\rho)_T$ can be expressed as fluctuations in E, H, V, and N, respectively. As such, they must always be positive. The positiveness of these quantities is translated in thermodynamic language as the condition of stability of the system. Thus, $C_V > 0$ and $C_p > 0$ are the conditions for thermal stability of a closed system at constant volume and pressure, respectively. $\kappa_T > 0$ expresses the mechanical stability of a closed system at constant temperature. Of particular importance, in the context of this book, is the material stability. A positive value of $(\partial\mu/\partial\rho)_T$ means that the chemical potential is always a monotonically increasing function of the density. At equilibrium, any fluctuation which causes an increase in the local density will necessarily increase the local chemical potential. This local fluctuation will be reversed by the flow of material from the higher to the lower chemical potential, hence restoring the system to its equilibrium state. In chapter 4, we shall also encounter fluctuations and cross-fluctuations in multicomponent systems.

1.4 The classical limit of statistical thermodynamics

In section 1.2, we introduced the quantum mechanical partition function in the T, V, N ensemble. In most applications of statistical thermodynamics to problems in chemistry and biochemistry, the classical limit of the quantum mechanical partition function is used. In this section, we present the so-called classical canonical partition function.

The canonical partition function introduced in section 1.2 is defined as

$$Q(T, V, N) = \sum_i \exp(-\beta E_i) = \sum_E W(E, V, N)\exp(-\beta E) \qquad (1.59)$$

where the first sum is over all possible states of the T, V, N system. In the second sum all states having the same energy E are grouped first, and then we sum over all the different energy levels. $W(E, V, N)$ is simply the degeneracy of the energy level E (given V and N), i.e., the number of states having the same energy E.

The classical analog of $Q(T, V, N)$ for a system of N simple particles (i.e., spherical particles having no internal structure) is

$$Q(T, V, N) = (1/h^{3N}N!) \int \cdots \int d\boldsymbol{p}^N d\boldsymbol{R}^N \exp(-\beta H). \tag{1.60}$$

Here, h is the Planck constant ($h = 6.625 \times 10^{-27}$ erg s) and H is the classical Hamiltonian of the system, given by

$$H(\boldsymbol{p}^N, \boldsymbol{R}^N) = \sum_{i=1}^{N} (\boldsymbol{p}_i^2/2m) + U_N(\boldsymbol{R}^N). \tag{1.61}$$

Here $\boldsymbol{p}_i$ is the momentum vector of the ith particle (presumed to possess only translational degrees of freedom) and m is the mass of each particle. The total potential energy of the system at the specified configuration $\boldsymbol{R}^N$ is denoted by $U_N(\boldsymbol{R}^N)$.

Note that the expression (1.60) is not purely classical since it contains two corrections of quantum mechanical origin: the Planck constant h and the $N!$. Therefore, Q defined in (1.60) is actually the classical limit of the quantum mechanical partition function in (1.59). The purely classical partition function consists of the integral expression on the rhs of (1.60) without the factor $(h^{3N}N!)$. This partition function fails to produce the correct form of the chemical potential or of the entropy of the system.

The integration over the momenta in (1.60) can be performed straightforwardly to obtain

$$\begin{aligned} h^{-3N} \int_{\infty}^{\infty} d\boldsymbol{p}^N \exp\left[-\beta \sum_{i=1}^{N} (\boldsymbol{p}_i^2/2m)\right] &= \left[h^{-1} \int_{-\infty}^{\infty} d\boldsymbol{p} \exp(-\beta \boldsymbol{p}^2/2m)\right]^{3N} \\ &= \left[h^{-1}(2m/\beta)^{1/2} \int_{-\infty}^{\infty} \exp(-x^2)\, dx\right]^{3N} \\ &= \left[(2\pi mkT)^{3/2}/h^3\right]^N = \Lambda^{-3N}. \end{aligned} \tag{1.62}$$

In (1.62) we have introduced the momentum partition function, defined by

$$\Lambda = \frac{h}{(2\pi mkT)^{1/2}}. \tag{1.63}$$

This is also referred to as the thermal de Broglie wavelength of the particles at temperature T. Another important quantity is the configurational partition function, defined by

$$Z_N = \int \cdots \int d\boldsymbol{R}^N \exp[-\beta U_N(\boldsymbol{R}^N)]. \tag{1.64}$$

The canonical partition function in (1.60) can be rewritten as

$$Q(T, V, N) = \frac{Z_N}{N!\Lambda^{3N}}. \tag{1.65}$$

The condition required for the applicability of the classical partition function, as given in (1.60), is

$$\rho\Lambda^3 \ll 1 \tag{1.66}$$

i.e., when either the density is low, or the mass of the particles is large, or the temperature is high. Indeed, for most systems of interest in this book, we shall assume the validity of the condition (1.66), hence the validity of (1.60).

For a system of N nonspherical particles, the partition function (1.60) is modified as follows

$$Q(T, V, N) = \frac{q^N}{(8\pi^2)^N\Lambda^{3N}N!}\int\cdots\int d\mathbf{X}^N \exp[-\beta U_N(\mathbf{X}^N)]. \tag{1.67}$$

The integration on the rhs of (1.67) extends over all possible locations and orientations of the N particles. We shall refer to the vector $\mathbf{X}^N = \mathbf{X}_1, \ldots, \mathbf{X}_N$ as the *configuration* of the system of the N particles. The factor q, referred to as the internal partition function, includes the rotational, vibrational, electronic, and nuclear partition functions of a single molecule. We shall always assume in this book that the internal partition functions are separable from the configurational partition function. Such an assumption cannot always be granted, especially when strong interactions between the particles can perturb the internal degrees of freedom of the particles involved.

In the classical T, V, N ensemble, the basic distribution function is the probability density for observing the configuration $\mathbf{X}^N$,

$$P(\mathbf{X}^N) = \frac{\exp[-\beta U_N(\mathbf{X}^N)]}{\int\cdots\int d\mathbf{X}^N \exp[-\beta U_N(\mathbf{X}^N)]}. \tag{1.68}$$

In the classical T, P, N ensemble, the basic distribution function is the probability density of finding a system with a volume V and a configuration $\mathbf{X}^N$, i.e.,

$$P(\mathbf{X}^N, \mathbf{V}) = \frac{\exp[-\beta \mathbf{P}\mathbf{V} - \beta \mathbf{U}_\mathbf{N}(\mathbf{X}^\mathbf{N})]}{\int d\mathbf{V}\int\cdots\int d\mathbf{X}^N \exp[-\beta \mathbf{P}\mathbf{V} - \beta \mathbf{U}_\mathbf{N}(\mathbf{X}^\mathbf{N})]}. \tag{1.69}$$

The integration over V extends from zero to infinity. The probability density of observing a system with volume V, independently of the configuration, is obtained from (1.69) by integrating over all configurations, i.e.,

$$P(V) = \int\cdots\int d\mathbf{X}^N P(\mathbf{X}^N, V). \tag{1.70}$$

The conditional distribution function defined by[†]

$$P(\mathbf{X}^N/V) = \frac{P(\mathbf{X}^N, V)}{P(V)}$$

$$= \frac{\exp[-\beta PV - \beta U_N(\mathbf{X}^N)]}{\int\cdots\int d\mathbf{X}^N \exp[-\beta PV - \beta U_N(\mathbf{X}^N)]}$$

$$= \frac{\exp[-\beta U_N(\mathbf{X}^N)]}{\int\cdots\int d\mathbf{X}^N \exp[-\beta U_N(\mathbf{X}^N)]} \tag{1.71}$$

is the probability density of finding a system in the configuration $\mathbf{X}^N$, given that the system has the volume V.

In the classical T, V, μ ensemble, the basic distribution function defined by

$$P(\mathbf{X}^N, N) = \frac{(q^N/N!)\exp[\beta\mu N - \beta U_N(\mathbf{X}^N)]}{\sum_{N=0}^{\infty}(q^N/N!)[\exp(\beta\mu N)]\int\cdots\int d\mathbf{X}^N \exp[-\beta U_N(\mathbf{X}^N)]} \tag{1.72}$$

is the probability density of observing a system with precisely N particles and the configuration $\mathbf{X}^N$. The probability of finding a system in the T, V, μ ensemble with exactly N particles is obtained from (1.72) by integrating over all possible configurations namely,

$$P(N) = \int\cdots\int d\mathbf{X}^N P(\mathbf{X}^N, N) \tag{1.73}$$

which can be written as

$$P(N) = \frac{Q(T, V, N)\exp(\beta\mu N)}{\Xi(T, V, \mu)}. \tag{1.74}$$

The conditional distribution function, defined by

$$P(\mathbf{X}^N/N) = \frac{P(\mathbf{X}^N, N)}{P(N)} = \frac{\exp[-\beta U_N(\mathbf{X}^N)]}{\int\cdots\int d\mathbf{X}^N \exp[-\beta U_N(\mathbf{X}^N)]}, \tag{1.75}$$

is the probability density of observing a system in the configuration $\mathbf{X}^N$, given that the system contains precisely N particles.

[†] We use the slash sign for the conditional probability. In some texts, the vertical bar is used instead.

1.5 The ideal gas and small deviation from ideality

Theoretically, an ideal gas is a hypothetical system of noninteracting molecules, i.e.,

$$U_N(\mathbf{X}^N) \equiv 0 \tag{1.76}$$

for any configuration $\mathbf{X}^N$. Of course, there is no real system that obeys equation (1.76).

In practice, the ideal-gas behavior is obtained in the limit of very low densities or pressure, where interactions between the (real) molecules are on the average negligible. One should be careful, however, to distinguish between these two conditions for ideality. The two systems are not identical, as we shall see later in the book.

Using (1.76) in the classical partition function (1.67), we immediately obtain

$$\begin{aligned} Q(T,V,N) &= \frac{q^N}{(8\pi^2)^N \Lambda^{3N} N!} \int \cdots \int d\mathbf{X}^N \\ &= \frac{q^N}{(8\pi^2)^N \Lambda^{3N} N!} \left[\int_V d\mathbf{R} \int_0^{2\pi} d\phi \int_0^{\pi} \sin\theta d\theta \int_0^{2\pi} d\psi \right]^N \\ &= \frac{q^N V^N}{\Lambda^{3N} N!}. \end{aligned} \tag{1.77}$$

For simple spherical particles, sometimes referred to as "structureless" particles, equation (1.77) reduces to

$$Q(T,V,N) = \frac{V^N}{\Lambda^{3N} N!}. \tag{1.78}$$

Note that q and Λ depend on the temperature and not on the volume V or on N. An important consequence of this is that the equation of state of an ideal gas is independent of the particular molecules constituting the system. To see this, we derive the expression for the pressure. Differentiating (1.77) with respect to volume, we obtain

$$P = kT \left(\frac{\partial \ln Q}{\partial V} \right)_{T,N} = \frac{kTN}{V} = \rho kT. \tag{1.79}$$

This equation of state is universal, in the sense that it does not depend on the properties of the specific molecules. This behavior is not shared by all thermodynamic quantities of the ideal gas. For instance, the chemical potential obtained by differentiation of (1.77) and using the Stirling approximation[†] is

$$\mu = -kT\left(\frac{\partial \ln Q}{\partial N}\right)_{T,V} = kT\ln(\Lambda^3 q^{-1}) + kT\ln\rho$$
$$= \mu^{0g}(T) + kT\ln\rho \tag{1.80}$$

where $\rho = N/V$ is the number density and $\mu^{0g}(T)$ is the standard chemical potential. The latter depends on the properties of the individual molecules in the system. Note that the value of $\mu^{0g}(T)$ depends on the choice of units of ρ. The quantity $\rho\Lambda^3$, however, is dimensionless. Hence, μ is independent of the choice of the concentration units.

Another useful expression is that for the entropy of an ideal gas, which can be obtained from (1.77):

$$S = k\ln Q + kT\left(\frac{\partial \ln Q}{\partial T}\right)_{V,N}$$
$$= \tfrac{5}{2}kN - Nk\ln(\rho\Lambda^3 q^{-1}) + kTN\frac{\partial \ln q}{\partial T}. \tag{1.81}$$

Clearly, the entropy in (1.81) depends on the properties of the specific gas. For simple particles, this reduces to the well-known Sackur–Tetrode equation for the entropy:

$$S = \tfrac{5}{2}kN - Nk\ln\rho\Lambda^3. \tag{1.82}$$

The dependence of both μ and S on the density ρ through $\ln\rho$ is confirmed by experiment. We note here that had we used the purely classical partition function [i.e., the integral excluding the factors $h^{3N}N!$ in (1.60)], we would not have obtained such a dependence on the density. This demonstrates the necessity of using the correction factors $h^{3N}N!$ even in the classical limit of the quantum mechanical partition function.

Similarly, the energy of an ideal-gas system of simple particles is obtained from (1.78) and (1.82), i.e.,

$$E = A + TS = kT\ln\rho\Lambda^3 - kTN + T(\tfrac{5}{2}kN - Nk\ln\rho\Lambda^3) = \tfrac{3}{2}kTN \tag{1.83}$$

which in this case is entirely due to the kinetic energy of particles.

The heat capacity for a system of simple particles is obtained directly from (1.83) as

$$C_V = (\partial E/\partial T)_V = \tfrac{3}{2}kN \tag{1.84}$$

[†] In this book, we always use the Stirling approximation in the form $\ln N! = N\ln N - N$. A better approximation for small values of N is $\ln N! = N\ln N - N + \frac{1}{2}\ln(2\pi N)$.

which may be viewed as originating from the accumulation of $k/2$ per translational degree of freedom of a particle. For molecules having also rotational degrees of freedom, we have

$$C_V = 3kN \tag{1.85}$$

which is built up of $\frac{3}{2}kN$ from the translational, and $\frac{3}{2}kN$ from the rotational degrees of freedom. If other internal degrees of freedom are present, there are additional contributions to C_V.

In all of the aforementioned discussions, we left unspecified the internal partition function of a single molecule. This, in general, includes contributions from the rotational, vibrational, and electronic states of the molecule. Assuming that these degrees of freedom are independent, the corresponding internal partition function may be factored into a product of the partition functions for each degree of freedom, namely,

$$q(T) = q_{\mathrm{r}}(T)q_{\mathrm{v}}(T)q_{\mathrm{e}}(T). \tag{1.86}$$

We shall never need to use the explicit form of the internal partition function in this book. Such knowledge is needed for the actual calculation, for instance, of the equilibrium constant of a chemical reaction.

The equation of state (1.79) has been derived theoretically for an ideal gas for which (1.76) was assumed. In reality, equation (1.79) is obtained when the density is very low, $\rho \approx 0$, such that intermolecular interactions, though existing, may be neglected.

We now present some corrections to the ideal-gas equation of state (1.79). Formally, we write βP as a power series in the density, presuming that such an expansion exists,

$$\begin{aligned}\beta P &= \rho\left(\frac{\partial(\beta P)}{\partial\rho}\right)_{T,\rho=0} + \tfrac{1}{2}\rho^2\left(\frac{\partial^2(\beta P)}{\partial\rho^2}\right)_{T,\rho=0} + \cdots \\ &= \rho + B_2(T)\rho^2 + B_3(T)\rho^3 + \cdots\end{aligned} \tag{1.87}$$

where the coefficients $B_k(T)$ are evaluated at $\rho = 0$, and hence are functions of the temperature only.†

One of the most remarkable results of statistical mechanics is that it provides explicit expressions for the coefficients in (1.87). The first-order coefficient is

$$\begin{aligned}B_2(T) &= -\frac{1}{2V(8\pi^2)^2}\int \{\exp[-\beta U(\boldsymbol{X}_1,\boldsymbol{X}_2)] - 1\}\, d\boldsymbol{X}_1 d\boldsymbol{X}_2 \\ &= -\frac{1}{2(8\pi^2)}\int \{\exp[-\beta U(\boldsymbol{X})] - 1\}\, d\boldsymbol{X}.\end{aligned} \tag{1.88}$$

† The coefficients $B_2(T)$, $B_3(T)$, etc., are sometimes denoted by B, C, D, etc.

This is known as the second virial coefficient. In the second step on the rhs of (1.88), we exploit the fact that $U(\boldsymbol{X}_1, \boldsymbol{X}_2)$ is actually a function of six coordinates, not twelve as implied in $\boldsymbol{X}_1$, $\boldsymbol{X}_2$; i.e., we can hold $\boldsymbol{X}_1$ fixed, say at the origin, and view the potential function $U(\boldsymbol{X}_1, \boldsymbol{X}_2)$ as depending on the relative locations and orientations of the second particle, which we denote by $\boldsymbol{X}$. Thus integrating over $\boldsymbol{X}_1$ produces a factor $V8\pi^2$ and the final form of $B_2(T)$ is obtained.

Note also that since the potential function $U(\boldsymbol{X})$ has a short range, say of a few molecular diameters, the integral over the entire volume is actually over only a very short distance from the particle that we held fixed at the origin. This is the reason why $B_2(T)$ is not a function of the volume.

Expression (1.88) can be further simplified when the pair potential is a function of the scalar distance $R = |\boldsymbol{R}_2 - \boldsymbol{R}_1|$. In this case, the integration over the orientations produce the factor $8\pi^2$ and the integration over the volume can be performed after transforming to polar coordinates to obtain

$$B_2(T) = -\tfrac{1}{2}\int_0^\infty \{\exp[(-\beta U(R))] - 1\}4\pi R^2 dR. \tag{1.89}$$

Note that we chose infinity as the upper limit of the integral. In practice, the integration extends to a finite distance of the order of a few molecular diameters, i.e., the effective range of the interaction potential. Beyond this limit, $U(R)$ is zero and therefore the integrand becomes zero as well. Hence, the extension of the range of integration does not affect the value of $B_2(T)$.

Of the virial coefficients, $B_2(T)$ is the most useful. The theory also provides expressions for the higher order corrections to the equation of state. We cite here the expression for the third virial coefficient,

$$\begin{aligned} B_3(T) = -\frac{1}{3(8\pi^2)^2}\int \{&\exp[-\beta U_3(\boldsymbol{X}_1, \boldsymbol{X}_2, \boldsymbol{X}_3)] \\ &- \exp[-\beta U(\boldsymbol{X}_1, \boldsymbol{X}_2) - \beta U(\boldsymbol{X}_2, \boldsymbol{X}_3)] \\ &- \exp[-\beta U(\boldsymbol{X}_1, \boldsymbol{X}_2) - \beta U(\boldsymbol{X}_1, \boldsymbol{X}_3)] \\ &- \exp[-\beta U(\boldsymbol{X}_1, \boldsymbol{X}_3) - \beta U(\boldsymbol{X}_2, \boldsymbol{X}_3)] \\ &+ \exp[-\beta U(\boldsymbol{X}_1, \boldsymbol{X}_2)] + \exp[-\beta U(\boldsymbol{X}_1, \boldsymbol{X}_3)] \\ &+ \exp[-\beta U(\boldsymbol{X}_2, \boldsymbol{X}_3)] - 1\} d\boldsymbol{X}_2 d\boldsymbol{X}_3. \end{aligned} \tag{1.90}$$

We see that this expression is fairly complicated. If the total potential energy is pairwise additive, in the sense that

$$U_3(\boldsymbol{X}_1, \boldsymbol{X}_2, \boldsymbol{X}_3) = U(\boldsymbol{X}_1, \boldsymbol{X}_2) + U(\boldsymbol{X}_1, \boldsymbol{X}_3) + U(\boldsymbol{X}_2, \boldsymbol{X}_3) \tag{1.91}$$

the integrand in (1.90) simplifies to[†]

$$B_3(T) = -\frac{1}{3(8\pi^2)^2}\int f(\boldsymbol{X}_1, \boldsymbol{X}_2,)f(\boldsymbol{X}_1, \boldsymbol{X}_3)f(\boldsymbol{X}_2, \boldsymbol{X}_3)d\boldsymbol{X}_2 d\boldsymbol{X}_3 \qquad (1.92)$$

where f, the so-called Mayer f-function, is defined by

$$f(\boldsymbol{X}_i, \boldsymbol{X}_j) = \exp[-\beta U(\boldsymbol{X}_i, \boldsymbol{X}_j)] - 1. \qquad (1.93)$$

Extending the same procedure for mixtures, say of two components, A and B will give us the second virial coefficient for a mixture. The first-order correction to the ideal-gas behavior of the mixture is

$$\beta P = \rho_A + \rho_B + B_{AA}\rho_A^2 + B_{BB}\rho_B^2 + 2B_{AB}\rho_A\rho_B + \cdots \qquad (1.94)$$

In terms of the total density $\rho_T = \rho_A + \rho_B$, and the mole fraction $x_A = \rho_A/\rho_T$ (1.94) can be written as

$$\beta P = \rho_T + [B_{AA}x_A^2 + B_{BB}x_B^2 + 2B_{AB}x_Ax_B]\rho_T^2 + \cdots \qquad (1.95)$$

where the term in the square brackets may be interpreted as the average second virial coefficient of the mixture. $B_{\alpha\beta}$ is related to $U_{\alpha\beta}$ by the same relation as B_2 to U in (1.88) or (1.89).

1.6 Suggested references on general thermodynamics and statistical mechanics

There are many good textbooks on thermodynamics: Denbigh (1966, 1981), Prigogine and Defay (1954) and Callen (1960).

Books on the elements of statistical thermodynamics: Hill (1960), McQuarrie (1976) and Ben-Naim (1992).

Advanced books on statistical thermodynamics: Hill (1956), Münster (1969,1974) and Hansen and McDonald (1976).

[†] Note that in both (1.90) and (1.92), integration over X_1 has been performed so that the integrands are not functions of X_1.

TWO

Molecular distribution functions

In this chapter, we introduce the concepts of molecular distribution function (MDF), in one- and multicomponent systems. The MDFs are the fundamental ingredients in the modern molecular theories of liquids and liquid mixtures. As we shall see, these quantities convey local information on the densities, correlation between densities at two points (or more) in the system, etc.

We start with detailed definitions of the singlet and the pair distribution functions. We then introduce the pair correlation function, a function which is the cornerstone in any molecular theory of liquids. Some of the salient features of these functions are illustrated both for one- and for multicomponent systems. Also, we introduce the concepts of the generalized molecular distribution functions. These were found useful in the application of the mixture model approach to liquid water and aqueous solutions.

In this chapter, we shall not discuss the methods of obtaining information on molecular distribution functions. There are essentially three sources of information: analyzing and interpreting x-ray and neutron diffraction patterns; solving integral equations; and simulation of the behavior of liquids on a computer. Most of the illustrations for this chapter were done by solving the Percus–Yevick equation. This method, along with some comments on the numerical solution, are described in Appendices B–F.

2.1 The singlet distribution function

We start with the simplest MDF, the *singlet* distribution function. The presentation here is done at great length, far more than is necessary, but, as we shall soon see, fully understanding the meaning of this quantity will be essential for the understanding the higher MDF as well as the generalized MDF.

In this and the following chapter, we shall always start with a one-component system, then generalize for multicomponent mixtures. This is done mainly for notational convenience. We also discuss rigid molecules, i.e., molecules without internal rotational degrees of freedom. The state of each molecule is fully described by the six-dimensional vector $\boldsymbol{X}$ consisting of three locational coordinates $\boldsymbol{R} = (x, y, z)$ and three orientational coordinates $\boldsymbol{\Omega} = (\phi, \theta, \psi)$.

We start with a system consisting of N rigid particles at a given temperature T, contained in volume V. The basic probability density for such a system is essentially the Boltzmann distribution

$$P(\boldsymbol{X}^N) = \frac{\exp[-\beta U_N(\boldsymbol{X}^N)]}{\int \cdots \int d\boldsymbol{X}^N \exp[-\beta U_N(\boldsymbol{X}^N)]}. \tag{2.1}$$

In general, an average of any function of the configuration, $F(\boldsymbol{X}^N)$, in the T, V, N ensemble, is defined by

$$F = \int \cdots \int d\boldsymbol{X}^N P(\boldsymbol{X}^N) F(\boldsymbol{X}^N). \tag{2.2}$$

In some cases, we shall also use either the symbol $\langle F \rangle$ or $\overline{F}$ for an *average* quantity. However, we shall refrain from using this notation whenever the meaning of that quantity as an average is evident.

As a first and very simple example, let us calculate the average number of particles in a region S within the system. (A particle is said to be in the region S whenever its *center* falls within that region.) Let $N(\boldsymbol{X}^N, S)$ be the number of particles in S, given that the system is at a particular configuration $\boldsymbol{X}^N$. One may imagine taking a snapshot of the system at some instant and *counting* the number of particles that happen to fall within S at that configuration. Hence, $N(\boldsymbol{X}^N, S)$ is also referred to as a *counting function.* A two-dimensional illustration is given in figure 2.1.

The *average number* of particles in S is, according to (2.2)

$$N(S) = \int \cdots \int d\boldsymbol{X}^N P(\boldsymbol{X}^N) N(\boldsymbol{X}^N, S). \tag{2.3}$$

This relation can be written in an alternative form which will turn out to be useful for later applications.

Let us define the *characteristic* function

$$A_i(\boldsymbol{R}_i, S) = \begin{cases} 1 & \text{if } \boldsymbol{R}_i \in S \\ 0 & \text{if } \boldsymbol{R}_i \notin S. \end{cases} \tag{2.4}$$

The symbol $\in$ means "belongs to." Hence, $A_i(\boldsymbol{R}_i, S)$ is unity whenever $\boldsymbol{R}_i$ is within S and zero elsewhere. The quantity $N(\boldsymbol{X}^N, S)$ can be expressed as

$$N(\boldsymbol{X}^N, S) = \sum_{i=1}^{N} A_i(\boldsymbol{R}_i, S). \tag{2.5}$$

Clearly, in order to *count* the number of particles within S, we have to check the location of each particle separately. Each particle whose center falls within S will

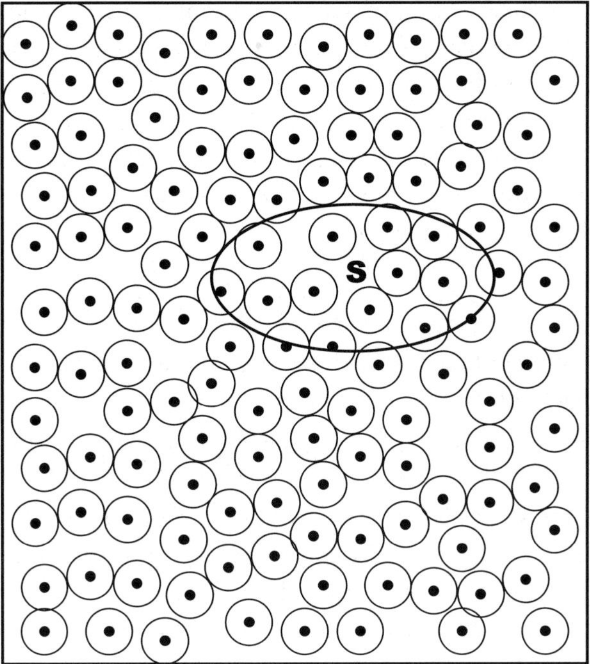

Figure 2.1. An arbitrary region S within the system of volume V. In the particular configuration shown here, the number of particles in S is 12.

contribute unity to the sum on the rhs of (2.5); hence, the sum *counts the exact number of particles in* S, given a specific configuration $\boldsymbol{X}^N$. Introducing (2.5) into (2.3), we obtain the *average* number of particles in S:

$$\begin{aligned} N(S) &= \int \cdots \int d\boldsymbol{X}^N P(\boldsymbol{X}^N) \sum_{i=1}^{N} A_i(\boldsymbol{R}_i, S) \\ &= \sum_{i=1}^{N} \int \cdots \int d\boldsymbol{X}^N P(\boldsymbol{X}^N) A_i(\boldsymbol{R}_i, S) \\ &= N \int \cdots \int d\boldsymbol{X}^N P(\boldsymbol{X}^N) A_1(\boldsymbol{R}_1, S). \end{aligned} \tag{2.6}$$

Since all the particles are *equivalent*, the sum over the index i produces N integrals having the same magnitude. We may therefore select one of these integrals, say $i = 1$, and replace the sum by N times that specific integral. The *mole fraction* of particles within S is defined as

$$x(S) = \frac{N(S)}{N} = \int \cdots \int d\boldsymbol{X}^N\, P(\boldsymbol{X}^N)\, A_1(\boldsymbol{R}_1, S). \tag{2.7}$$

$x(S)$ is the *average* fraction of particles found in S. This quantity may also be assigned a probabilistic meaning that is often useful. To see this, we recall that the function $A_1(\mathbf{R}_1, S)$ used in (2.7) has the effect of reducing the *range of integration* from V to a restricted range which fulfills the condition: "$\mathbf{R}_1$ being located in S." Symbolically, this can be written as

$$\int_V \cdots \int_V d\mathbf{X}^N P(\mathbf{X}^N) A_1(\mathbf{R}_1, S) = \int_{\mathbf{R}_1 \in S} \cdots \int d\mathbf{X}^N P(\mathbf{X}^N) = P_1(S). \qquad (2.8)$$

Thus, the integration over the entire volume V is reduced to the region for which $\mathbf{R}_1 \in S$.

We recall that $P(\mathbf{X}^N)$ is the probability density of the occurrence of the event $\mathbf{X}^N$, i.e., that the N particles are found at the specific configuration $\mathbf{X}_1, \ldots, \mathbf{X}_N$. Therefore, integration over *all the events* $\mathbf{X}^N$ for which the condition $\mathbf{R}_1 \in S$ is fulfilled gives the probability of the occurrence of the condition, i.e., $P_1(S)$ is the probability that a specific particle, say number 1, will be found in S. From (2.7) and (2.8) we arrive at an important relation:

$$x(S) = P_1(S), \qquad (2.9)$$

which states that the *mole fraction* of particles in S equals the *probability* that a *specific particle*, say 1, will be found in S. [Of course, we could have chosen in (2.9) any other specific particle other than particle 1.]

We now introduce the singlet molecular distribution function, which is obtained from $N(S)$ in the limit of a very small region S. First we note that $A_i(\mathbf{R}_i, S)$ can also be written as

$$A_i(\mathbf{R}_i, S) = \int_S \delta(\mathbf{R}_i - \mathbf{R}')\, d\mathbf{R}', \qquad (2.10)$$

where $\delta(\mathbf{R}_i - \mathbf{R}')$ is the Dirac delta function. The integral over $\delta(\mathbf{R}_i - \mathbf{R}')$ is unity if $\mathbf{R}_i \in S$, and zero otherwise.

When S is an infinitesimally small region $d\mathbf{R}'$, we have

$$A_i(\mathbf{R}_i, d\mathbf{R}') = \delta(\mathbf{R}_i - \mathbf{R}')\, d\mathbf{R}'. \qquad (2.11)$$

Hence, from (2.6) we obtain the average quantity

$$N(d\mathbf{R}') = d\mathbf{R}' \int \cdots \int d\mathbf{X}^N P(\mathbf{X}^N) \sum_{i=1}^{N} \delta(\mathbf{R}_i - \mathbf{R}'). \qquad (2.12)$$

The average local (number) density of particles in the element of volume at $d\mathbf{R}'$ at $\mathbf{R}'$ is now defined by

$$\rho^{(1)}(\mathbf{R}') = \frac{N(d\mathbf{R}')}{d\mathbf{R}'} = \int \cdots \int d\mathbf{X}^N P(\mathbf{X}^N) \sum_{i=1}^{N} \delta(\mathbf{R}_i - \mathbf{R}'). \qquad (2.13)$$

Table 2.1

Event	Probability of the event
Particle 1 in $d\boldsymbol{R}'$	$P^{(1)}(\boldsymbol{R}')\, d\boldsymbol{R}'$
Particle 2 in $d\boldsymbol{R}'$	$P^{(1)}(\boldsymbol{R}')\, d\boldsymbol{R}'$
⋮	⋮
Particle N in $d\boldsymbol{R}'$	$P^{(1)}(\boldsymbol{R}')\, d\boldsymbol{R}'$

Note that $d\boldsymbol{R}'$ is an element of *volume* $dx'dy'dz'$ at $\boldsymbol{R}'$. The quantity $\rho^{(1)}(\boldsymbol{R}')$ is referred to as the *singlet molecular distribution function.*

The meaning of $\rho^{(1)}(\boldsymbol{R}')$ as a *local density* will prevail in all our applications. However, in some cases one may also assign to $\rho^{(1)}(\boldsymbol{R}')$ the meaning of probability density. This must be done with some caution, as will be shown below. First, we rewrite (2.13) in the form

$$\rho^{(1)}(\boldsymbol{R}') = N\int\cdots\int d\boldsymbol{X}^N P(\boldsymbol{X}^N)\delta(\boldsymbol{R}_1 - \boldsymbol{R}') = NP^{(1)}(\boldsymbol{R}'). \tag{2.14}$$

The interpretation of $P^{(1)}(\boldsymbol{R}')d\boldsymbol{R}'$ follows from the same argument as in the case of $P_1(S)$ in (2.8). This is the *probability* of finding a *specific* particle, say 1, in $d\boldsymbol{R}'$ at $\boldsymbol{R}'$. Hence, $P^{(1)}(\boldsymbol{R}')$ is often referred to as the *specific singlet distribution function.*

The next question is: "What is the probability of finding *any* particle in $d\boldsymbol{R}'$?" To answer this question, we consider the events listed in Table 2.1.

Since all particles are equivalent, we have exactly the same probability for each of the events listed on the left-hand side (lhs).

The event "any particle in $d\boldsymbol{R}'$" means either "particle 1 in $d\boldsymbol{R}'$" or "particle 2 in $d\boldsymbol{R}'$", ..., or "particle N in $d\boldsymbol{R}'$." In probability language, this event is called the *union* of all the events as listed above, and is written symbolically as

$$\{\text{any particle in } d\boldsymbol{R}'\} = \bigcup_{i=1}^{N}\{\text{particle } i \text{ in } d\boldsymbol{R}'\}. \tag{2.15}$$

It is at this point that care must be exercised in writing the probability of the event on the lhs of (2.15). In general, there exists no simple relation between the probability of a union of events and the probabilities of the individual events. However, if we choose $d\boldsymbol{R}'$ to be small enough so that no more than a single particle may be found in $d\boldsymbol{R}'$ at any given time, then all the events listed above become disjoint (i.e., occurrence of one event precludes the possibility of simultaneous occurrence of any other event). In this case, we have the additivity relation for the probability of the union of the events, namely:

$$\begin{aligned}\Pr\{\text{any particle in } d\mathbf{R}'\} &= \sum_{i=1} \Pr\{\text{particle } i \text{ in } d\mathbf{R}'\} \\ &= \sum_{i=1} P^{(1)}(\mathbf{R}')\, d\mathbf{R}' \\ &= NP^{(1)}(\mathbf{R}')\, d\mathbf{R}' \\ &= \rho^{(1)}(\mathbf{R}')\, d\mathbf{R}'. \end{aligned} \tag{2.16}$$

Relation (2.16) provides the probabilistic meaning of the quantity $\rho^{(1)}(\mathbf{R}')d\mathbf{R}'$, which is contingent upon the choice of a *sufficiently small element of volume* $d\mathbf{R}'$. The quantity $\rho^{(1)}(\mathbf{R}')$ is referred to as the *generic singlet* distribution function[†]. Clearly, the generic singlet distribution function is the physically meaningful quantity. We *can* measure the average number of particles in a given element of volume. We *cannot* measure the probability of finding a *specific* particle in a given element of volume.

Caution must also be exercised when using the probabilistic meaning of $\rho^{(1)}(\mathbf{R}')d\mathbf{R}'$. For instance, the *probability* of finding a specific particle, say 1, in a region S is obtained from the specific singlet distribution function simply by integration:

$$P_1(S) = \int_S P^{(1)}(\mathbf{R}')\, d\mathbf{R}'. \tag{2.17}$$

This interpretation follows from the fact that the events "particle 1 in $d\mathbf{R}'$" and "particle 1 in $d\mathbf{R}''$" are disjoint events (i.e., a *specific* particle cannot be in two different elements $d\mathbf{R}'$ and $d\mathbf{R}''$ simultaneously). Hence, the probability of the union is obtained as the sum (or integral) of the probabilities of the individual events.

This property is not shared by the *generic* singlet distribution function, and the integral

$$\int_S \rho^{(1)}(\mathbf{R}')\, d\mathbf{R}' \tag{2.18}$$

does not have the meaning of the probability of the event "any particle in S." The reason is that the events "a particle in $d\mathbf{R}'$" and "a particle in $d\mathbf{R}''$" are not disjoint events; hence, one cannot obtain the probability of their union in a

[†] The adjectives "specific" and "generic" were introduced by Gibbs. Since the particles of a given species are indistinguishable, only the *generic* MDF has physical meaning. However, the *specific* MDF is an important step in the definition of MDFs. One first "labels" the particles to obtain the *specific MDF*, then "un-labels" them to obtain the *generic* MDF.

simple fashion. It is for this reason that the meaning of $\rho^{(1)}(\boldsymbol{R}')$ as *a local density* at $\boldsymbol{R}'$ should be preferred. If $\rho^{(1)}(\boldsymbol{R}')\, d\boldsymbol{R}'$ is viewed as the *average* number of particles in $d\boldsymbol{R}'$, then clearly (2.18) is the average number of particles in S. The meaning of $\rho^{(1)}(\boldsymbol{R}')d\boldsymbol{R}'$ as an *average* number of particles is preserved upon integration; the probabilistic meaning is not. A particular example of (2.18) occurs when S is chosen as the total volume of the system, i.e.,

$$\int_V \rho^{(1)}(\boldsymbol{R}')\, d\boldsymbol{R}' = N \int_V P^{(1)}(\boldsymbol{R}')d\boldsymbol{R}' = N. \tag{2.19}$$

The last equality follows from the normalization of $\rho^{(1)}(\boldsymbol{R}')$; i.e., the probability of finding particle 1 in *any* place in V is unity. The normalization condition (2.19) can also be obtained directly from (2.13).

In a homogeneous fluid, we expect that $\rho^{(1)}(\boldsymbol{R}')$ will have the same value at any point $\boldsymbol{R}'$ within the system. (This is true apart from a very small region near the surface of the system, which we always neglect in considering macroscopic systems.) Therefore, we write

$$\rho^{(1)}(\boldsymbol{R}') = \text{const.} \tag{2.20}$$

and, from (2.19) and (2.20), we obtain

$$\text{const.} \times \int_V d\boldsymbol{R}' = N. \tag{2.21}$$

Hence

$$\rho^{(1)}(\boldsymbol{R}') = \frac{N}{V} = \rho. \tag{2.22}$$

The last relation is almost a self-evident result for homogenous systems. It states that the local density at any point $\boldsymbol{R}'$ is equal to the bulk density ρ. That is, of course, not true in an inhomogeneous system.

In a similar fashion, we can define the singlet distribution function for location and orientation, which by analogy to (2.14) is defined as

$$\begin{aligned}\rho^{(1)}(\boldsymbol{X}') &= \int \cdots \int d\boldsymbol{X}^N P(\boldsymbol{X}^N) \sum_{i=1}^{N} \delta(\boldsymbol{X}_i - \boldsymbol{X}') \\ &= N \int \cdots \int d\boldsymbol{X}^N P(\boldsymbol{X}^N) \delta(\boldsymbol{X}_1 - \boldsymbol{X}') \\ &= NP^{(1)}(\boldsymbol{X}').\end{aligned} \tag{2.23}$$

Here $P^{(1)}(\boldsymbol{X}')$ is the probability density of finding a *specific* particle at a given configuration $\boldsymbol{X}'$.

Again, in a homogeneous and isotropic fluid, we expect that

$$\rho^{(1)}(\boldsymbol{X}') = \text{const.} \tag{2.24}$$

Using the normalization condition

$$\int \rho^{(1)}(\boldsymbol{X}')d\boldsymbol{X}' = N \int P^{(1)}(\boldsymbol{X}')d\boldsymbol{X}' = N \tag{2.25}$$

we get

$$\rho^{(1)}(\boldsymbol{X}') = \frac{N}{V8\pi^2} = \frac{\rho}{8\pi^2}. \tag{2.26}$$

The connection between $\rho^{(1)}(\boldsymbol{R}')$ and $\rho^{(1)}(\boldsymbol{X}')$ is obtained simply by integration over all the orientations:

$$\rho^{(1)}(\boldsymbol{R}') = \int \rho^{(1)}(\boldsymbol{X}')\, d\Omega' = \rho. \tag{2.27}$$

2.2 The pair distribution function

In this section, we introduce the pair distribution function. We first present its meaning as a *probability density* and then show how it can be reinterpreted as an *average* quantity. Again, the starting point is the basic probability density $P(\boldsymbol{X}^N)$, (2.1), in the T, V, N ensemble. The *specific* pair distribution function is defined as the probability density of finding particle 1 at $\boldsymbol{X}'$ and particle 2 at $\boldsymbol{X}''$. This can be obtained from $P(\boldsymbol{X}^N)$ by integrating over all the configurations of the remaining $N-2$ molecules†:

$$P^{(2)}(\boldsymbol{X}', \boldsymbol{X}'') = \int \cdots \int d\boldsymbol{X}_3 \ldots d\boldsymbol{X}_N P(\boldsymbol{X}', \boldsymbol{X}'', \boldsymbol{X}_3, \ldots, \boldsymbol{X}_N). \tag{2.28}$$

Clearly, $P^{(2)}(\boldsymbol{X}', \boldsymbol{X}'')\, d\boldsymbol{X}'\, d\boldsymbol{X}''$ is the probability of finding a specific particle, say 1, in $d\boldsymbol{X}'$ at $\boldsymbol{X}'$ and *another* specific particle, say 2, in $d\boldsymbol{X}''$ at $\boldsymbol{X}''$. The same probability applies for any specific pair of two different particles.

As in the case of the singlet MDF, here we also start with the *specific* pair distribution function defined in (2.28). To get the generic pair distribution function, consider the list of events and their corresponding probabilities in table 2.2. Note that the probabilities of all the events on the left-hand column of table 2.2 are equal.

† We use primed vectors like $\boldsymbol{X}'$ and $\boldsymbol{X}'', \ldots$ to distinguish them from the vectors $\boldsymbol{X}_3$, $\boldsymbol{X}_4, \ldots$ whenever each of the two sets of vectors has a different "status." For instance, in (2.28) the primed vectors are fixed in the integrand. Such a distinction is not essential, although it may help to avoid confusion.

Table 2.2

Event	Probability
Particle 1 in $d\boldsymbol{X}'$ and particle 2 in $d\boldsymbol{X}''$	$P^{(2)}(\boldsymbol{X}', \boldsymbol{X}'')d\boldsymbol{X}'d\boldsymbol{X}''$
Particle 1 in $d\boldsymbol{X}'$ and particle 3 in $d\boldsymbol{X}''$	$P^{(2)}(\boldsymbol{X}', \boldsymbol{X}'')d\boldsymbol{X}'d\boldsymbol{X}''$
⋮	
Particle 1 in $d\boldsymbol{X}'$ and particle N in $d\boldsymbol{X}''$	$P^{(2)}(\boldsymbol{X}', \boldsymbol{X}'')d\boldsymbol{X}'d\boldsymbol{X}''$
Particle 2 in $d\boldsymbol{X}'$ and particle 1 in $d\boldsymbol{X}''$	$P^{(2)}(\boldsymbol{X}',\boldsymbol{X}'')\, d\boldsymbol{X}'\, d\boldsymbol{X}''$
⋮	
Particle N in $d\boldsymbol{X}'$ and particle $N-1$ in $d\boldsymbol{X}''$	$P^{(2)}(\boldsymbol{X}', \boldsymbol{X}'')\, d\boldsymbol{X}'\, d\boldsymbol{X}''$

The event:

$$\{\text{any particle in } d\boldsymbol{X}' \text{ and any other particle in } d\boldsymbol{X}''\} \tag{2.29}$$

is clearly the union of all the $N(N-1)$ events listed in table 2.2. However, the probability of the event (2.29) is the sum of all probabilities of the events on the left-hand column of table 2.2 *only* if the latter are disjoint. This condition can be realized when the elements of volume $d\boldsymbol{R}'$ and $d\boldsymbol{R}''$ (contained in $d\boldsymbol{X}'$ and $d\boldsymbol{X}''$, respectively) are small enough so that no more than one of the events in table 2.2 may occur at any given time.

We now define the *generic* pair distribution function as

$$\begin{aligned}
\rho^{(2)}(\boldsymbol{X}',\boldsymbol{X}'')d\boldsymbol{X}'d\boldsymbol{X}'' &= \Pr\{\text{a particle in } d\boldsymbol{X}' \text{ and a different particle in } d\boldsymbol{X}''\} \\
&= \sum_{i\neq j} \Pr\{\text{particle } i \text{ in } d\boldsymbol{X}' \text{ and another particle } j \text{ in } d\boldsymbol{X}''\} \\
&= \sum_{i\neq j} P^{(2)}(\boldsymbol{X}',\boldsymbol{X}'')d\boldsymbol{X}'d\boldsymbol{X}'' \\
&= N(N-1)P^{(2)}(\boldsymbol{X}',\boldsymbol{X}'')d\boldsymbol{X}'d\boldsymbol{X}''.
\end{aligned} \tag{2.30}$$

The last equality in (2.30) follows from the equivalence of all the $N(N-1)$ pairs of specific and different particles. Using the definition of $P^{(2)}(\boldsymbol{X}', \boldsymbol{X}'')$ in (2.28), we can transform the definition of $\rho^{(2)}(\boldsymbol{X}', \boldsymbol{X}'')$ into an expression which may be interpreted as an average quantity:

$$\begin{aligned}
&\rho^{(2)}(\boldsymbol{X}',\boldsymbol{X}'')d\boldsymbol{X}'d\boldsymbol{X}'' \\
&\quad= N(N-1)d\boldsymbol{X}'d\boldsymbol{X}'' \int\cdots\int d\boldsymbol{X}_3\ldots d\boldsymbol{X}_N P(\boldsymbol{X}',\boldsymbol{X}'',\boldsymbol{X}_3,\ldots,\boldsymbol{X}_N) \\
&\quad= N(N-1)d\boldsymbol{X}'d\boldsymbol{X}'' \int\cdots\int d\boldsymbol{X}_1\ldots d\boldsymbol{X}_N P(\boldsymbol{X}_1,\ldots,\boldsymbol{X}_N)\delta(\boldsymbol{X}_1-\boldsymbol{X}')\delta(\boldsymbol{X}_2-\boldsymbol{X}''.)
\end{aligned}$$

$$= d\boldsymbol{X}' d\boldsymbol{X}'' \int \cdots \int d\boldsymbol{X}^N P(\boldsymbol{X}^N) \sum_{\substack{i=1 \\ i \neq j}}^{N} \sum_{j=1}^{N} \delta(\boldsymbol{X}_i - \boldsymbol{X}') \delta(\boldsymbol{X}_j - \boldsymbol{X}''). \qquad (2.31)$$

In the second form of the rhs of (2.31), we employ the basic property of the Dirac delta function, so that integration is now extended over all the vectors $\boldsymbol{X}_1, \ldots, \boldsymbol{X}_N$. In the third form we have used the equivalence of the N particles, as we have done in (2.30), to get an average of the quantity

$$d\boldsymbol{X}' d\boldsymbol{X}'' \sum_{\substack{i=1 \\ i \neq j}}^{N} \sum_{j=1}^{N} \delta(\boldsymbol{X}_i - \boldsymbol{X}') \, \delta(\boldsymbol{X}_j - \boldsymbol{X}''). \qquad (2.32)$$

This can be viewed as a *counting function*, i.e., for any specific configuration $\boldsymbol{X}^N$, this quantity counts the number of *pairs* of particles occupying the elements $d\boldsymbol{X}'$ and $d\boldsymbol{X}''$. Hence, the integral (2.31) is the *average number* of pairs occupying $d\boldsymbol{X}'$ and $d\boldsymbol{X}''$. The normalization of $\rho^{(2)}(\boldsymbol{X}', \boldsymbol{X}'')$ follows directly from (2.31):

$$\int \int d\boldsymbol{X}' d\boldsymbol{X}'' \rho^{(2)}(\boldsymbol{X}', \boldsymbol{X}'') = N(N-1) \qquad (2.33)$$

which is the *exact* number of pairs in V. As in the previous section, we note that the meaning of $\rho^{(2)}(\boldsymbol{X}', \boldsymbol{X}'')$ as an average quantity is preserved upon integration over any region S. This is not the case, however, when its probabilistic meaning is adopted.

For instance, the quantity

$$\int_S \int_S d\boldsymbol{X}' d\boldsymbol{X}'' \rho^{(2)}(\boldsymbol{X}', \boldsymbol{X}'') \qquad (2.34)$$

is the *average number* of pairs occupying the region S. This quantity is, in general, not a probability.

It is also useful to introduce the locational (or spatial) pair distribution function, defined by

$$\rho^{(2)}(\boldsymbol{R}', \boldsymbol{R}'') = \int \int d\boldsymbol{\Omega}' d\boldsymbol{\Omega}'' \rho^{(2)}(\boldsymbol{X}', \boldsymbol{X}''), \qquad (2.35)$$

where integration is carried out over the orientations of the two particles. Here, $\rho^{(2)}(\boldsymbol{R}', \boldsymbol{R}'')\, d\boldsymbol{R}'\, d\boldsymbol{R}''$ is the average number of pairs occupying $d\boldsymbol{R}'$ and $d\boldsymbol{R}''$ or, alternatively, for infinitesimal elements $d\boldsymbol{R}'$ and $d\boldsymbol{R}''$, the probability of finding one particle in $d\boldsymbol{R}'$ at $\boldsymbol{R}'$ and a second particle in $d\boldsymbol{R}''$ at $\boldsymbol{R}''$. It is sometimes convenient to denote the quantity defined in (2.35) by $\overline{\rho}^{(2)}(\boldsymbol{R}', \boldsymbol{R}'')$, to distinguish it from the different function $\rho^{(2)}(\boldsymbol{X}', \boldsymbol{X}'')$. However, since we specify the arguments of the functions there should be no reason for confusion as to this notation.

2.3 The pair correlation function

We now introduce the most important and most useful function in the theory of liquids: the *pair correlation* function. Consider the two elements of volume $d\boldsymbol{X}'$ and $d\boldsymbol{X}''$ and the *intersection* of the two events:

$$\{\text{a particle in } d\boldsymbol{X}'\} \text{ and } \{\text{a particle in } d\boldsymbol{X}''\}. \tag{2.36}$$

The combined event written in (2.36) means that the first *and* the second events occur i.e., this is the intersection of the two events.

Two events are called independent whenever the probability of their intersection is equal to the product of the probabilities of the two events. In general, the two separate events given in (2.36) are not independent; the occurrence of one of them may influence the likelihood, or the probability, of the occurrence of the other. For instance, if the separation $R = |\boldsymbol{R}'' - \boldsymbol{R}'|$ between the two elements is very small (compared to the molecular diameter of the particles), then the occurrence of one event strongly affects the chances of the occurrence of the second.

In a fluid, we expect that if the separation R between two particles is very large, then the two events in (2.36) become independent. Therefore, we can write for the probability of their intersection

$$\begin{aligned}\rho^{(2)}(\boldsymbol{X}', \boldsymbol{X}'')d\boldsymbol{X}'d\boldsymbol{X}'' &= \Pr\{\text{a particle in } d\boldsymbol{X}'\} \text{ and } \{\text{a particle in } d\boldsymbol{X}''\}\\ &= \Pr\{\text{a particle in } d\boldsymbol{X}'\} \times \Pr\{\text{a particle in } d\boldsymbol{X}''\}\\ &= \rho^{(1)}(\boldsymbol{X}')d\boldsymbol{X}'\ \rho^{(1)}(\boldsymbol{X}'')d\boldsymbol{X}'', \quad \text{for } R \to \infty,\end{aligned} \tag{2.37}$$

or in short,

$$\rho^{(2)}(\boldsymbol{X}', \boldsymbol{X}'') = \rho^{(1)}(\boldsymbol{X}')\rho^{(1)}(\boldsymbol{X}'') = (\rho/8\pi^2)^2, \quad R \to \infty. \tag{2.38}$$

The last equality is valid for a homogeneous and isotropic fluid. If (2.38) holds, it is often said that the local densities at $\boldsymbol{X}'$ and $\boldsymbol{X}''$ are uncorrelated. (The limit $R \to \infty$ should be understood as large enough compared with the molecular diameter, but still within the boundaries of the system.)

For any finite distance R, factoring of $\rho^{(2)}(\boldsymbol{X}', \boldsymbol{X}'')$ into a product is, in general, not valid. We now introduce the pair correlation function $g(\boldsymbol{X}', \boldsymbol{X}'')$

which measures the extent of deviation from (2.38) and is defined by[†]

$$\begin{aligned}\rho^{(2)}(\boldsymbol{X}',\boldsymbol{X}'') &= \rho^{(1)}(\boldsymbol{X}')\rho^{(1)}(\boldsymbol{X}'')g(\boldsymbol{X}',\boldsymbol{X}'')\\ &= (\rho/8\pi^2)^2 g(\boldsymbol{X}',\boldsymbol{X}''). \end{aligned} \tag{2.39}$$

The second equality holds for a homogeneous and isotropic fluid. A related quantity is the locational pair correlation function, defined in terms of the locational pair distribution function, i.e.,

$$\rho^{(2)}(\boldsymbol{R}',\boldsymbol{R}'') = \rho^2 g(\boldsymbol{R}',\boldsymbol{R}''). \tag{2.40}$$

The relation between $g(\boldsymbol{R}',\boldsymbol{R}'')$ and $g(\boldsymbol{X}',\boldsymbol{X}'')$ follows from (2.35), (2.39) and (2.40):

$$g(\boldsymbol{R}',\boldsymbol{R}'') = \frac{1}{(8\pi^2)^2}\int\int d\boldsymbol{\Omega}'\, d\boldsymbol{\Omega}'' g(\boldsymbol{X}',\boldsymbol{X}''), \tag{2.41}$$

which can be viewed as the average of $g(\boldsymbol{X}', \boldsymbol{X}'')$ over all the orientations of the two particles. Note that this average is taken with the probability distribution $d\boldsymbol{\Omega}' d\boldsymbol{\Omega}''/(8\pi^2)^2$. This is the probability of finding one particle in orientation $d\boldsymbol{\Omega}'$ and a second particle in $d\boldsymbol{\Omega}''$ when they are at infinite separation from each other. At any finite separation, the probability of finding one particle in $d\boldsymbol{\Omega}'$ and the second in $d\boldsymbol{\Omega}''$ given the locations of $\boldsymbol{R}'$ and $\boldsymbol{R}''$ is

$$\begin{aligned}\Pr(\boldsymbol{\Omega}',\boldsymbol{\Omega}''/\ \boldsymbol{R}',\boldsymbol{R}'')\ d\boldsymbol{\Omega}'\ d\boldsymbol{\Omega}'' &= \frac{\rho^{(2)}(\boldsymbol{X}',\boldsymbol{X}'')d\boldsymbol{\Omega}'\ d\boldsymbol{\Omega}''}{\int \rho^{(2)}(\boldsymbol{X}',\boldsymbol{X}'')d\boldsymbol{\Omega}'\ d\boldsymbol{\Omega}''}\\ &= \frac{g(\boldsymbol{X}',\boldsymbol{X}'')d\boldsymbol{\Omega}'\ d\boldsymbol{\Omega}''}{(8\pi^2)^2 g(\boldsymbol{R}',\boldsymbol{R}'')}. \end{aligned} \tag{2.42}$$

It is only for $|\boldsymbol{R}'' - \boldsymbol{R}'| \to \infty$ that this probability distribution becomes $d\boldsymbol{\Omega}'$ $d\boldsymbol{\Omega}''/(8\pi^2)^2$.

In this book, we shall only be interested in *homogeneous* and *isotropic* fluids. In such a case, there is a redundancy in specifying the full configuration of the pair of particles by 12 coordinates $(\boldsymbol{X}', \boldsymbol{X}'')$. It is clear that for any configuration of the pair $\boldsymbol{X}', \boldsymbol{X}''$, the correlation $g(\boldsymbol{X}', \boldsymbol{X}'')$ is invariant to translation and rotation of the pair as a unit, keeping the relative configuration of one particle toward the other fixed. Therefore, we can reduce to six the number of independent variables necessary for the full description of the pair correlation function. For instance, we may choose the location of one particle at the origin of the coordinate system, $\boldsymbol{R}' = \boldsymbol{0}$, and fix its orientation, say, at $\phi' = \theta' = \psi' = 0$. Hence, the pair correlation function is a function only of the six variables $\boldsymbol{X}'' = \boldsymbol{R}''$, $\boldsymbol{\Omega}''$.

[†] The correlation function as defined here differs from the correlation defined in probability theory. In probability theory, it is defined as the *difference* between the probability of the intersection of the two events and the product of the probabilities of each of the events. It is also normalized in such a way that its range of variation is between -1 and $+1$.

Similarly, the function $g(\boldsymbol{R}', \boldsymbol{R}'')$ is a function only of the scalar distance $R = |\boldsymbol{R}'' - \boldsymbol{R}'|$. For instance, $\boldsymbol{R}'$ may be chosen at the origin $\boldsymbol{R}' = \boldsymbol{0}$, and because of the isotropy of the fluid, the relative orientation of the second particle is of no importance. Therefore, only the separation R is left as the independent variable. The function $g(R)$, i.e., the pair correlation function expressed explicitly as a function of the distance R, is often referred to as the *radial distribution* function. This function plays a central role in the theory of fluids.

The generalization to multicomponent systems is quite straightforward. Instead of one pair correlation $g(\boldsymbol{X}', \boldsymbol{X}'')$, we shall have pair correlation functions for each pair of species $\alpha\beta$. For instance, if A and B are spherical particles, then we have three different pair correlation functions $g_{AA}(R)$, $g_{AB}(R) = g_{BA}(R)$ and $g_{BB}(R)$. We shall describe these in more detail in section 2.9.

2.4 Conditional probability and conditional density

We now turn to a somewhat different interpretation of the pair distribution function. We define the *conditional probability* of observing a particle in $d\boldsymbol{X}''$ at $\boldsymbol{X}''$, given a particle at $\boldsymbol{X}'$, by

$$\begin{aligned} \rho(\boldsymbol{X}''/\boldsymbol{X}')\ d\boldsymbol{X}'' &= \frac{\rho^{(2)}(\boldsymbol{X}', \boldsymbol{X}'')\ d\boldsymbol{X}'\ d\boldsymbol{X}''}{\rho^{(1)}(\boldsymbol{X}')\ d\boldsymbol{X}'} \\ &= \rho^{(1)}(\boldsymbol{X}'')g(\boldsymbol{X}', \boldsymbol{X}'')\ d\boldsymbol{X}'' \end{aligned} \tag{2.43}$$

The last equality follows from the definition of $g(\boldsymbol{X}', \boldsymbol{X}'')$ in (2.39). Note that the probability of finding a particle at an *exact* configuration $\boldsymbol{X}''$ is zero, which is the reason for taking an infinitesimal element of volume at $\boldsymbol{X}''$. On the other hand, the *conditional probability* may be defined for an *exact condition*: "given a particle at $\boldsymbol{X}'$." This may be seen formally from (2.43), where $d\boldsymbol{X}'$ cancels out once we form the ratio of the two distribution functions. Hence, one can actually take the limit $d\boldsymbol{X}' \to 0$ in the definition of the conditional probability. What remains is a conditional probability of finding a particle at $\boldsymbol{X}''$, given a particle at exact configuration $\boldsymbol{X}'$.

We recall that the quantity $\rho^{(1)}(\boldsymbol{X}'')\ d\boldsymbol{X}''$ is the *local density* of particles at $\boldsymbol{X}''$. We now show that the quantity defined in (2.43) is the *conditional local density* at $\boldsymbol{X}''$, given a particle at $\boldsymbol{X}'$. In other words, we place a particle at $\boldsymbol{X}'$ and view the rest of the $N-1$ particles as a system subjected to the field of force produced by the particle at $\boldsymbol{X}'$. Clearly, the new system is no longer homogeneous, nor isotropic. Therefore, the local density may be different at each point of the system. To show this we first define the *binding energy*, B_1, of one particle, say 1, to the rest of the system by

$$\begin{aligned} U_N(\boldsymbol{X}_1, \ldots, \boldsymbol{X}_N) &= U_{N-1}(\boldsymbol{X}_2, \ldots, \boldsymbol{X}_N) + \sum_{j=2}^{N} U_{1,j}(\boldsymbol{X}_1, \boldsymbol{X}_j) \\ &= U_{N-1} + B_1. \end{aligned} \tag{2.44}$$

In (2.44), we have split the total potential energy of the system of N particles into two parts: the potential energy of the interaction among the $N-1$ particles and the interaction of one particle, chosen as particle 1, with the $N-1$ particles. Once we fix the configuration of particle 1 at $\boldsymbol{X}_1$, the rest of the system can be viewed as a system in an "external" field defined by B_1.

From the definitions (2.1), (2.23), (2.31) and (2.43), we get

$$\begin{aligned} \rho(\boldsymbol{X}''/\boldsymbol{X}') &= \frac{N(N-1)\int\cdots\int d\boldsymbol{X}^N \exp[-\beta U_N(\boldsymbol{X}^N)]\delta(\boldsymbol{X}_1-\boldsymbol{X}')\delta(\boldsymbol{X}_2-\boldsymbol{X}'')}{N\int\cdots\int d\boldsymbol{X}^N \exp[-\beta U_N(\boldsymbol{X}^N)]\delta(\boldsymbol{X}_1-\boldsymbol{X}')} \\ &= \frac{(N-1)\int\cdots\int d\boldsymbol{X}_2\ldots d\boldsymbol{X}_N \exp[-\beta U_N(\boldsymbol{X}',\boldsymbol{X}_2,\ldots,\boldsymbol{X}_N)]\delta(\boldsymbol{X}_2-\boldsymbol{X}'')}{\int\cdots\int d\boldsymbol{X}_2\ldots d\boldsymbol{X}_N \exp[-\beta U_N(\boldsymbol{X}',\boldsymbol{X}_2,\ldots,\boldsymbol{X}_N)]} \\ &= (N-1)\int\cdots\int d\boldsymbol{X}_2\ldots d\boldsymbol{X}_N\, P^*(\boldsymbol{X}',\boldsymbol{X}_2,\ldots,\boldsymbol{X}_N)\delta(\boldsymbol{X}_2-\boldsymbol{X}'') \end{aligned} \tag{2.45}$$

where $P^*(\boldsymbol{X}', \boldsymbol{X}_2, \ldots, \boldsymbol{X}_N)$ is the basic probability density of a system of $N-1$ particles in an "external" field produced by a particle fixed at $\boldsymbol{X}'$, i.e.,

$$P^*(\boldsymbol{X}',\boldsymbol{X}_2,\ldots,\boldsymbol{X}_N) = \frac{\exp(-\beta U_{N-1} - \beta B_1)}{\int\cdots\int d\boldsymbol{X}_2\ldots d\boldsymbol{X}_N \exp(-\beta U_{N-1} - \beta B_1)}. \tag{2.46}$$

We now observe that relation (2.45) has the same structure as relation (2.23) but with two differences. First, (2.45) refers to a system of $N-1$ instead of N particles. Second, the system of $N-1$ particles is in an "external" field. Hence, (2.45) is interpreted as the *local density at* $\boldsymbol{X}''$ of a system of $N-1$ particles placed in the external field B_1. This is an example of a conditional singlet molecular distribution function which is not constant everywhere.

Similarly, for the *locational pair* correlation function, we have the relation

$$\rho(\boldsymbol{R}''/\boldsymbol{R}') = \rho g(\boldsymbol{R}', \boldsymbol{R}''), \tag{2.47}$$

where $\rho(\boldsymbol{R}''/\boldsymbol{R}')$ is the conditional average density at $\boldsymbol{R}''$ given a particle at $\boldsymbol{R}'$. In the last relation, the pair correlation function measures the deviation of the local density at $\boldsymbol{R}''$, given a particle at $\boldsymbol{R}'$ from the bulk density ρ. In Appendix F we present another expression for the correlation function in terms of local fluctuation in the density. Note again that in a multicomponent system, we have several different conditional densities, e.g., the conditional density of A at a distance R from an A-particle, the conditional density of A at a distance R from a B-particle, etc.

2.5 Some general features of the radial distribution function

In this section, we illustrate the general features of the radial distribution function (RDF), $g(R)$, for a system of simple spherical particles. From the definitions (2.31) and (2.39) (applied to spherical particles), we get

$$g(\mathbf{R}', \mathbf{R}'') = \frac{N(N-1)}{\rho^2} \frac{\int \cdots \int d\mathbf{R}_3 \ldots d\mathbf{R}_N \exp[-\beta U_N(\mathbf{R}', \mathbf{R}'', \mathbf{R}_3, \ldots, \mathbf{R}_N)]}{\int \cdots \int d\mathbf{R}_1 \ldots d\mathbf{R}_N \exp[-\beta U_N(\mathbf{R}_1, \ldots, \mathbf{R}_N)]}. \tag{2.48}$$

A useful expression, which we shall need only for demonstrative purposes, is the density expansion of $g(\mathbf{R}', \mathbf{R}'')$, which reads†

$$g(\mathbf{R}', \mathbf{R}'') = \exp[-\beta U(\mathbf{R}', \mathbf{R}'')]\{1 + B(\mathbf{R}', \mathbf{R}'')\rho + C(\mathbf{R}', \mathbf{R}'')\rho^2 + \cdots\} \tag{2.49}$$

where the coefficients $B(\mathbf{R}', \mathbf{R}'')$, $C(\mathbf{R}', \mathbf{R}'')$, etc., are given in terms of integrals over the so-called Mayer f-function, defined by

$$f(\mathbf{R}', \mathbf{R}'') = \exp[-\beta U(\mathbf{R}', \mathbf{R}'')] - 1. \tag{2.50}$$

For instance, the expression for $B(\mathbf{R}', \mathbf{R}'')$ is

$$B(\mathbf{R}', \mathbf{R}'') = \int_V f(\mathbf{R}', \mathbf{R}_3) f(\mathbf{R}'', \mathbf{R}_3)\, d\mathbf{R}_3. \tag{2.51}$$

We now turn to some specific cases.

2.5.1 Theoretical ideal gas

A *theoretical* ideal gas is defined as a system of strictly noninteracting particles. The RDF for such a system can be obtained directly from definition (2.48). With $U_N \equiv 0$ for *all configurations*, the integrations in (2.48) become trivial and we get

$$g(\mathbf{R}', \mathbf{R}'') = \frac{N(N-1)}{\rho^2} \frac{\int \cdots \int d\mathbf{R}_3 \ldots d\mathbf{R}_N}{\int \cdots \int d\mathbf{R}_1 \ldots d\mathbf{R}_N} = \frac{N(N-1) V^{N-2}}{\rho^2 V^N} \tag{2.52}$$

† See, for example, Hill (1956). We shall not need this expansion in ρ of the pair correlation function. However, it should be noted that this expansion is derived in an open system, i.e., using the grand partition function. In a closed system, we always have an additional term of the order of N^{-1}. See Appendix G for details.

or equivalently

$$g(R) = 1 - \frac{1}{N}. \tag{2.53}$$

As is expected, $g(R)$ is *practically* unity for any value of R. This behavior reflects the basic property of an ideal gas i.e., the absence of correlation follows from the absence of interaction. The term N^{-1} is typical of a closed system[†]. At the thermodynamic limit $N \to \infty$, $V \to \infty$, $N/V = \text{const.}$, this term, for most purposes, may be dropped. Of course, in order to get the correct normalization of $g(R)$, one should use the exact relation (2.53), i.e.,

$$\rho \int_V g(\mathbf{R}', \mathbf{R}'') d\mathbf{R}'' = \rho \int_V [1 - (1/N)] d\mathbf{R}'' = N - 1, \tag{2.54}$$

which is exactly the total number of particles in the system, excluding the one fixed at $\mathbf{R}'$.

It should be clear that the pair correlation function has, in general, two contributions. One is due to interaction, which in this case is unity. The second arises from the *closure* condition with respect to N. Placing a particle at a fixed position changes the conditional density of particles everywhere in the system from N/V into $(N-1)/V$. Hence, the pair correlation due to this effect is

$$g(R) = \frac{(N-1)/V}{N/V} = 1 - \frac{1}{N}.$$

More on this aspect of the pair correlation can be found in Appendix G.

2.5.2 **Very dilute gas**

For any *real* gas at very low densities, $\rho \to 0$, we may neglect all powers of ρ in the density expansion of $g(R)$, in which case we get, from (2.49)[‡]

$$g(R) = \exp[-\beta U(R)], \quad \rho \to 0, \tag{2.55}$$

where $U(R)$ is the pair potential operating between two particles. Relation (2.55) is essentially the Boltzmann distribution law. Since at low densities encounters in which more than two particles are involved are very rare, the pair distribution function is determined solely by the pair potential.

A direct way of obtaining (2.55) from the definition (2.48) (and not through the density expansion) is to consider the case of a system containing only two particles.

[†] In an open system, $g(R)$ is everywhere unity for a theoretical ideal gas. For more details, see Appendix G.

[‡] Again, we note that since (2.49) is derived for the open system, also (2.55) is valid for an open system.

Letting $N=2$ in (2.48), we get

$$g(R) = \frac{2}{\rho^2}\frac{\exp[-\beta U(R)]}{Z_2}, \tag{2.56}$$

where Z_2 is the configurational partition function for a system of two particles in V.

Since we choose $U(R) \to 0$ as $R \to \infty$, we can use (2.56) to form the ratio

$$\frac{g(R)}{g(\infty)} = \exp[-\beta U(R)]. \tag{2.57}$$

Assuming that as $R \to \infty$, $g(\infty)$ is practical unity[†], we get from (2.57)

$$g(R) = \exp[-\beta U(R)], \tag{2.58}$$

which is the same as (2.55). Note that (2.55) and (2.58) have been obtained for two apparently different systems ($\rho \to 0$ on one hand and $N=2$ on the other). The identical results for $g(R)$ in two cases reflects the fact that at very low densities, only interactions between pairs determine the behavior of $g(R)$.

The form of $g(R)$ as $\rho \to 0$ for a system of hard spheres (HS) and Lennard-Jones (LJ) particles is depicted in figure 2.2. It is seen that for HS particles as $\rho \to 0$, correlation exists only for $R < \sigma$. For $R > \sigma$, the function $g(R)$ is identically unity. For LJ particles, we observe a single peak in $g(R)$ at the same point for which $U(R)$ has a minimum, namely at $R = 2^{1/6}\,\sigma$. There are two features of the behavior of $g(R)$ which are common to any gas. First, at large distance $R \to \infty$, $g(R) \to 1$; this is normally attained for R on the order of a few molecular diameters. Second, for $R < \sigma$, $g(R) \to 0$, where σ is a length referred to as the molecular diameter of the particles. For LJ particles $U(R = \sigma) = 0$.

It should be noted that for any gas with any intermolecular interactions, when $\rho \to 0$, we obtain the ideal-gas behavior. For instance, the equation of state has the typical and well-known form. One should distinguish between the ideal-gas behavior of a *real gas* as $\rho \to 0$, and a theoretical ideal gas which is a model system, where no interactions exist. Such a system does not exist; however, the equation of state of such a model system is the same as the equation of state of a real system as $\rho \to 0$.

In this section, we have seen that in the limit $\rho \to 0$, the pair correlation is (2.55). This is different from the theoretical ideal gas case obtained in section (2.5.1). There, the form of $g(R)$ is valid for *any* density provided that all

[†] By $R \to \infty$ we mean here a very large distance compared with the molecular diameter of the particles, but still within the macroscopic system of volume V. The assumption that $g(\infty) = 1$ is valid for an open system. In a closed system, we have an additional N^{-1} term. This is negligible whenever we are interested in $g(R)$ itself. It becomes important when we integrate over the entire volume of the systems. See also Appendix G.

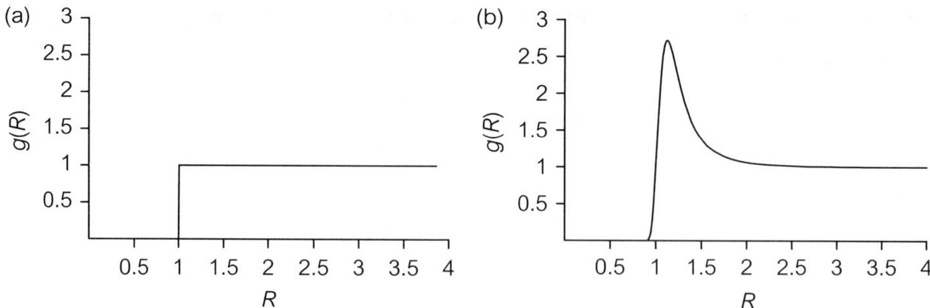

Figure 2.2. The form of the pair correlation function $g(R)$ at very low densities ($\rho \to 0$): (a) for hard spheres with $\sigma = 1$; (b) for Lennard-Jones particles with parameters $\sigma = 1$ and $\varepsilon/kT = 0.5$.

intermolecular interactions are strictly zero. In (2.55) we have the limit of a *ratio* $\rho^{(1)}(\mathbf{R}'/\mathbf{R}'')$ and $\rho^{(1)}(\mathbf{R}')$. Both of these densities tend to zero at $\rho \to 0$, but their ratio is finite at this limit.

2.5.3 **Slighty dense gas**

In the context of this section, a slightly dense gas is a gas properly described by the first-order expansion in the density, i.e., up to the linear term in (2.49). Before analyzing the content of the coefficient $B(\mathbf{R}', \mathbf{R}'')$ in the expansion of $g(\mathbf{R})$, let us demonstrate its origin by considering a system of exactly three particles. Putting $N = 3$ in (2.48), we get

$$g(\mathbf{R}', \mathbf{R}'') = \frac{6}{\rho^2} \frac{\int d\mathbf{R}_3 \exp[-\beta U(\mathbf{R}', \mathbf{R}'', \mathbf{R}_3)]}{Z_3} \tag{2.59}$$

where Z_3 is the configurational partition function for a system of three particles.

Assuming pairwise additivity of the potential energy U_3, and using the definition of the function f in (2.50), we can transform (2.59) into

$$g(\mathbf{R}', \mathbf{R}'') = \frac{6}{\rho^2} \exp[-\beta U(\mathbf{R}', \mathbf{R}'')] \times \frac{\int d\mathbf{R}_3 [f(\mathbf{R}', \mathbf{R}_3) f(\mathbf{R}_3, \mathbf{R}'') + f(\mathbf{R}', \mathbf{R}_3) + f(\mathbf{R}'', \mathbf{R}_3) + 1]}{Z_3}. \tag{2.60}$$

Noting again that $U(\mathbf{R}', \mathbf{R}'') = 0$ for $R = |\mathbf{R}'' - \mathbf{R}'| \to \infty$, we form the ratio

$$\frac{g(R)}{g(\infty)} = \exp[-\beta U(R)] \times \frac{\int d\mathbf{R}_3 f(\mathbf{R}', \mathbf{R}_3) f(\mathbf{R}_3, \mathbf{R}'') + 2 \int d\mathbf{R}_3 f(\mathbf{R}', \mathbf{R}_3) + V}{\lim_{R \to \infty} [\int d\mathbf{R}_3 f(\mathbf{R}', \mathbf{R}_3) f(\mathbf{R}_3, \mathbf{R}'') + 2 \int d\mathbf{R}_3 f(\mathbf{R}', \mathbf{R}_3) + V]}. \tag{2.61}$$

Clearly, the two integrals over $f(\boldsymbol{R}', \boldsymbol{R}_3)$ and $f(\boldsymbol{R}'', \boldsymbol{R}_3)$ are equal and independent of the separation R. Denoting by

$$C = \int_V d\boldsymbol{R}_3 \, f(\boldsymbol{R}', \boldsymbol{R}_3) = \int_V d\boldsymbol{R}_3 \, f(\boldsymbol{R}'', \boldsymbol{R}_3) \tag{2.62}$$

and noting that since $f(R)$ is a short-range function of R, the integral in (2.62) does not depend on V, for macroscopic V. On the other hand, we have the limiting behavior

$$\lim_{R\to\infty} \int d\boldsymbol{R}_3 \, f(\boldsymbol{R}', \boldsymbol{R}_3) f(\boldsymbol{R}_3, \boldsymbol{R}'') = 0, \tag{2.63}$$

which follows from the fact that two factors in the integrand contribute to the integral only if $\boldsymbol{R}_3$ is close simultaneously to both $\boldsymbol{R}'$ and $\boldsymbol{R}''$, a situation that cannot be attained if $R = |\boldsymbol{R}'' - \boldsymbol{R}'| \to \infty$.

Using (2.62), (2.63), and (2.51), we can now rewrite (2.61) as

$$\frac{g(R)}{g(\infty)} = \exp[-\beta U(R)] \frac{B(\boldsymbol{R}', \boldsymbol{R}'') + 2C + V}{2C + V}. \tag{2.64}$$

Since C is constant, it may be neglected, as compared with V, in the thermodynamic limit. Also, assuming that $g(\infty)$ is practically unity,† we get the final form of $g(R)$ for this case:

$$g(R) = \exp[-\beta U(R)][1 + (1/V)B(\boldsymbol{R}', \boldsymbol{R}'')], \quad R = |\boldsymbol{R}'' - \boldsymbol{R}'|. \tag{2.65}$$

Note that $1/V$, appearing in (2.65), replaces the density ρ in (2.49). In fact, the quantity $1/V$ may be interpreted as the density of "free particles" (i.e., the particles besides the two fixed at $\boldsymbol{R}'$, $\boldsymbol{R}''$) for the case $N = 3$.

The derivation of (2.65) illustrates the origin of the coefficient $B(\boldsymbol{R}', \boldsymbol{R}'')$, which in principle results from the simultaneous interaction of three particles [compare this result with (2.58)]. This is actually the meaning of the term "slightly dense gas." Whereas in a very dilute gas we take account of interactions between *pairs* only, here we also consider the effect of interactions among *three particles*, but not more. For hard spheres (HS), we can calculate $B(\boldsymbol{R}', \boldsymbol{R}'')$ exactly; in this case we have

$$f(R) = \begin{cases} -1 & \text{for } R < \sigma \\ 0 & \text{for } R > \sigma \end{cases} \tag{2.66}$$

Thus, the only contribution to the integral in (2.51) comes from regions in which both $f(\boldsymbol{R}', \boldsymbol{R}_3)$ and $f(\boldsymbol{R}'', \boldsymbol{R}_3)$ are equal to -1. This occurs for $R < 2\sigma$. The integrand vanishes when either $|\boldsymbol{R}' - \boldsymbol{R}_3| > \sigma$ or $|\boldsymbol{R}'' - \boldsymbol{R}_3| > \sigma$. Furthermore, for $|\boldsymbol{R}'' - \boldsymbol{R}'| < \sigma$, the exponential factor in (2.65), $\exp[-\beta U(\boldsymbol{R}', \boldsymbol{R}'')]$,

† Note again that this is strictly true for an open system. See Appendix G.

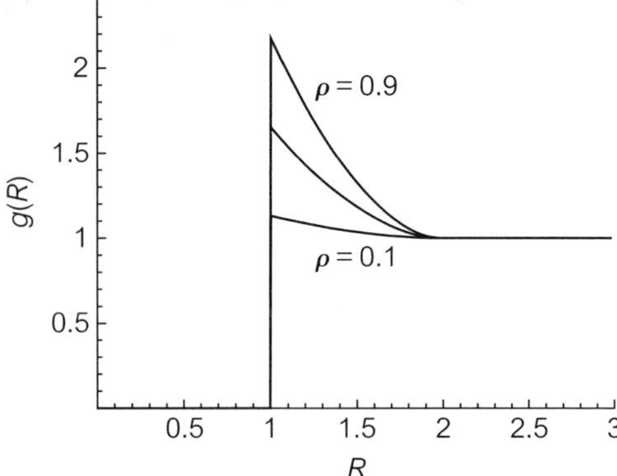

Figure 2.3. The form of $g(R)$ for hard-sphere particles ($\sigma = 1$), using the first-order expansion in the density (equation 2.68). The three curves correspond to $\rho = 0.1$ (lower), $\rho = 0.4$ (intermediate), and $\rho = 0.9$ (upper).

vanishes. Thus, the only region of interest is $\sigma \leq R \leq 2\sigma$. Since the value of the integrand in the region where it is nonzero equals $(-1) \times (-1) = 1$, the integration in (2.51) reduces to the geometric problem of computing the volume of the intersection of the two spheres of radius σ. The solution to this problem is well known.† The result is

$$B(R) = \frac{4\pi\sigma^3}{3}\left[1 - \frac{3}{4}\frac{R}{\sigma} + \frac{1}{16}\left(\frac{R}{\sigma}\right)^3\right]. \tag{2.67}$$

Using (2.67), we can now rewrite explicitly the form of the radial distribution function for hard spheres at "slightly dense" concentration:

$$g(R) = \begin{cases} 0 & \text{for } R < \sigma \\ 1 + \rho\dfrac{4\pi\sigma^3}{3}\left[1 - \dfrac{3}{4}\dfrac{R}{\sigma} + \dfrac{1}{16}\left(\dfrac{R}{\sigma}\right)^3\right] & \text{for } \sigma < R < 2\sigma \\ 1 & \text{for } R > 2\sigma. \end{cases} \tag{2.68}$$

The form of this function is depicted in figure 2.3.

2.5.4 **Lennard-Jones particles at moderately high densities**

Lennard-Jones (LJ) particles are model particles, the behavior of which resembles the behavior of real, simple spherical particles such as argon. In this section, we present some further information on the behavior of $g(R)$ and its dependence on density and on temperature. The LJ particles are defined by means of their pair potential as

$$U_{LJ}(R) = 4\varepsilon\left[\left(\frac{\sigma}{R}\right)^{12} - \left(\frac{\sigma}{R}\right)^{6}\right]. \tag{2.69}$$

Figure 2.4 demonstrates the variation of $g(R)$ as we increase the density. The dimensionless densities $\rho\sigma^3$ are recorded next to each curve. At very low

† See, for instance, Ben-Naim (1992), page 279.

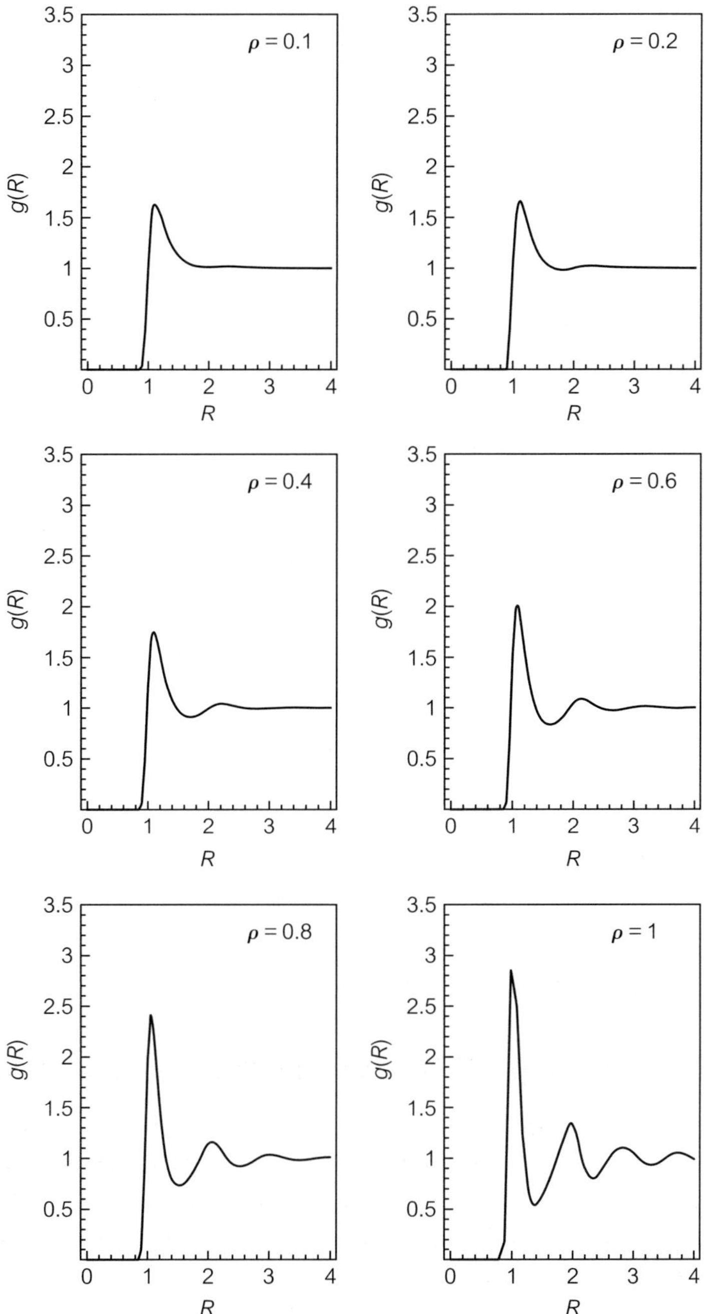

Figure 2.4. Dependence of the pair correlation function $g(R)$ for the LJ particles on the number density. The density ρ is indicated next to each curve in the dimensionless quantity $\rho\sigma^3$. We choose $\sigma = 1$ and $\varepsilon/kT = 0.5$ in the LJ potential. All the illustrations of $g(R)$ for this book were obtained by numerical solution of the Percus–Yevick equation. See Appendix E for more details.

densities, there is a single peak, corresponding to the minimum in the potential function (2.69). The minimum of $U_{LJ}(R)$ is at $R_{min} = 2^{\frac{1}{6}}\sigma$. Hence, the first maximum of $g(R)$ at low densities is also at R_{min}. At successively higher densities, new peaks develop which become more and more pronounced as the density increases. The location of the first peak is essentially unchanged, though its height increases steadily. The locations of the new peaks occur nearly at integral multiples of σ, i.e., at $R \approx \sigma, 2\sigma, 3\sigma, \ldots$ This feature reflects the propensity of the spherical molecules to pack, at least locally, in concentric and nearly equidistant spheres about a given molecule. This is a very fundamental property of simple fluids and deserves further attention.

Consider a random configuration of spherical particles in the fluid. An illustration in two dimensions is depicted in figure 2.5. Now consider a spherical shell of width $d\sigma$ and radius σ, and count the average number of particles in this element of volume. If the center of the spherical shell has been chosen at random, as on the rhs of figure 2.5, we should find that on the average, the number of particles is $\rho 4\pi\sigma^2 d\sigma$[†]. On the other hand, if we choose the center of a spherical shell so that it coincides with the center of the particle, then on the average, we find more particles in this element of volume. The drawing on the left illustrates this case for one configuration. One sees that, in this example, there are more particles in the element of volume on the left as compared with the elements of volume on the right. Similarly, we could have drawn spherical shells of width $d\sigma$ at 2σ and again have found excess particles in the element of volume, the origin of which has been chosen at the center of the particle. The excess of particles at the distances of about σ, 2σ, 3σ, etc., from the center of a particle is manifested in the various peaks of the function $g(R)$. Clearly, this effect decays rapidly as the distance from the center increases. We see from figure 2.4 that $g(R)$ is almost unity for $R \geq 4\sigma$. This means that correlation between the local densities at two points R' and R'' extends over a relatively short range, of a few molecular diameters only.

At short distances, say in the range of $\sigma \leq R \leq 5\sigma$, in spite of the random distribution of the particles, there is a sort of order as revealed by the form of the RDF. This order is often referred to as the local structure of the *liquid*. The *local* character of this structure should be noted. It contrasts with the long-range order typical of the *solid* state.

From the definition of $g(R)$, it follows that the average number of particles in a spherical shell of radius R (from the center of a given particle) and width dR is

$$N(dR) = \rho g(R)\ 4\pi R^2\ dR. \tag{2.70}$$

[†] This is for the three-dimensional case. In the two-dimensional case illustrated in the figure, the average number of particles in a ring of width $d\sigma$ is $\rho 2\pi\sigma d\sigma$.

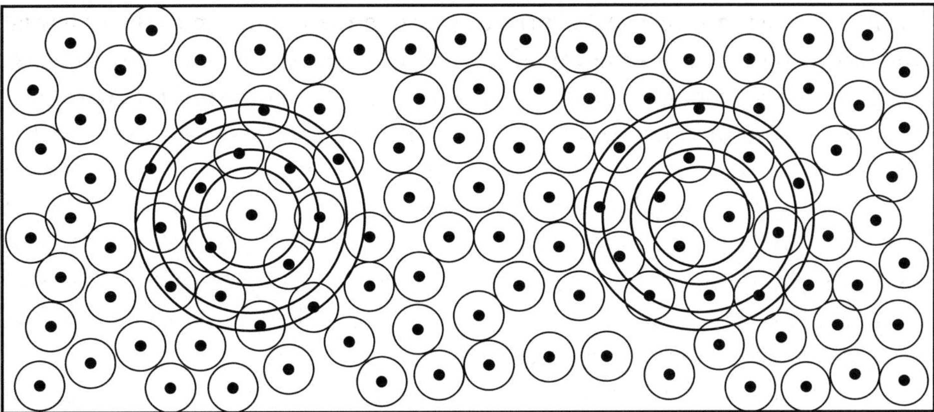

Figure 2.5. A random distribution of spheres in two dimensions. Two spherical shells of width $d\sigma$ with radius σ and 2σ are drawn (the diameter of the spheres is σ). On the left, the center of the spherical shell coincides with the center of one particle, whereas on the right, the center of the spherical shell has been chosen at a random point. It is clearly observed that two shells on the left are filled by centers of particles to a larger extent than the corresponding shells on the right. The average excess of particles in these shells, drawn from the center of a given particle, is manifested by the various peaks of $g(R)$.

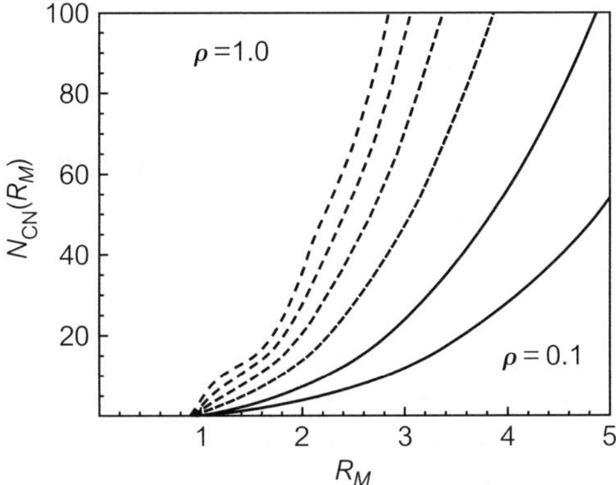

Figure 2.6. The average coordination number N_{CN} as a function of R_M (equation 2.71) for different densities $\rho\sigma^3$ and $\varepsilon/kT = 0.5$. The curves from the lowest upwards correspond to $\rho = 0.1$, 0.2, 0.4, 0.6, 0.8 and 1.

Hence, the average number of particles in a sphere of radius R_M (excluding the particle at the center) is

$$N_{CN}\ (R_M) = \rho \int_0^{R_M} g(R)\, 4\pi R^2\, dR. \tag{2.71}$$

The quantity $N_{CN}(R_M)$ may be referred to as the *coordination* number of particles, computed for the particular sphere of radius R_M. A choice of

$\sigma \leq R_M < 1.5\sigma$ will give a coordination number that conforms to the common meaning of the concept of the *first* coordination number. Figure 2.6 illustrates the dependence of the coordination number on R_M for LJ particles, for different densities $\rho\sigma^3$ (and constant $\varepsilon/kT = 0.5$). At large values of R_M the function

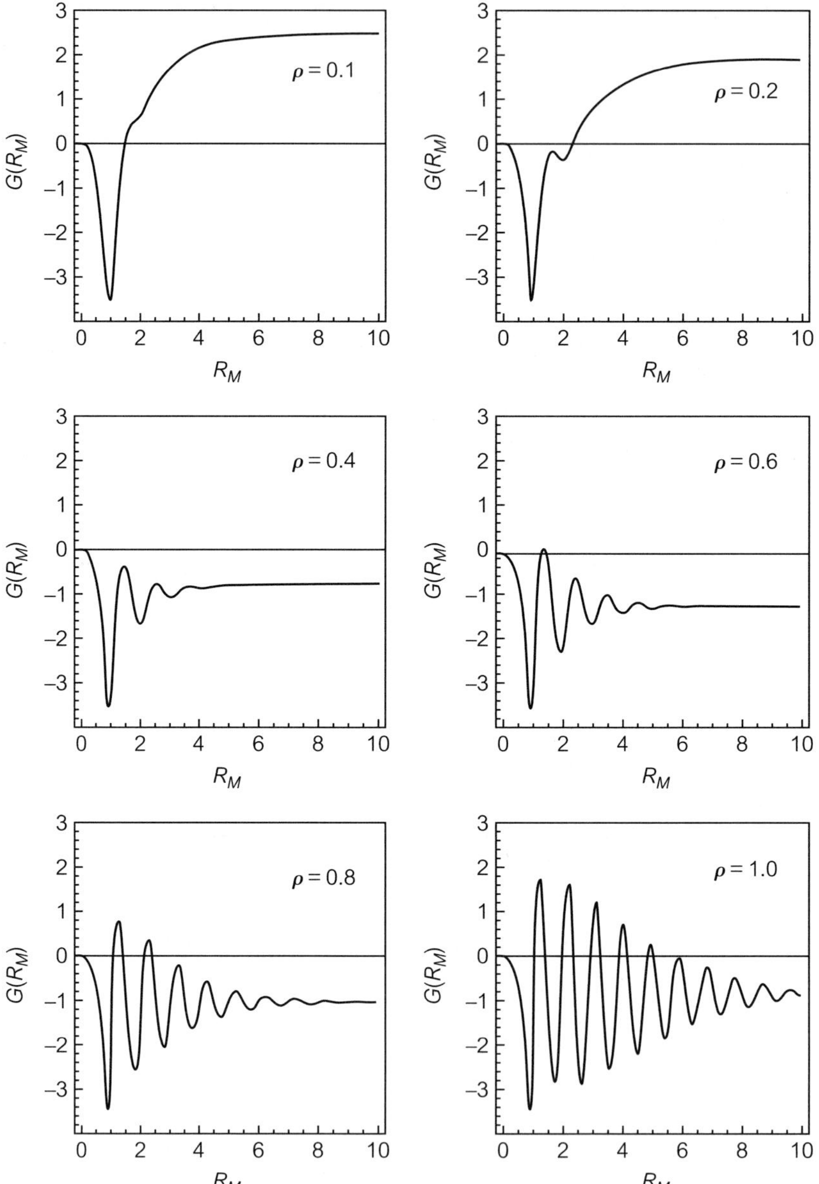

Figure 2.7. The functions $G(R_M)$ defined in (2.72) as a function of R_M for the same system and the same densities as in figure 2.4.

takes the form $\rho \frac{4}{3}\pi R^3$. Figure 2.7 shows the integrals

$$G(R_M) = \int_0^{R_M} [g(R) - 1]\, 4\pi R^2\, dR \tag{2.72}$$

at several densities, corresponding to the values in figure 2.4. The quantity $\rho G(R_M)$ is the excess[†] in the average number of particles in a spherical volume of radius R_M centered at a given particle, relative to the average number of particles in a random sphere of the same radius. Note that all curves start at zero at $R_M = 0$. At large R_M of the order of a few molecular diameters, the function $G(R_M)$ tends to a constant value.

The limit

$$G = \lim_{R_M \to \infty} G(R_M) = \int_0^{\infty} [g(R) - 1]\, 4\pi R^2\, dR \tag{2.73}$$

is the so-called Kirkwood–Buff integral. We shall encounter these integrals very frequently throughout this book.

Figure 2.8 shows one of the functions $G(R_M)$ and the corresponding pair correlation functions. Note that the maxima and minima of the function

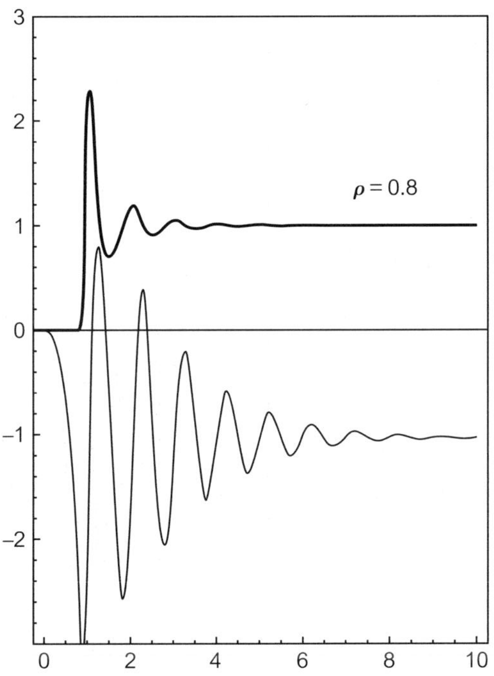

Figure 2.8. The combined curves of $g(R)$, upper curve, and $G(R_M)$, lower curve, for the case of $\rho = 0.8$ and $\varepsilon/kT = 0.5$.

[†] A negative excess is considered a deficiency.

$G(R_M)$ correspond to points at which $g(R) = 1$. Note also that the oscillations in $G(R_M)$ are quite pronounced even at distances where $g(R)$ looks almost flat on the scale of this figure.

In figure 2.9 we show the variation of $g(R)$ with ε/kT for a given density $\rho = 0.8$. The values of ε in units of kT are indicated near each curve. Clearly, one

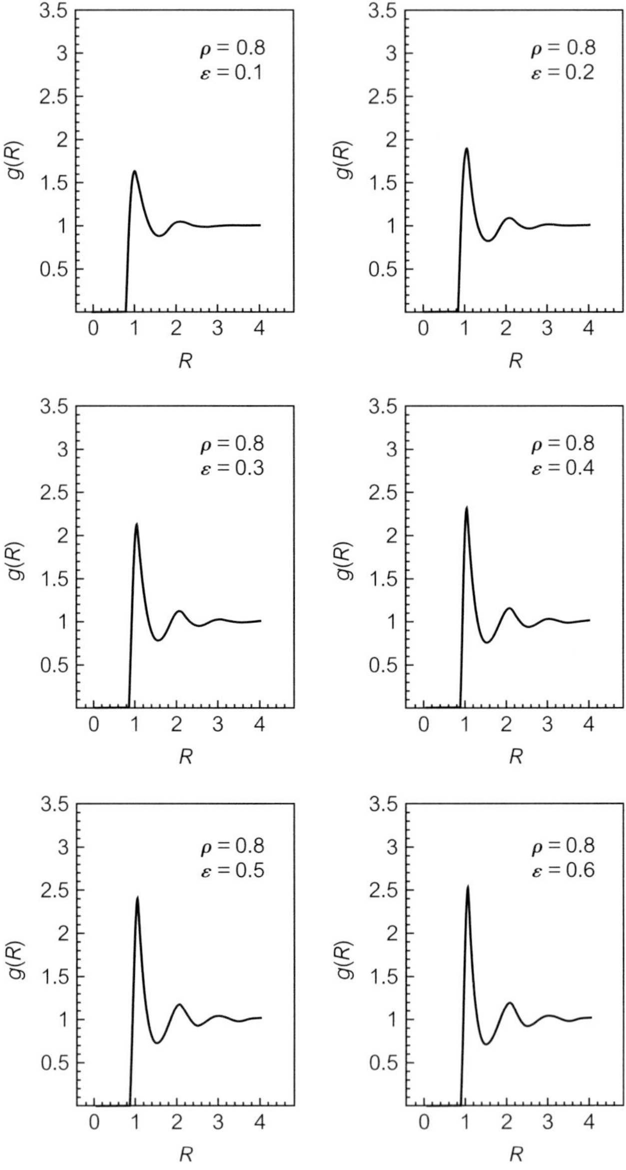

Figure 2.9. Variation of $g(R)$ with ε (in units of kT) for a specific density $\rho = 0.8$.

can interpret the variation of $g(R)$ either as a result of changing ε (in units of kT), or as changing the temperature T (in units of ε/k).

Finally we show two illustrations of $g(R)$ for real liquids, first, figure 2.10 for liquid argon (drawn as a function of the reduced distance $R^* = R/3.5$). Clearly the general behavior is similar to the LJ fluid. It is also shown in the figure that the theoretical curve, obtained from the solution of the Percus–Yevick equation, is almost indistinguishable from the experiment curve.

The second, figure 2.11, shows $g(R)$ for H_2O and D_2O at 4 °C. Note that the two curves are almost indistinguishable on the scale of the figure. In water we see a second peak of $g(R)$ at 4.5 Å, which indicates a high degree of "structure" in this liquid. For a normal, spherical particles of diameter 2.8 Å we would have expected a second peak at about 5.6 Å.

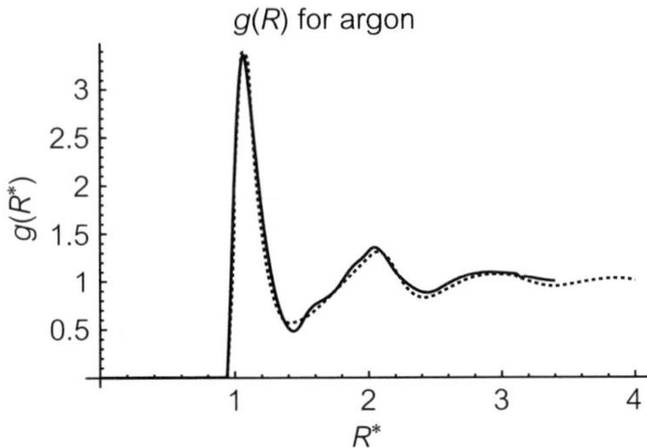

Figure 2.10 The pair correlation function for $g(R^*)$ for liquid argon (at 84.25 K and 0.71 atm) with $R^* = R/3.5$. The dotted curve is experimental values provided by N.S. Gingrich (to which the author is very grateful). The solid curve is a solution of the Percus–Yevick equation with parameters $\sigma = 3.5\text{Å}$, $\varepsilon/kT = 1.39$ and $\rho\sigma^3 = 0.85$ (for details see Appendix E). The theoretical and experimental curves are almost indistinguishable.

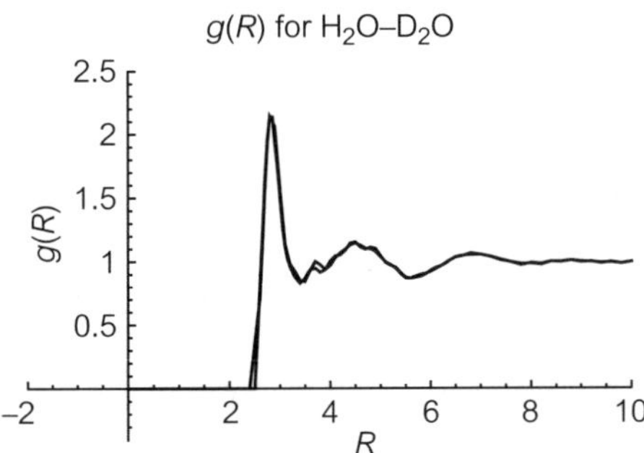

Figure 2.11 The pair correlation function for H_2O and D_2O at 4 °C and 1 atm, as a function of R (in Å). The two curves are almost indistinguishable at this scale (based on data provided by A.N. Narten to which the author is very grateful).

2.6 Molecular distribution functions in the grand canonical ensemble

In the previous section, we introduced the MDF in the canonical ensemble, i.e., the MDF in a *closed* system with fixed values of T, V, N. Similarly, one can define the MDF in any other ensemble, such as the T, P, N ensemble. Of particular interest, for this book, are the MDFs in the grand canonical ensemble, i.e., the MDF pertaining to an open system characterized by the variables T, V, μ. The fundamental probability in the grand canonical ensemble is

$$P(N) = \frac{Q(T, V, N) \exp[\beta\mu N]}{\Xi(T, V, \mu)}, \tag{2.74}$$

where $Q\ (T,\ V,\ N)$ and $\Xi\ (T,\ V,\ \mu)$ are the canonical and the grand canonical partition functions in the two ensembles, respectively. $P(N)$ is the probability of finding a system in the T, V, μ ensemble with exactly N particles.

The conditional nth-order MDF of finding the configuration $\boldsymbol{X}^n$, given that the system has N particles, is[†]

$$\rho^{(n)}\ (\boldsymbol{X}^n/N) = \frac{N!}{(N-n)!} \frac{\int \cdots \int d\boldsymbol{X}_{n+1} \ldots d\boldsymbol{X}_N \exp[-\beta U_N(\boldsymbol{X}^N)]}{\int \cdots \int d\boldsymbol{X}^N \exp[-\beta U_N(\boldsymbol{X}^N)]}. \tag{2.75}$$

This quantity is defined for $n \leq N$ only. The nth-order MDF in the T, V, μ ensemble is defined as the average of (2.75) with the weight given in (2.74), i.e.,

$$\begin{aligned}
\overline{\rho^{(n)}\ (\boldsymbol{X}^n)} &= \sum_{N \geq n} P(N)\ \rho^{(n)}\ (\boldsymbol{X}^n/N) \\
&= \frac{1}{\Xi} \sum_{N \geq n} \frac{N!}{(N-n)!} \\
&\times \frac{Q(T, V, N) \exp(\beta\mu N) \int \cdots \int d\boldsymbol{X}_{n+1} \ldots d\boldsymbol{X}_N \exp[-\beta U_N\ (\boldsymbol{X}^N)]}{Z_N}.
\end{aligned} \tag{2.76}$$

The bar over $\rho^{(n)}\ (\boldsymbol{X}^N)$ denotes the average in the T, V, μ ensemble[‡]. Recalling that the canonical partition function is

$$Q(T, V, N) = (q^N/N!) Z_N \tag{2.77}$$

and denoting by λ the absolute activity which is related to the chemical potential by

$$\lambda = \exp(\beta\mu), \tag{2.78}$$

[†] For more details on this and other expressions in this section, see Ben-Naim (1992).
[‡] We use either an over-bar or the brackets $\langle\ \rangle$ to denote averages in the open system.

we can rewrite (2.76) as

$$\overline{\rho^{(n)}\ (X^n)} = \frac{1}{\Xi} \sum_{N \geq n} \frac{(\lambda q)^N}{(N-n)!} \int \cdots \int d\boldsymbol{X}_{n+1} \ldots d\boldsymbol{X}_N \exp[-\beta U_N\ (\boldsymbol{X}^N)]. \tag{2.79}$$

The normalization condition for $\overline{\rho^{(n)}\ (\boldsymbol{X}^n)}$ is obtained from (2.76) by integrating over all the configurations $\boldsymbol{X}^n$:

$$\int \cdots \int d\boldsymbol{X}^n\ \overline{\rho^{(n)}\ (\boldsymbol{X}^n)} = \sum_{N \geq n} P(N)\ \frac{N!}{(N-n)!} = \left\langle \frac{N!}{(N-n)!} \right\rangle. \tag{2.80}$$

Two simple important cases are the following. For $n = 1$, we have

$$\int d\boldsymbol{X}_1\ \overline{\rho^{(1)}\ (\boldsymbol{X}_1)} = \langle N!/(N-1)! \rangle = \langle N \rangle, \tag{2.81}$$

which is simply the average number of particles in a system in the T, V, μ ensemble (compare this with (2.19) in the T, V, N ensemble). Using essentially the same arguments as in section 2.1, we get for a homogeneous and isotropic system

$$\overline{\rho^{(1)}\ (\boldsymbol{X})} = \frac{\langle N \rangle}{8\pi^2 V} = \frac{\rho}{8\pi^2}, \tag{2.82}$$

which is the same as in (2.26) but with the replacement of the exact N by the average $\langle N \rangle$.

For $n = 2$, we get from (2.80)

$$\begin{aligned} \int \int d\boldsymbol{X}_1 d\boldsymbol{X}_2\ \overline{\rho^{(2)}\ (\boldsymbol{X}_1, \boldsymbol{X}_2)} &= \langle N!/(N-2)! \rangle = \langle N(N-1) \rangle \\ &= \langle N^2 \rangle - \langle N \rangle. \end{aligned} \tag{2.83}$$

As in the T, V, N ensemble, one may introduce correlation functions in the T, V, μ ensemble. Of particular importance is the pair correlation function defined by

$$\overline{\rho^{(2)}\ (\boldsymbol{X}_1, \boldsymbol{X}_2)} = \overline{\rho^{(1)}\ (\boldsymbol{X}_1)}\ \overline{\rho^{(1)}\ (\boldsymbol{X}_2)}\ \overline{g\ (\boldsymbol{X}_1, \boldsymbol{X}_2)}. \tag{2.84}$$

One important property of $\overline{g(\boldsymbol{X}_1, \boldsymbol{X}_2)}$, defined in the T, V, μ ensemble, is its limiting behavior at low densities, i.e.,

$$\overline{g(\boldsymbol{X}_1, \boldsymbol{X}_2)} \xrightarrow{\rho \to 0} \exp[-\beta U(\boldsymbol{X}_1, \boldsymbol{X}_2)] \tag{2.85}$$

which is strictly true without additional terms on the order of $\langle N \rangle^{-1}$. Also for the (theoretical) ideal gas, where $U(\boldsymbol{X}_1, \boldsymbol{X}_2) = 0$, (2.85) reduces to

$$\overline{g(\boldsymbol{X}_1, \boldsymbol{X}_2)} = 1. \tag{2.86}$$

See also Appendix G.

2.7 **Generalized molecular distribution functions**

We present here a few examples of generalized molecular distribution functions MDFs (see Ben-Naim 1973a). Of particular interest is the singlet GMDF. These have been found very useful to establish a firm basis for the mixture model approach to any liquid (Ben-Naim 1972a, b, and 1973b, 1974), and in particular to aqueous solutions. It also provides some new relationships between MDFs and thermodynamic quantities. These will be presented in the next chapter.

The general procedure of defining the generalized MDF is the following. We recall the general definition of the *n*th-order MDF, say in the *T, V, N* ensemble, which for a system of spherical particles is written in the following two equivalent forms:

$$
\begin{aligned}
\rho^{(n)}(\mathbf{S}_1,\ldots,\mathbf{S}_n) &= \frac{N!}{(N-n)!}\int\cdots\int d\mathbf{R}_{n+1}\ldots d\mathbf{R}_N\, P(\mathbf{S}_1,\ldots,\mathbf{S}_n;\mathbf{R}_{n+1},\ldots,\mathbf{R}_N) \\
&= \sum_{\substack{i_1=1 \\ i_1\neq i_2\neq\cdots i_n}}^{N} \cdots \sum_{i_n=1}^{N} \int\cdots\int d\mathbf{R}^N\, P(\mathbf{R}^N) \\
&\quad \times [\delta(\mathbf{R}_{i_1}-\mathbf{S}_1)\cdots\delta(\mathbf{R}_{i_n}-\mathbf{S}_n)].
\end{aligned}
\tag{2.87}
$$

Here, $P(\mathbf{R}^N)$ is the basic probability density in the *T, V, N* ensemble. In the first form on the rhs of (2.87), we have made the distinction between *fixed* variables $\mathbf{S}_1,\ldots,\mathbf{S}_n$ and dummy variables $\mathbf{R}_{n+1},\ldots,\mathbf{R}_N$. The latter undergo integration. The second form on the rhs has the form of an average quantity of the function in the squared brackets. We first recognize that the squared brackets in the integrand comprise a stipulation on the range of integration, i.e., they serve to extract from the entire configurational space only those configurations (or regions) for which the vector $\mathbf{R}_{i_1}$ attains the value $\mathbf{S}_1,\ldots$ and the vector $\mathbf{R}_{i_n}$ attains the value $\mathbf{S}_n$.

2.7.1 **The singlet generalized molecular distribution function**

In this section, we present a special case of the generalization procedure outlined above. Consider the ordinary singlet MDF:

$$
\begin{aligned}
N_L^{(1)}(\mathbf{S}_1)\, d\mathbf{S}_1 &= d\mathbf{S}_1 \int\cdots\int d\mathbf{R}^N\; P(\mathbf{R}^N) \sum_{i=1}^{N} \delta(\mathbf{R}_i-\mathbf{S}_1) \\
&= N\, d\mathbf{S}_1 \int\cdots\int d\mathbf{R}^N P(\mathbf{R}^N)\, \delta(\mathbf{R}_1-\mathbf{S}_1).
\end{aligned}
\tag{2.88}
$$

Here, $N_L^{(1)}(\boldsymbol{S}_1)\, d\boldsymbol{S}_1$ is the average number† of particles occupying the element of volume $d\boldsymbol{S}_1$. For the present treatment, we limit our discussion to spherical molecules only. As we have already stressed in section 2.1, the quantity defined in (2.88) can be assigned two different meanings. The first follows from the first form on the rhs of (2.88), which is an average quantity in the T, V, N ensemble. The second form on the rhs of (2.88) provides the probability of finding particle 1 in the element of volume $d\boldsymbol{S}_1$. Clearly, this probability is given by $N_L^{(1)}(\boldsymbol{S}_1)d\boldsymbol{S}_1/N$.

Let us now rewrite (2.88) in a somewhat more complicated way. For each configuration $\boldsymbol{R}^N$, we define the *property* of the particle i as

$$L_i(\boldsymbol{R}^N) = \boldsymbol{R}_i. \tag{2.89}$$

The property of the i-th particle, defined in (2.89), is the *location* of particle i, giving a configuration $\boldsymbol{R}^N$ which is simply $\boldsymbol{R}_i$. This is the reason for using the letter L in the definition of the function $L_i\,(\boldsymbol{R}^N)$.

Next, we define the *counting function* of the property L by

$$N_L^{(1)}(\boldsymbol{R}^N, \boldsymbol{S}_1)d\boldsymbol{S}_1 = \sum_{i=1}^{N} \delta[L_i(\boldsymbol{R}^N) - \boldsymbol{S}_1]d\boldsymbol{S}_1. \tag{2.90}$$

This is the number of particles whose property L attains a value within $d\boldsymbol{S}_1$ at $\boldsymbol{S}_1$, given the configuration $\boldsymbol{R}^N$. The average number (here in the T, V, N ensemble) of such particles is

$$\begin{aligned} N_L^{(1)}(\boldsymbol{S}_1)d\boldsymbol{S}_1 &= \left\langle N_L^{(1)}(\boldsymbol{R}^N, \boldsymbol{S}_1)\right\rangle d\boldsymbol{S}_1 \\ &= d\boldsymbol{S}_1 \int\cdots\int d\boldsymbol{R}^N P(\boldsymbol{R}^N) \sum_{i=1}^{N} \delta[L_i(\boldsymbol{R}^N) - \boldsymbol{S}_1] \end{aligned} \tag{2.91}$$

which is the same as (2.88).

We present a few illustrative examples of properties that may replace L in (2.90) and (2.91), and which are of interest in the theory of liquids and solutions.

2.7.2 **Coordination number**

A simple property which has been the subject of many investigations is the coordination number (CN). We recall that the *average* coordination number can be obtained from the pair distribution function (section 2.5). Here, we are interested in more detailed information on the *distribution* of CN.

† Here we use the letter N rather than ρ for the density of particles. This is done in order to unify the system of notation for the continuous as well as discrete cases that are treated in this section.

Let R_C be a fixed number, to serve as the radius of the first coordination shell. If σ is the effective diameter of the particles of the system, a reasonable choice of R_C for our purposes could be $\sigma \leq R_C \leq 1.5\sigma$. This range for R_C is in conformity with the meaning of the concept of the radius of the first coordination sphere around a given particle. In what follows, we assume that R_C had been fixed, and we omit it from the notation.

The property to be considered here is the CN of the particle i at a given configuration $\boldsymbol{R}^N$ of the system. This is defined by

$$C_I(\boldsymbol{R}^N) = \sum_{j=1, j \neq i}^{N} H(|\boldsymbol{R}_j - \boldsymbol{R}_i| - R_C), \tag{2.92}$$

where $H(x)$ is a unit step function, defined as

$$H(x) = \begin{cases} 0 & \text{if } x > 0 \\ 1 & \text{if } x \leq 0. \end{cases} \tag{2.93}$$

Each term in (2.92) contributes unity whenever $|\boldsymbol{R}_j - \boldsymbol{R}_i| < R_C$, i.e., whenever the center of particle j falls within the first coordination sphere of particle i. Hence, $C_i(\boldsymbol{R}^N)$ is the number of particles $(j \neq i)$ that falls in the coordination sphere of a particle i for a given configuration $\boldsymbol{R}^N$. Next, we define the *counting* function for this property by

$$N_C^{(1)}(\boldsymbol{R}^N, K) = \sum_{i=1}^{N} \delta[C_i(\boldsymbol{R}^N) - K] \tag{2.94}$$

Here, we have used the notation $\delta(x - K)$ for the Kronecker delta function, instead of the more common notation $\delta_{x,K}$, for the sake of unity of notation. The meaning of δ as a Dirac or Kronecker delta should be clear from the context. In the sum of (2.94) we scan all the particles $(i = 1,2,\ldots,N)$ of the system at a given configuration $\boldsymbol{R}^N$. Each particle whose CN is exactly K contributes unity to the sum (2.94), and zero otherwise. Hence, the sum in (2.94) *counts* all particles whose CN is exactly K for the particular configuration $\boldsymbol{R}^N$. The average number of such particles is

$$\begin{aligned} N_C^{(1)}(K) &= \left\langle N_C^{(1)}(\boldsymbol{R}^N, K) \right\rangle \\ &= N \int \cdots \int d\boldsymbol{R}^N P(\boldsymbol{R}^N)\, \delta[C_1(\boldsymbol{R}^N) - K]. \end{aligned} \tag{2.95}$$

We can also define the following quantity:

$$x_C(K) = \frac{N_C^{(1)}(K)}{N} = \int \cdots \int d\boldsymbol{R}^N P(\boldsymbol{R}^N)\, \delta[C_1(\boldsymbol{R}^N) - K]. \tag{2.96}$$

From the definition of $N_C^{(1)}(K)$ in (2.95), it follows that $x_C(K)$ is the mole fraction of particles whose coordination number is equal to K. On the other hand, the second form on the rhs of (2.96) provides the probabilistic meaning of $x_C(K)$; i.e., this is the probability that a specific particle, say 1, will be found with CN equal to K. The quantity $x_C(K)$ can be viewed as a component of a vector

$$\boldsymbol{x}_C = (x_C(0), x_C(1), \ldots,). \tag{2.97}$$

This vector gives the "composition" of the system with respect to the CN, i.e., each component is the mole fraction of particles with a given CN. The average CN of particles in the system is given by†

$$\langle K \rangle = \sum_{K=0}^{\infty} K x_C(K). \tag{2.98}$$

We also use this example to demonstrate that changes in the condition can be achieved easily. For instance, with the same property (CN), we can ask for the average number of particles whose CN is less than or equal to, say, five. This is obtained from (2.95):

$$N_C^{(1)}(K \leq 5) = \sum_{K=0}^{5} N_C^{(1)}(K). \tag{2.99}$$

The CN, as defined above, may be viewed as a property conveying the *local density* around the particles. Another quantity conveying a similar meaning will be introduced in section 2.7.4.

2.7.3 **Binding energy**

An example of a continuously varying property is the binding energy (BE). This is defined for particle i and for the configuration $\boldsymbol{R}^N$ as follows:

$$\begin{aligned} B_i(\boldsymbol{R}^N) &= U_N(\boldsymbol{R}_1, \ldots, \boldsymbol{R}_{i-1}, \boldsymbol{R}_i, \boldsymbol{R}_{i+1}, \ldots, \boldsymbol{R}_N) \\ &\quad - U_{N-1}(\boldsymbol{R}_1, \ldots, \boldsymbol{R}_{i-1}, \boldsymbol{R}_{i+1}, \ldots, \boldsymbol{R}_N). \end{aligned} \tag{2.100}$$

This is the work required to bring a particle from an infinite distance with respect to all the other particles, to the position $\boldsymbol{R}_i$. For a system of pairwise additive potentials, (2.100) is simply the sum

$$B_i(\boldsymbol{R}^N) = \sum_{j=1, j\neq i}^{N} U(\boldsymbol{R}_i, \boldsymbol{R}_j). \tag{2.101}$$

† Note that $\langle K \rangle$ as defined in (2.98) coincides with the definition of the average CN given in section 2.5 provided that we choose R_C in this section to be the same as R_M in section 2.5.

The *counting* function corresponding to this property is

$$N_B^{(1)}(\boldsymbol{R}^N, v)\, dv = dv \sum_{i=1}^{N} \delta[B_i(\boldsymbol{R}^N) - v], \tag{2.102}$$

which is the number of particles having BE between v and $v + dv$ for the specified configuration $\boldsymbol{R}^N$. Note that since v is a continuous variable, the δ-function in (2.102) is the Dirac delta function. The average number of particles having BE between v and $v + dv$ is thus

$$N_B^{(1)}(v)\, dv = dv \left\langle \sum_{i=1}^{N} \delta[B_i(\boldsymbol{R}^N) - v] \right\rangle. \tag{2.103}$$

The corresponding mole fraction is

$$x_B(v)\, dv = \frac{N_B^{(1)}(v)\, dv}{N}, \tag{2.104}$$

With the normalization condition

$$\int_{-\infty}^{\infty} x_B(v)\, dv = 1. \tag{2.105}$$

The function $x_B(v)$ is referred to as the *distribution* of BE. By analogy with the vector (2.97) which has discrete components, we often write $\boldsymbol{x}_B$ for the *whole* distribution function, the components of which are $x_B(v)$. For simple spherical particles, the function $x_B(v)$ has one maximum at $2\langle U_N \rangle / N$. For more complex liquids such as water, this function has more "structure," reflecting the possibility of the different structural environments of a molecule in the liquid (for more details, see Ben-Naim 1974).

2.7.4 **Volume of the Voronoi polyhedron**

Another continuous-type local property of interest in the study of liquids is the Voronoi polyhedron (VP), or the Dirichlet region, defined as follows. Consider a specific configuration $\boldsymbol{R}^N$ and a particular particle i. Let us draw all the segments $l_{ij}(j = 1, \ldots, N, j \neq i)$ connecting the centers of particles i and j. Let P_{ij} be the plane perpendicular to and bisecting the line l_{ij}. Each plane P_{ij} divides the entire space into two parts. Denote by V_{ij} that part of space that includes the point $\boldsymbol{R}_i$. The VP of particle i for the configuration $\boldsymbol{R}^N$ is defined as the intersection of all the V_{ij} $(j = 1, \ldots, N, j \neq i)$:

$$(\mathrm{VP})_i = \bigcap_{j=1, j \neq i}^{N} V_{ij}(\boldsymbol{R}_i, \boldsymbol{R}_j). \tag{2.106}$$

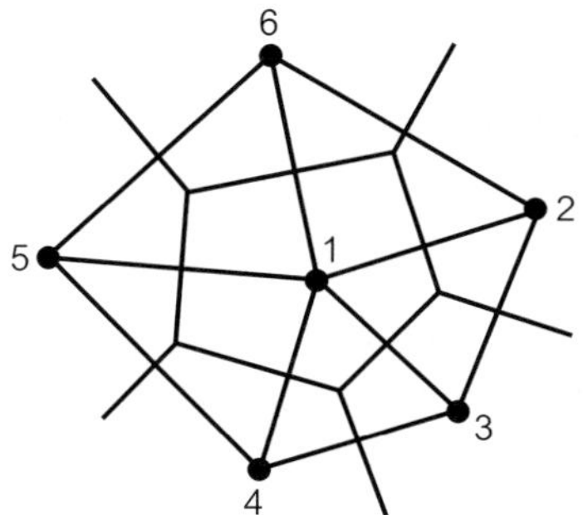

Figure 2.12. Construction of the Voronoi polygon of particle 1 in a two-dimensional system of particles.

A two-dimensional illustration of the construction of a VP is shown in figure 2.12. It is clear from the definition that the region $(\mathrm{VP})_i$ includes all the points in space that are "nearer" to $\boldsymbol{R}_i$ than to any $R_j (j \neq i)$. Furthermore, each VP contains the center of one and only one particle.

The concept of VP can be used to generate a few local properties†; the one we shall be using is the *volume* of the VP, which we denote by

$$\psi_i(\boldsymbol{R}^N) = \text{volume of } (\mathrm{VP})_i. \tag{2.107}$$

The counting function for this property is

$$N_\psi^{(1)}(\boldsymbol{R}^N, \phi)\, d\phi = d\phi \sum_{i=1}^{N} \delta[\psi_i(\boldsymbol{R}^N) - \phi], \tag{2.108}$$

and its average is

$$N_\psi^{(1)}(\phi)\, d\phi = d\phi \left\langle \sum_{i=1}^{N} \delta[\psi_i(\boldsymbol{R}^N) - \phi] \right\rangle. \tag{2.109}$$

$N_\psi^{(1)}(\phi) d\phi$ is the average number of particles whose VP has a volume between ϕ and $\phi + d\phi$. The VP of a particle *i*, in a system at a specific configuration $\boldsymbol{R}^N$, conveys a measure of the contribution of this particle to the total volume of the system at this specific configuration. See also section 3.6 for the relation between the *volume* of the system and the partial molar volume of the "species" of particles having a specific volume of VP. Clearly, the larger the volume of the VP, the smaller the local density around the particle.

† Note that the *form* of the VP is also a property which can be considered in the context of this section. Other properties of interest are the number of faces of the VP, the surface area of the VP, etc. The distribution functions defined in this section involve random variables whose values are real numbers. If we choose the *form* of the VP as a random variable, then its range of variation is the space of geometric figures and not real numbers.

2.7.5 Combination of properties

One way of generating new properties is by combination of properties. For instance, the counting function of BE *and* the volume of the VP is defined as

$$N^{(1)}_{B,\psi}(\boldsymbol{R}^N, v, \phi)\, dv\, d\phi = dv\, d\phi \sum_{i=1}^{N} \delta[B_i(\boldsymbol{R}^N) - v]\delta[\psi_i(\boldsymbol{R}^N) - \phi] \quad (2.110)$$

which counts the number of particles having BE between v and $v + dv$ *and* the volume of the VP between ϕ and $\phi + d\phi$. The average number of such particles is

$$N^{(1)}_{B,\psi}(v, \phi)\, dv\, d\phi = dv\, d\phi \left\langle \sum_{i=1}^{N} \delta[B_i(\boldsymbol{R}^N) - v]\delta[\psi_i(\boldsymbol{R}^N) - \phi] \right\rangle. \quad (2.111)$$

Note that although we have combined *two* properties, we still have a *singlet* generalized MDF. A related singlet generalized MDF which conveys similar information to that in (2.111), but is simpler for computational purposes, is constructed by the combination of BE and CN, i.e.

$$N^{(1)}_{B,C}(v, K)\, dv = dv \left\langle \sum_{i=1}^{N} \delta[B_i(\boldsymbol{R}^N) - v]\delta[C_i(\boldsymbol{R}^N) - K]. \right\rangle. \quad (2.112)$$

In (2.112) the firs δ on the rhs is a Dirac delta function, whereas the second is a Kronecker delta function.

The general procedure of defining generalized MDFs is now clear. We first define a *property* which is a function definable on the configurational space, and then introduce its distribution function in the appropriate ensemble. Examples of some of these may be found in Ben-Naim (1973a and 1974).

2.8 Potential of mean force

The potential of mean force (PMF) is an important quantity related to the pair correlation function. In this section, we show that PMF as defined below, equation (2.113), is the work involved (the Helmholtz energy in the *T, V, N* ensemble or the Gibbs energy in the *T, P, N* ensemble) in bringing two selected particles from infinite separation to the final configuration $\boldsymbol{X}'$, $\boldsymbol{X}''$. We shall also show that the gradient of this function is the average force exerted on one particle at $\boldsymbol{X}'$, given a second particle at $\boldsymbol{X}''$, averaged over all configurations of the particles in the system. We shall start with the definition of the PMF for a one-component system consisting of N spherical particles in the T, V, N ensemble.

$$W(\boldsymbol{R}', \boldsymbol{R}'') = -kT \ln g(\boldsymbol{R}', \boldsymbol{R}''). \tag{2.113}$$

Using the definition of the pair correlation function in (2.48) we have

$$\exp[-\beta W(\boldsymbol{R}', \boldsymbol{R}'')] = \frac{N(N-1)}{\rho^2} \frac{\int \cdots \int d\boldsymbol{R}_3 \ldots d\boldsymbol{R}_N \exp[-\beta U_N(\boldsymbol{R}', \boldsymbol{R}'', \boldsymbol{R}_3, \ldots, \boldsymbol{R}_N)]}{Z_N}. \tag{2.114}$$

We now take the gradient of $W(\boldsymbol{R}', \boldsymbol{R}'')$ with respect to the vector $\boldsymbol{R}'$, and get

$$-\beta \boldsymbol{\nabla}' W(\boldsymbol{R}', \boldsymbol{R}'') = \boldsymbol{\nabla}' \left\{ \ln \int \cdots \int d\boldsymbol{R}_3 \ldots d\boldsymbol{R}_N \exp[-\beta U_N(\boldsymbol{R}', \boldsymbol{R}'', \boldsymbol{R}_3, \ldots, \boldsymbol{R}_N)] \right\}. \tag{2.115}$$

The symbol $\boldsymbol{\nabla}'$ stands for the gradient with respect to the vector $\boldsymbol{R}' = (x', y', z')$, i.e.,

$$\boldsymbol{\nabla}' = \left(\frac{\partial}{\partial x'}, \frac{\partial}{\partial y'}, \frac{\partial}{\partial z'} \right). \tag{2.116}$$

We also assume that the total potential energy is pairwise additive. Hence, we write

$$U_N(\boldsymbol{R}', \boldsymbol{R}'', \boldsymbol{R}_3, \ldots, \boldsymbol{R}_N) = U_{N-2}(\boldsymbol{R}_3, \ldots, \boldsymbol{R}_N) + \sum_{i=3}^{N} [U(\boldsymbol{R}_i, \boldsymbol{R}') + U(\boldsymbol{R}_i, \boldsymbol{R}'')] + U(\boldsymbol{R}', \boldsymbol{R}''). \tag{2.117}$$

The gradient of U_N with respect to $\boldsymbol{R}'$ in (2.117) is

$$\boldsymbol{\nabla}' U_N(\boldsymbol{R}', \boldsymbol{R}'', \boldsymbol{R}_3, \ldots, \boldsymbol{R}_N) = \sum_{i=3}^{N} \boldsymbol{\nabla}' U(\boldsymbol{R}_i, \boldsymbol{R}') + \boldsymbol{\nabla}' U(\boldsymbol{R}', \boldsymbol{R}''). \tag{2.118}$$

Taking the gradient of W in (2.115), we get

$$-\boldsymbol{\nabla}' W(\boldsymbol{R}', \boldsymbol{R}'') = \frac{\int \cdots \int d\boldsymbol{R}_3 \ldots d\boldsymbol{R}_N \exp(-\beta U_N) \left[-\sum_{i=3}^{N} \boldsymbol{\nabla}' U(\boldsymbol{R}_i, \boldsymbol{R}') - \boldsymbol{\nabla}' U(\boldsymbol{R}', \boldsymbol{R}'') \right]}{\int \cdots \int d\boldsymbol{R}_3 \cdots d\boldsymbol{R}_N \exp(-\beta U_N)}. \tag{2.119}$$

Note that the integration in the numerator of (2.119) is over $\mathbf{R}_3,\ldots,\mathbf{R}_N$, and the quantity $\nabla' U(\mathbf{R}', \mathbf{R}'')$ is independent of these variables. We also introduce the conditional probability density of finding the $N-2$ particles at a specified configuration $\mathbf{R}_3,\ldots,\mathbf{R}_N$ given that the two particles are at $\mathbf{R}'$ and $\mathbf{R}''$, namely

$$\begin{aligned} P(\mathbf{R}_3,\ldots,\mathbf{R}_N/\mathbf{R}',\mathbf{R}'') &= \frac{P(\mathbf{R}',\mathbf{R}'',\mathbf{R}_3,\ldots,\mathbf{R}_N)}{P(\mathbf{R}',\mathbf{R}'')} \\ &= \frac{\exp[-\beta U_N(\mathbf{R}',\mathbf{R}'',\mathbf{R}_3,\ldots,\mathbf{R}_N]}{Z_N} \\ &\quad \times \frac{Z_N}{\int\cdots\int d\mathbf{R}_3\ldots d\mathbf{R}_N \exp[-\beta U_N(\mathbf{R}',\mathbf{R}'',\mathbf{R}_3,\ldots,\mathbf{R}_N)]} \\ &= \frac{\exp[-\beta U_N(\mathbf{R}',\mathbf{R}'',\mathbf{R}_3,\ldots,\mathbf{R}_N)]}{\int\cdots\int d\mathbf{R}_3\ldots d\mathbf{R}_N \exp[-\beta U_N(\mathbf{R}',\mathbf{R}'',\mathbf{R}_3,\ldots,\mathbf{R}_N)]}. \end{aligned} \tag{2.120}$$

Using (2.120) we rewrite (2.119) as

$$\begin{aligned} -\nabla' W(\mathbf{R}',\mathbf{R}'') &= -\nabla' U(\mathbf{R}',\mathbf{R}'') + \int\cdots\int d\mathbf{R}_3\ldots d\mathbf{R}_N P(\mathbf{R}_3,\ldots,\mathbf{R}_N/\mathbf{R}',\mathbf{R}'') \\ &\quad \times \sum_{i=3}^{N}[-\nabla' U(\mathbf{R}_i,\mathbf{R}')] \\ &= -\nabla' U(\mathbf{R}',\mathbf{R}'') + \left\langle -\sum_{i=3}^{N} \nabla' U(\mathbf{R}_i,\mathbf{R}')\right\rangle^{(N-2)}. \end{aligned} \tag{2.121}$$

In (2.121), we expressed $-\nabla W(\mathbf{R}', \mathbf{R}'')$ as a sum of two terms. The first term is simply the *direct* force exerted on the particle at $\mathbf{R}'$ when the second particle is at $\mathbf{R}''$. This is the same force operating between the two particles in vacuum. The second term is the *conditional* average force (note that the average has been calculated using the conditional probability density 2.120) exerted on the particle $\mathbf{R}'$ by all the other particles present in the system. It is an average over all the configurations of the $N-2$ particles given that the two particles are at $\mathbf{R}'$ and $\mathbf{R}''$. The latter may be referred to as the *indirect* force operating on the particle at $\mathbf{R}'$, which originates from all the other particles excluding the one at $\mathbf{R}''$. The foregoing discussion justifies the designation of $W(\mathbf{R}', \mathbf{R}'')$ as the *potential* of *mean force.* Its gradient gives the *average* force, including direct and indirect contributions, operating on the particle at $\mathbf{R}'$.

We can further simplify (2.121) by noting that the sum over i produces $N-2$ equal terms, i.e.,

$$\int \cdots \int d\boldsymbol{R}_3 \ldots d\boldsymbol{R}_N P(\boldsymbol{R}_3, \ldots, \boldsymbol{R}_N/\boldsymbol{R}', \boldsymbol{R}'') \sum_{i=3}^{N} \boldsymbol{\nabla}' U(\boldsymbol{R}_i, \boldsymbol{R}')$$

$$= (N-2) \int \cdots \int d\boldsymbol{R}_3 \ldots d\boldsymbol{R}_N P(\boldsymbol{R}_3, \ldots, \boldsymbol{R}_N/\boldsymbol{R}', \boldsymbol{R}'') \boldsymbol{\nabla}' U(\boldsymbol{R}_3, \boldsymbol{R}')$$

$$= (N-2) \int d\boldsymbol{R}_3 \boldsymbol{\nabla}' U(\boldsymbol{R}_3, \boldsymbol{R}') \int \cdots \int d\boldsymbol{R}_4 \ldots d\boldsymbol{R}_N P(\boldsymbol{R}_3, \ldots, \boldsymbol{R}_N/\boldsymbol{R}', \boldsymbol{R}'')$$

$$= (N-2) \int d\boldsymbol{R}_3 \boldsymbol{\nabla}' U(\boldsymbol{R}_3, \boldsymbol{R}') P(\boldsymbol{R}_3/\boldsymbol{R}', \boldsymbol{R}'')$$

$$= \int d\boldsymbol{R}\, [\boldsymbol{\nabla}' U(\boldsymbol{R}, \boldsymbol{R}')] \rho(\boldsymbol{R}/\boldsymbol{R}', \boldsymbol{R}''). \qquad (2.122)$$

The quantity $\rho(\boldsymbol{R}/\boldsymbol{R}', \boldsymbol{R}'')$, introduced in (2.122), is the *conditional* density at a point $\boldsymbol{R}$, *given* two particles at $\boldsymbol{R}'$ and $\boldsymbol{R}''$. This is a straightforward generalization of the conditional density introduced in section 2.4. The total force acting on 1 can now be written as

$$\boldsymbol{F}_1 = -\boldsymbol{\nabla}' U(\boldsymbol{R}', \boldsymbol{R}'') - \int d\boldsymbol{R}\, [\boldsymbol{\nabla}' U(\boldsymbol{R}, \boldsymbol{R}')] \rho(\boldsymbol{R}/\boldsymbol{R}', \boldsymbol{R}''). \qquad (2.123)$$

This form is useful in the study of forces applied to solutes or to groups in proteins, in aqueous solutions. The first term is referred to as the direct force and the second term as the solvent-induced force.

The form of the function $W(R)$, with $R = |\boldsymbol{R}'' - \boldsymbol{R}'|$, for LJ particles, and its density dependence are depicted in figure 2.13. At very low densities, the potential average force is identical to the pair potential; this follows from the negligible effect of all the other particles present in the system. At higher densities, the function $W(R)$ shows successive maxima and minima [corresponding to the minima and maxima of $g(R)$]. The interesting point worth noting is that the *indirect* force at, say, $R > \sigma$ can be either attractive or repulsive even in the region where the *direct* force is purely attractive.[†] We now derive an important relation between the PMF and the change of the Helmholtz energy.

Consider a system of N simple spherical particles in a volume V at temperature T. The Helmholtz energy for such a system is

$$\exp[-\beta A(T, V, N)] = (1/N!\Lambda^{3N}) \int \cdots \int d\boldsymbol{R}^N \exp[-\beta U(\boldsymbol{R}^N)]. \qquad (2.124)$$

Now consider a slightly modified system in which two specific particles, say 1 and 2, have been fixed at the points $\boldsymbol{R}'$ and $\boldsymbol{R}''$, respectively. The Helmholtz

[†] In fact this oscillatory behavior is manifested even by a system of hard spheres for which the *direct* force is zero beyond $R > \sigma$.

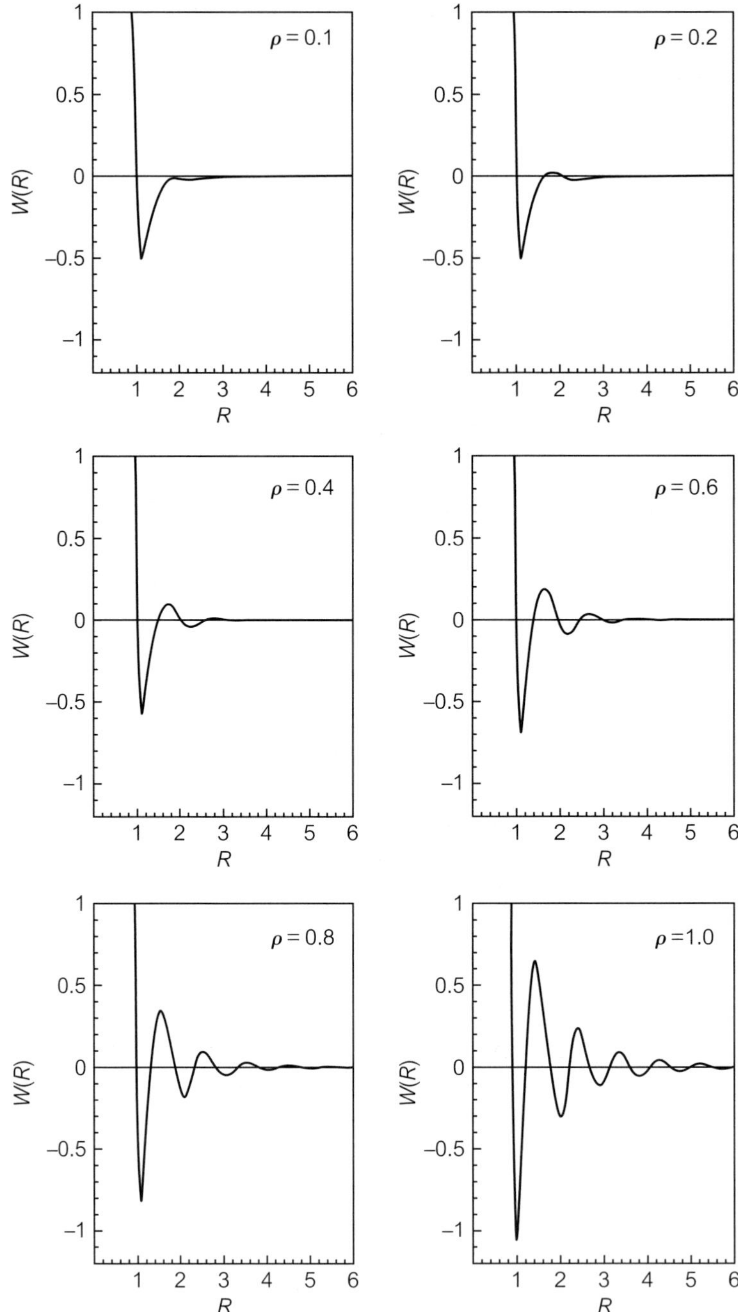

Figure 2.13. The potential of mean force $W(R)$ for the same system and the same densities as in figure 2.4.

energy for such a system is denoted by $A(\boldsymbol{R}', \boldsymbol{R}'')$ and we have

$$\exp[-\beta A(\boldsymbol{R}', \boldsymbol{R}'')] = \frac{1}{(N-2)!\Lambda^{3(N-2)}} \times \int \cdots \int d\boldsymbol{R}_3 \ldots d\boldsymbol{R}_N \, \exp[-\beta U_N(\boldsymbol{R}', \boldsymbol{R}'', \boldsymbol{R}_3, \ldots, \boldsymbol{R}_N)]. \tag{2.125}$$

Let us denote by $A(R)$, with $R = |\boldsymbol{R}'' - \boldsymbol{R}'|$, the Helmholtz energy of such a system when the separation between the two particles is R, and form the difference

$$\Delta A(R) = A(R) - A(\infty). \tag{2.126}$$

This is the work required to bring the two particles from fixed positions at infinite separation to fixed positions where the separation is R. The process is carried out at constant volume and temperature. From (2.113), (2.114), (2.125) and (2.126), we get

$$\begin{aligned}\exp[-\beta \Delta A(R)] &= \frac{\int \cdots \int d\boldsymbol{R}_3 \ldots d\boldsymbol{R}_N \exp[-\beta U_N(\boldsymbol{R}', \boldsymbol{R}'', \boldsymbol{R}_3, \ldots, \boldsymbol{R}_N)]}{\lim_{R\to\infty} \int \cdots \int d\boldsymbol{R}_3 \ldots d\boldsymbol{R}_N \exp[-\beta U_N(\boldsymbol{R}', \boldsymbol{R}'', \boldsymbol{R}_3, \ldots, \boldsymbol{R}_N)]} \\ &= g(R) = \exp\{-\beta[W(R) - W(\infty)]\}. \end{aligned} \tag{2.127}$$

This is an important and useful result. The correlation between two particles at distance R is related to the work required (here for constant T, V) to bring the two particles from infinite separation to a distance R. Since $g(R)$ is proportional to the probability density of finding the two particles at a distance R, we can conclude that the probability of finding the event "two particles at R" is related to the work required to create that event. This is a particular example of a much more general relation between the probability of observing an event and the work required to create that event.

In this section, we used the T, V, N ensemble to obtain relation (2.127). A similar relation can be obtained for any other ensemble. Of particular importance is the analog of (2.127) in the T, P, N ensemble. It has the same form but the events occur in a T, P, N system and instead of the Helmholtz energy change, we need to use the Gibbs energy change.

2.9 Molecular distribution functions in mixtures

Molecular distribution functions (MDFs) in mixtures are defined in a similar way as in the case of the one-component system; the only complication is notational. For two-component systems, we use the shorthand notation for the configuration of the entire system of N_A particles of type A and N_B particles of type B:

$$\boldsymbol{X}^{N_A+N_B} = \boldsymbol{X}_1, \boldsymbol{X}_2, \ldots, \boldsymbol{X}_{N_A}, \boldsymbol{X}_{N_A+1}, \boldsymbol{X}_{N_A+2}, \ldots, \boldsymbol{X}_{N_A+N_B}. \tag{2.128}$$

The total interaction energy for a specific configuration is

$$U_{N_A, N_B}(\boldsymbol{X}^{N_A}, \boldsymbol{X}^{N_B}) = \frac{1}{2}\sum_{i \neq j} U_{AA}(\boldsymbol{X}_i, \boldsymbol{X}_j) + \frac{1}{2}\sum_{i \neq j} U_{BB}(\boldsymbol{X}_i, \boldsymbol{X}_j) + \sum_{i=1}^{N_A}\sum_{j=1}^{N_B} U_{AB}(\boldsymbol{X}_i, \boldsymbol{X}_j). \tag{2.129}$$

Here we have assumed pairwise additivity of the total potential energy and adopted the convention that the order of arguments in the parentheses corresponds to the order of species as indicated by subscript of U. Thus $\boldsymbol{X}_i$ in the first sum on the rhs of (2.129) is the configuration of the ith molecule ($i = 1, 2, \ldots, N_A$) of species A, whereas $\boldsymbol{X}_j$, in the last term on the rhs of (2.129), stands for the configuration of the jth molecule ($j = 1, 2, \ldots, N_B$) of species B.

The basic probability density in the canonical ensemble is

$$P(\boldsymbol{X}^{N_A+N_B}) = P(\boldsymbol{X}^{N_A}, \boldsymbol{X}^{N_B}) = \frac{\exp[-\beta U_{N_A, N_B}(\boldsymbol{X}^{N_A}, \boldsymbol{X}^{N_B})]}{\int \cdots \int d\boldsymbol{X}^{N_A}\, d\boldsymbol{X}^{N_B}\, \exp[-\beta U_{N_A, N_B}(\boldsymbol{X}^{N_A}, \boldsymbol{X}^{N_B})]}. \tag{2.130}$$

The singlet distribution function for the species A is defined in complete analogy with the definition in the pure case (section 2.1),

$$\rho_A^{(1)}(\boldsymbol{X}') = \int \cdots \int d\boldsymbol{X}^{N_A+N_B}\, P(\boldsymbol{X}^{N_A+N_B}) \sum_{i=1}^{N_A} \delta(\boldsymbol{X}_i^A - \boldsymbol{X}') = N_A \int \cdots \int d\boldsymbol{X}^{N_A+N_B}\, P(\boldsymbol{X}^{N_A+N_B})\, \delta(\boldsymbol{X}_1^A - \boldsymbol{X}') \tag{2.131}$$

and similarly

$$\rho_B^{(1)}(\boldsymbol{X}') = N_B \int \cdots \int d\boldsymbol{X}^{N_A+N_B} P(\boldsymbol{X}^{N_A+N_B})\, \delta(\boldsymbol{X}_i^B - \boldsymbol{X}'). \tag{2.132}$$

Clearly, $\rho_A^{(1)}(\boldsymbol{X}')$ is the probability of finding any molecule of type A in a small region $d\boldsymbol{X}'$ at $\boldsymbol{X}'$. Similar interpretation applies to $\rho_B^{(1)}(\boldsymbol{X}')$.

As in the case of a one-component system, $\rho_A^{(1)}(\boldsymbol{X}')$ is also the average density of A molecules in the configuration $\boldsymbol{X}'$. In a homogenous and isotropic fluid, we have (see section 2.1 for more details)

$$\rho_A^{(1)}(\boldsymbol{X}') = \frac{N_A}{8\pi^2 V} \tag{2.133}$$

$$\rho_B^{(1)}(\boldsymbol{X}') = \frac{N_B}{8\pi^2 V}. \tag{2.134}$$

The average local density of A molecules at $\boldsymbol{R}'$ is defined by

$$\rho_A^{(1)}(\boldsymbol{R}') = \int d\boldsymbol{\Omega}' \rho_A^{(1)}(\boldsymbol{X}') = \frac{N_A}{V} = \rho_A \tag{2.135}$$

and a similar definition applies to $\rho_B^{(1)}(\boldsymbol{R}')$.

The probability density of finding an A-particle at a specific orientation Ω' (independently of its location) is

$$\rho_A^{(1)}(\boldsymbol{\Omega}') = \int \rho_A^{(1)}(\boldsymbol{X}')\, d\boldsymbol{R}' = \frac{N_A}{8\pi^2} \tag{2.136}$$

and a similar definition applies to B. Note that $\rho_A^{(1)}(\boldsymbol{R}')$ and $\rho_A^{(1)}(\boldsymbol{\Omega}')$ are the marginal probability densities, derived from $\rho_A^{(1)}(\boldsymbol{X}')$.

In a similar fashion, one defines the pair distribution functions for the four different pairs AA, AB, BA and BB. For instance,

$$\begin{aligned}\rho_{AA}^{(2)}(\boldsymbol{X}',\boldsymbol{X}'') &= \int \cdots \int d\boldsymbol{X}^{N_A+N_B} P(\boldsymbol{X}^{N_A+N_B}) \sum_{\substack{i=1\\ i\neq j}}^{N_A} \sum_{j=1}^{N_A} \delta(\boldsymbol{X}_i^A - \boldsymbol{X}')\delta(\boldsymbol{X}_j^A - \boldsymbol{X}'') \\ &= N_A(N_A-1) \int \cdots \int d\boldsymbol{X}^{N_A+N_B} P(\boldsymbol{X}^{N_A+N_B})\delta(\boldsymbol{X}_1^A - \boldsymbol{X}')\delta(\boldsymbol{X}_2^A - \boldsymbol{X}'').\end{aligned} \tag{2.137}$$

Similarly, for different species,

$$\begin{aligned}\rho_{AB}^{(2)}(\boldsymbol{X}',\boldsymbol{X}'') &= \int \cdots \int d\boldsymbol{X}^{N_A+N_B}\, P(\boldsymbol{X}^{N_A+N_B}) \sum_{i=1}^{N_A} \sum_{j=1}^{N_B}\, \delta(\boldsymbol{X}_i^A - \boldsymbol{X}')\delta(\boldsymbol{X}_j^B - \boldsymbol{X}'') \\ &= N_A N_B \int \cdots \int d\boldsymbol{X}^{N_A+N_B}\, P(\boldsymbol{X}^{N_A+N_B})\delta(\boldsymbol{X}_1^A - \boldsymbol{X}')\, \delta(\boldsymbol{X}_1^B - \boldsymbol{X}'').\end{aligned} \tag{2.138}$$

The pair correlation functions $g_{\alpha\beta}(\boldsymbol{X}', \boldsymbol{X}'')$ where α and β can either be A or B, are defined by

$$\rho_{\alpha\beta}^{(2)}(\boldsymbol{X}',\boldsymbol{X}'') = \rho_\alpha^{(1)}(\boldsymbol{X}')\rho_\beta^{(1)}(\boldsymbol{X}'')\, g_{\alpha\beta}(\boldsymbol{X}',\boldsymbol{X}'') \tag{2.139}$$

and the spatial pair correlation functions by

$$g_{\alpha\beta}(\boldsymbol{R}', \boldsymbol{R}'') = \frac{\int d\boldsymbol{\Omega}' \int d\boldsymbol{\Omega}'' g_{\alpha\beta}(\boldsymbol{X}', \boldsymbol{X}'')}{(8\pi^2)^2}. \tag{2.140}$$

As in a one-component system, the functions $g_{\alpha\beta}(\boldsymbol{R}', \boldsymbol{R}'')$ depend only on the scalar distance $R = |\boldsymbol{R}'' - \boldsymbol{R}'|$. Hence, for the spatial pair correlation function, we have

$$g_{AB}(R) = g_{BA}(R). \tag{2.141}$$

The conditional distribution functions are defined by

$$\rho_{A/B}(\boldsymbol{X}'/\boldsymbol{X}'') = \rho_{AB}^{(2)}(\boldsymbol{X}', \boldsymbol{X}'')/\rho_B^{(1)}(\boldsymbol{X}'') = \rho_A^{(1)}(\boldsymbol{X}')\ g_{AB}(\boldsymbol{X}', \boldsymbol{X}''). \tag{2.142}$$

As in the one-component case, $\rho_{A/B}(\boldsymbol{X}'/\boldsymbol{X}'')$ may be interpreted as the density (or probability density) of finding A in a small region at $\boldsymbol{X}'$, given that a B-particle is at an exactly fixed configuration $\boldsymbol{X}''$ (see also section 2.4). Note also that the probability interpretation of the singlet distribution function holds only in a very small region around $\boldsymbol{X}'$; the density interpretation holds true for any region including the entire range of configuration.

The normalization conditions for the pair distribution functions in a closed system are

$$\int \rho_{AA}^{(2)}(\boldsymbol{X}', \boldsymbol{X}'')\ d\boldsymbol{X}' d\boldsymbol{X}'' = N_A(N_A - 1) \tag{2.143}$$

$$\int \rho_{AB}^{(2)}(\boldsymbol{X}', \boldsymbol{X}'')\ d\boldsymbol{X}' d\boldsymbol{X}'' = N_A N_B. \tag{2.144}$$

The first is simply a statement that the total number of $A-A$ pairs is $N_A(N_A - 1)$. The second refers to the total number of $A-B$ pairs which is $N_A N_B$. Note that since we are in a closed system, these numbers are exact. We shall see in chapter 4 the analogs of these equations in an open system.

As in the case of the one-component system, we also expect here that as the distance becomes very large, the pair distribution function becomes a product of the corresponding singlet distribution functions.

We now turn to discuss some features of the pair correlation functions that are typical to mixtures of two (or more) components. We have seen in section 2.5 that for spherical particles, the pair correlation has peaks at roughly σ, 2σ, 3σ, etc., where σ is the effective diameter of the particles. However, it is not exactly at multiples of σ, first because the minimum of the pair potential is at $2^{\frac{1}{6}}\ \sigma$ and not at σ, and second because of the randomness of the packing of

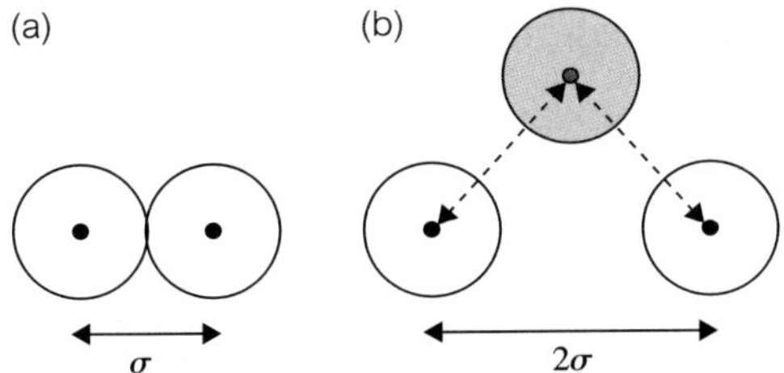

Figure 2.14. (a) At a distance of about $R = \sigma$ the correlation is determined mainly by the *direct* interaction between A and B (clear circles). (b) At $R > \sigma$, the direct interaction between A and B is weak. The correlation between A and B (clear circles) is mediated by the surrounding molecules (shaded circle), which interact with both A and B.

spheres in the liquid state. In mixtures, say of A and B, the location of the first maximum of $g_{AB}(R)$ is expected to be at about σ_{AB}, where σ_{AB} is defined as

$$\sigma_{AB} = \tfrac{1}{2}(\sigma_{AA} + \sigma_{BB}) \tag{2.145}$$

Note that this is the exact distance of closest approach for two hard sphere particles. For Lennard-Jones particles σ_{AB} is *defined* in (2.145). This is *practically* the distance of closest approach for A and B. The occurrence of the *first* peak at σ_{AB} is due to the dominance of the *direct* interaction between A and B at this distance. This is true for one-component as well as for multi-component systems. However, the other peaks of $g(R)$ are not determined by the direct interactions. Normally at a distance of about 2σ and beyond, $U(R)$ is very weak and what determines the location of the second, third, etc., peaks is *not* the direct interactions but the *indirect* correlation mediated by the surroundings of the pair A, B. The difference between these two cases is schematically depicted in figure 2.14.

We now turn to examine some features of the pair correlation functions of the mixture of A and B. Let A and B be two simple spherical molecules interacting through pair potentials which we denote by $U_{AA}(R)$, $U_{AB}(R)$, and $U_{BB}(R)$. For simplicity, assume Lennard-Jones particles

$$U_{AA}(R) = 4\varepsilon_{AA}\left[\left(\frac{\sigma_{AA}}{R}\right)^{12} - \left(\frac{\sigma_{AA}}{R}\right)^{6}\right] \tag{2.146}$$

$$U_{BB}(R) = 4\varepsilon_{BB}\left[\left(\frac{\sigma_{BB}}{R}\right)^{12} - \left(\frac{\sigma_{BB}}{R}\right)^{6}\right] \tag{2.147}$$

$$U_{AB}(R) = U_{BA}(R) = 4\varepsilon_{AB}\left[\left(\frac{\sigma_{AB}}{R}\right)^{12} - \left(\frac{\sigma_{AB}}{R}\right)^{6}\right]. \tag{2.148}$$

We also assume the combination rules

$$\sigma_{AB} = \sigma_{BA} = (\sigma_{AA} + \sigma_{BB})/2 \tag{2.149}$$

$$\varepsilon_{AB} = \varepsilon_{BA} = (\varepsilon_{AA}\varepsilon_{BB})^{\frac{1}{2}}. \tag{2.150}$$

Before proceeding to mixtures at high densities, it is instructive to recall the density dependence of $g(R)$ for a one-component system (see section 2.5). We have noticed that the second, third, etc., peaks of $g(R)$ develop as the density increases. The illustrations in sections 2.5 were calculated for Lennard-Jones particles with $\sigma = 1.0$ and increasing (number) density ρ. It is clear, however, that the important parameter determining the form of $g(R)$ is the *dimensionless* quantity $\rho\sigma^3$ (assuming for the moment that ε/kT is fixed). This can be illustrated schematically with the help of figure 2.15. In the two boxes, we have the same *number density*, whereas the volume density (qualitatively the "actual" volume occupied by the particles) defined below is quite different. Clearly, the behavior of these two systems will differ markedly even when A and B are hard spheres differing only in their diameters. Thus we expect that the form of $g(R)$ will be quite different for these two systems. The reason is that although the average *separation* between the particles is the *same* in *a* and *b*, the average *interaction* between the particles is quite different in *a* and *b*. In this illustration, the particles in *a* are most of the time within the range of the intermolecular interactions, whereas in *b*, the particles are far apart relative to the range of interactions; hence the effects of intermolecular interactions are negligible.

Now consider mixtures of A and B (with $\sigma_{AA} \gg \sigma_{BB}$) at different compositions but at constant *total number density* ρ. If we study the dependence of say $g_{AB}(R)$ on the mole fraction of x_A, we shall find that at $x_A \approx 1$, $g_{AB}(R)$ behaves

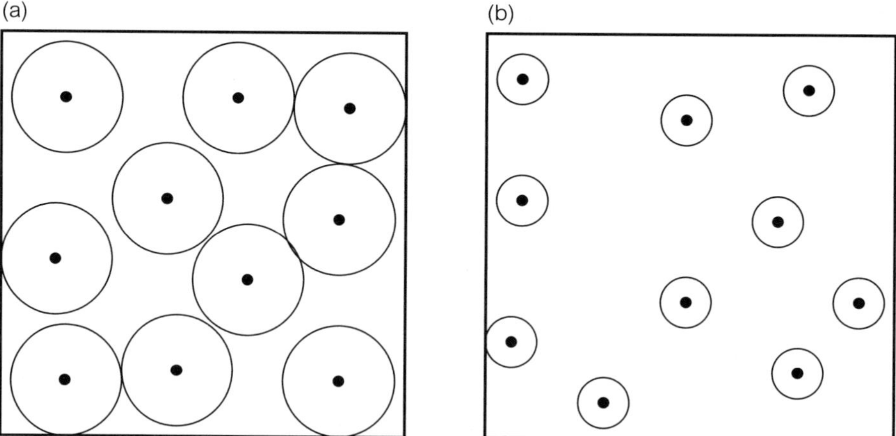

Figure 2.15. Two systems with the same number density but differing in the volume densities.

as in the case of a high-density fluid, whereas at $x_A \approx 0$, we shall observe the behavior of the low-density fluid. In order to highlight those effects that are specific to the properties of the mixtures, it is advantageous to study the behavior of the pair correlation function when the total "volume density" is constant. The latter is defined as follows. In a one-component system of particles with effective diameter σ, the ratio of the volume occupied by the particles to the total volume of the system is

$$\eta = \frac{N}{V} \frac{4\pi(\sigma/2)^3}{3} = \frac{\rho\pi\sigma^3}{6}. \tag{2.151}$$

Similarly, the total volume density of a mixture of two components A and B is defined by

$$\eta = \tfrac{1}{6}\pi(\rho_A \sigma_{AA}^3 + \rho_B \sigma_{BB}^3) = \tfrac{1}{6}\pi\rho(x_A \sigma_{AA}^3 + x_B \sigma_{BB}^3) \tag{2.152}$$

In the second equation on the rhs of (2.152), we have expressed η in terms of the total (number) density and the mole fractions.

We shall now illustrate some of the most salient features of the behavior of the various pair correlation functions in systems of Lennard-Jones particles obeying relations (2.146)–(2.148) with the parameters

$$\sigma_{AA} = 1.0 \quad \sigma_{BB} = 1.5$$

$$\frac{\varepsilon_{AA}}{kT} = \frac{\varepsilon_{BB}}{kT} = 0.5 \quad \eta = 0.45. \tag{2.153}$$

Note that the volume density of closed pack spheres is about $\eta_{cp} \approx 0.74$. The choice of $\eta = 0.45$, which is about 6/10 of the maximum density, was chosen for convenience. In fact even at these densities converging of the Percus–Yevick equation is quite slow (see also Appendix E).

We shall discuss separately three regions of compositions.

(1) Systems that are dominated by the presence of A's, between any pair of particles, i.e., $x_A \approx 1$.

Figure 2.16 shows the three pair correlation functions for a system with composition $x_A = 0.99$. Here, $g_{AA}(R)$ is almost identical to the pair correlation function for pure A. The peaks occur at about σ_{AA}, $2\sigma_{AA}$, $3\sigma_{AA}$, and $4\sigma_{AA}$. Since $\eta = 0.45$ in (2.153) corresponds to quite a high density, we have four pronounced peaks. The function $g_{AB}(R)$ has the first peak at about σ_{AB}. (The value of σ_{AB} is $(\sigma_{AA} + \sigma_{BB})/2 = 1.25$, but due to errors in the numerical computation and the fact that the minimum of U_{AB} is at $2^{\frac{1}{6}}\sigma_{AB}$, we actually obtain the first maximum at about $R = 1.3$.) Similarly the first peak of $g_{BB}(R)$ is at about $\sigma_{BB} = 1.5$. The second, third, and fourth peaks of $g_{AB}(R)$ are determined not by

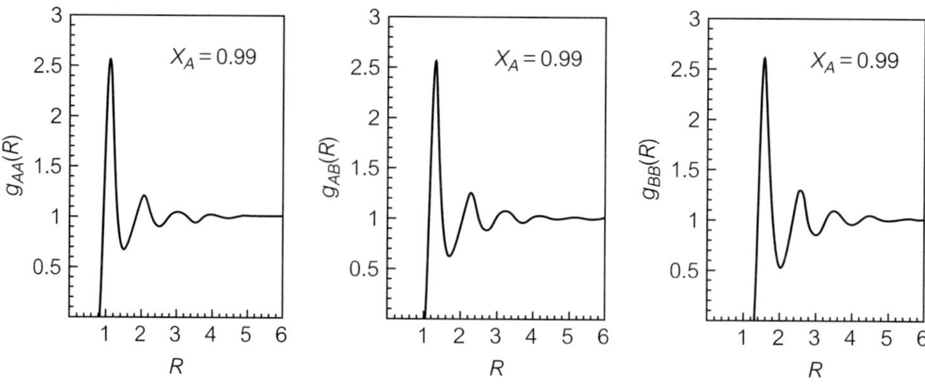

Figure 2.16. The three pair correlation functions g_{AA}, $g_{BB} = g_{BA}$ and g_{BB} for a system with parameters as in equation (2.153) and $\eta = 0.45$ and $x_A = 0.99$.

multiples of σ_{AB}, but by the addition of σ_{AA}.† That is, the maxima are at $R \approx \sigma_{AB}$, $\sigma_{AB} + \sigma_{AA}$, $\sigma_{AB} + 2\sigma_{AA}$, etc. This is a characteristic feature of a dilute solution of B in A, where the spacing between the maxima is determined by σ_{AA}, i.e., the diameter of the dominating species. The molecular reason for this is very simple. The spacing between, say, the first and second peaks is determined by the *size* of the molecule that will most probably fill the space between the two molecules under observation. Because of the prevalence of A molecules in this case, they are the most likely to fill the space between A and B. The situation is depicted schematically in Figure 2.17 where we show the most likely filling of space between a pair of molecules for the case $x_A \approx 1$, i.e., for a very dilute solution of B in A. The first row in Figure 2.17 shows the approximate locations of the first three peaks of $g_{AA}(R)$; other rows correspond successively to $g_{AB}(R) = g_{BA}(R)$ and $g_{BB}(R)$.

For $x_A \approx 1$, the component A may be referred to as the *solvent* and B as the solute. For any pair of species $\alpha\beta$, we can pick up two specific particles (one of species α and the other of species β) and refer to these particles as a "dimer." From the first row of figure 2.17, we see that the most probable configurations of the dimers occur either when the separation is $\sigma_{\alpha\beta}$ or when they are "solvent separated," i.e., when the distances are $R \approx \sigma_{\alpha\beta} + n\sigma_{AA}$, where $n = 1, 2, 3$, for the second, third, and fourth peaks. Note that because of the approximate nature of the computations, the curves $g_{AB}(R)$ and $g_{BA}(R)$ may come out a little different; however, theoretically they should be identical and in our computation they are nearly identical and may not be distinguished on the scale of figure 2.16.

† The second peak of $g_{AB}(R)$ is clearly related to $\sigma_{AB} + \sigma_{AA}$ and the third to $\sigma_{AB} + 2\sigma_{AA}$. If we had chosen $\sigma_{AA} = 1.0$ and $\sigma_{BB} = 2.0$ then we could not have distinguished between $\sigma_{AB} + 2\sigma_{AA}$ and $\sigma_{AB} + \sigma_{BB}$. It is for this reason that we have chosen the values of $\sigma_{AA} = 1.0$ and $\sigma_{BB} = 1.5$ which could lead to less ambiguity in the interpretation of the first two peaks.

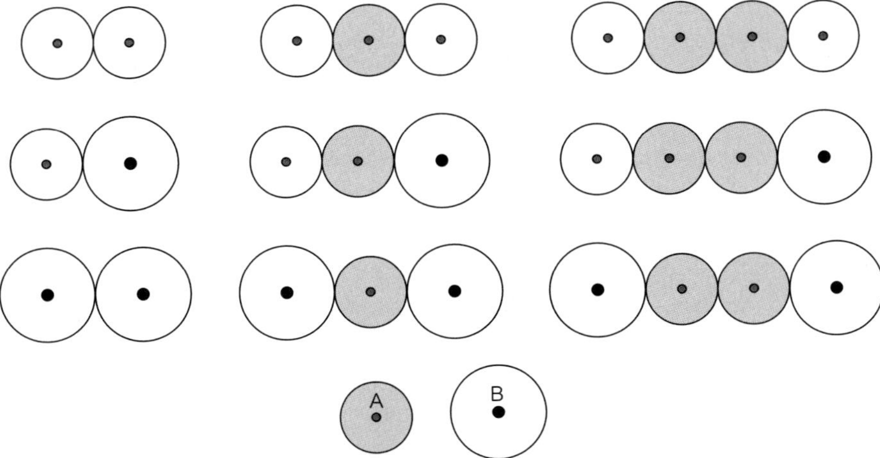

Figure 2.17. Configurations corresponding to the first three peaks of $g_{\alpha\beta}(R)$ for a system of B diluted in A (e.g., $x_A = 0.99$) corresponding to figure 2.16. The two unshaded particles are the ones under observation, i.e., these are the particles for which $g_{\alpha\beta}(R)$ is considered. The shaded particles here, which are invariably of species A, are the ones that fill the spaces between the observed particles. The locations of the expected peaks of $g_{\alpha\beta}(R)$ can be estimated with the help of this figure with $\sigma_A = 1.0$ and $\sigma_B = 1.5$.

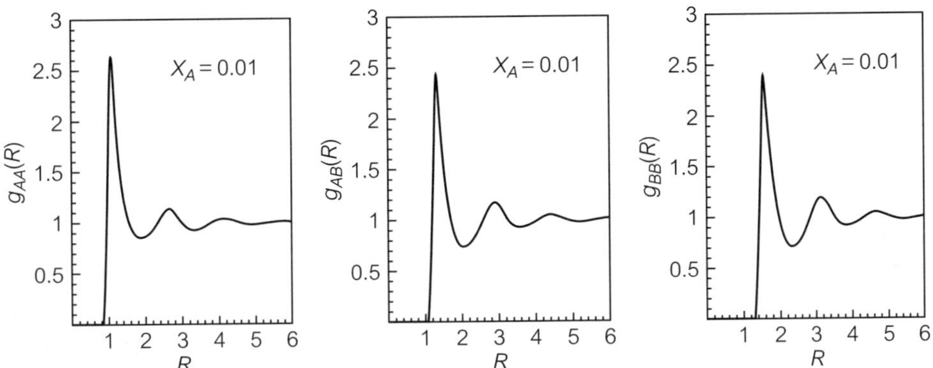

Figure 2.18. The three pair correlation functions g_{AA}, $g_{AB} = g_{BA}$ and g_{BB} for the same system as in figure 2.16 and $\eta = 0.45$ and $x_A = 0.01$.

(2) System dominated by the presence of B's, between any pair of particles, i.e., $x_A \approx 0$.

This is the other extreme case where $x_A \approx 0$ or $x_B \approx 1$. Figure 2.18 shows the pair correlation functions for this case. Here A is diluted in B and the separation between the peaks is determined by σ_{BB}, since now it is B that dominate the space between any pair of particles. Thus the first peak of $g_{AA}(R)$ appears at σ_{AA} as expected. But the second and third peaks are roughly at $R \approx \sigma_{AA} + n\sigma_{BB}$, $n = 1, 2, 3$ for the second, third, and fourth peaks.

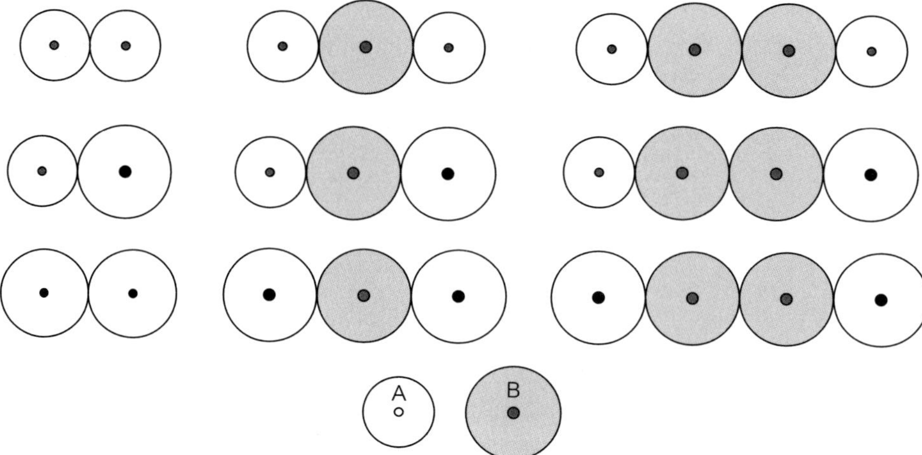

Figure 2.19. Same as figure 2.17 but for the case $x_A = 0.01$. The particles that fill the space between the pair $\alpha\beta$ in $g_{\alpha\beta}(R)$ are now B particles.

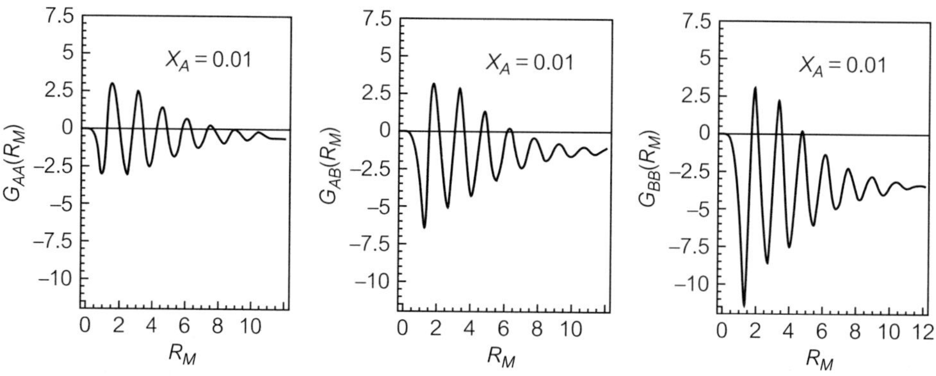

Figure 2.20. The functions $G_{\alpha\beta}(R_M)$ for the same system as that of figure 2.18.

Figure 2.19 shows the configurations corresponding to first three peaks of $g_{\alpha\beta}(R)$ for the system of A diluted in B. Note that in this case it is the B particles that fill the space between the pair of particles for which $g_{\alpha\beta}(R)$ is under consideration.

Figures 2.20 and 2.21 show the functions $G_{\alpha\beta}(R_M)$ and the potential of mean force $W_{\alpha\beta}(R)$ for the same system as in figure 2.18.

(3) Systems of intermediate composition; $x_A = 0.64$.

Figure 2.22 shows the pair correlation functions $g_{\alpha\beta}(R)$ for the composition $x_A = 0.64$. The most remarkable feature of these curves is the almost complete

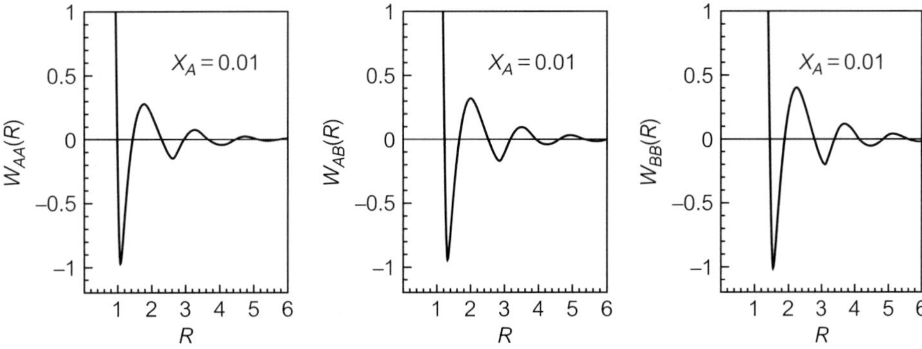

Figure 2.21. The potential of mean force $W_{\alpha\beta}$ (R) for the same system as in figure 2.18.

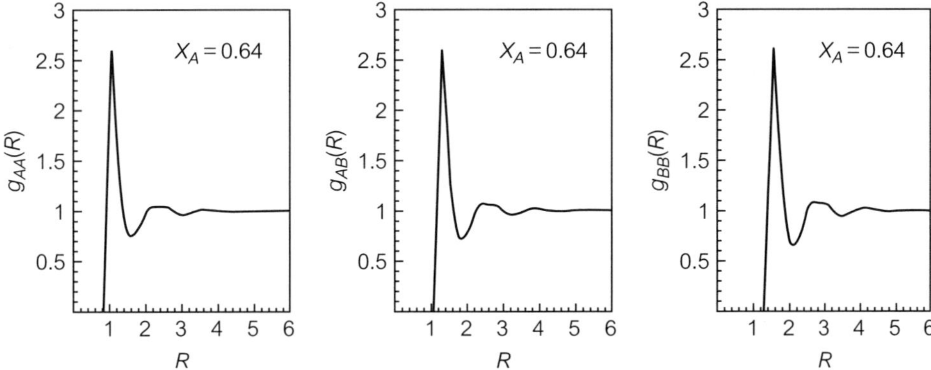

Figure 2.22. The three pair correlation functions g_{AA}, $g_{AB}=g_{BA}$ and g_{BB} for the same system as in figure 2.16 but with $\eta=0.45$ and $x_A=0.64$. Note the relatively flat region where a second peak is expected.

disappearance of the third and fourth peaks. The second peak is less pronounced than in the case of either $x_A=0.99$, or $x_A=0.01$. Since there is no component that is dominant in this case, we cannot describe the most likely configuration as we did in figure 2.17 and 2.19.

It is interesting to note the composition dependence of g_{AA} (R) in the region $1.2 \leq R \leq 3.0$. The most important point to be noted is the way the location of the second peak changes from about $\sigma_{AA}+\sigma_{AA}$ at $x_A=0.99$ (A being the "solvent") to about $\sigma_{AA}+\sigma_{BB}$ at $x_A=0.01$ (B being the "solvent"). The second peak has its maximal value of about 1.2 for the case $x_A=0.99$. It gradually decreases when the composition changes until about $x_A=0.64$. The curve becomes flat in the region between $\sigma_{AA}+\sigma_{AA}$ and $\sigma_{AA}+\sigma_{BB}$. When x_A decreases further, a new peak starts to develop at $\sigma_{AA}+\sigma_{BB}$, which reaches its highest value of about 1.18 at $x_A=0.01$. Figure 2.23a shows g_{AA} (R) for the three compositions, $x_A=0.01$, 0.64 and 0.99, and figure 2.23b shows the

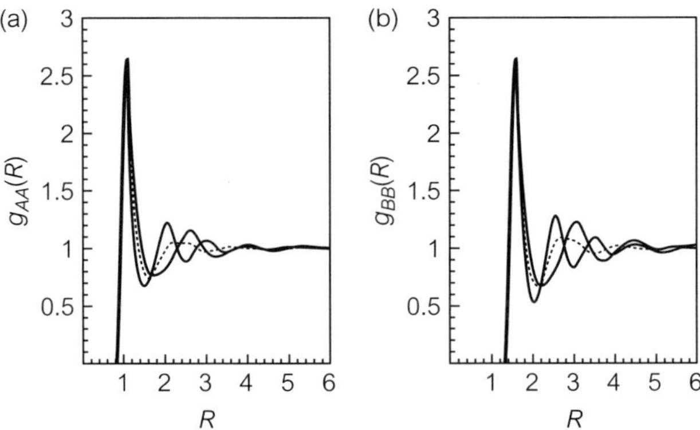

Figure 2.23. (a) The function $g_{AA}(R)$ for three compositions: $x_A = 0.01$, 0.99 and 0.64 (dashed). Other parameters are as in figure 2.18. (b) The function $g_{BB}(R)$ for the three compositions: $x_A = 0.01$, 0.99 and 0.64 (dashed). Other parameters are as in figure 2.18.

behavior of $g_{BB}(R)$ for the same three compositions. A more detailed variation of $g_{AA}(R)$ as a function of x_A has been described by Ben-Naim (1992).

We stress that the fading away of the second peak of $g_{AA}(R)$ as the composition changes is not a result of the decrease in the density of the system. We recall that in a one-component system all the peaks of $g(R)$, except the first one, will vanish as $\rho \to 0$. The same is true in the mixture if we let $\rho_A + \rho_B \to 0$. In both cases the disappearance of successive peaks in $g_{\alpha\beta}(R)$ is simply a result of the fact that as $\rho \to 0$ the availability of the particles to occupy the space between the "tagged dimer" become vanishingly small. The phenomenon we have observed in the mixture at a relatively high volume density ($\eta = 0.45$) is not a result of the scarcity of particles in the system but a result of the competition between the species A and B, to occupy the space between the two selected particles.

We recall that the location of the second peak is determined principally by the size of the particles that fill the space between the two selected particles. For $x_A = 0.99$ it is most likely that the space will be filled by A molecules. Similarly, for $x_A = 0.01$, it is most probable that the B molecules will be filling the space. The strong peak at $2\sigma_{AA}$ in the first case and at $\sigma_{AA} + \sigma_{BB}$ in the second case reflects the high degree of certainty with which the system chooses the species for filling the space between any pair of selected particles. As the mole fraction of A decreases, the B molecules become competitive with A for the "privilege" of filling the space. At about $x_A \approx 0.64$, B is in a state of emulating A (in the sense of filling the space). The situation is schematically shown in figure 2.24.

The fact that this occurs at $x_A \approx 0.64$ and not, say at $x_A \approx 0.5$ is a result of the difference in σ of the two components. Since B is "larger" than A, its prevalence as volume occupant is effective at $x_B \approx 0.36 < 0.5$. The fading of the second

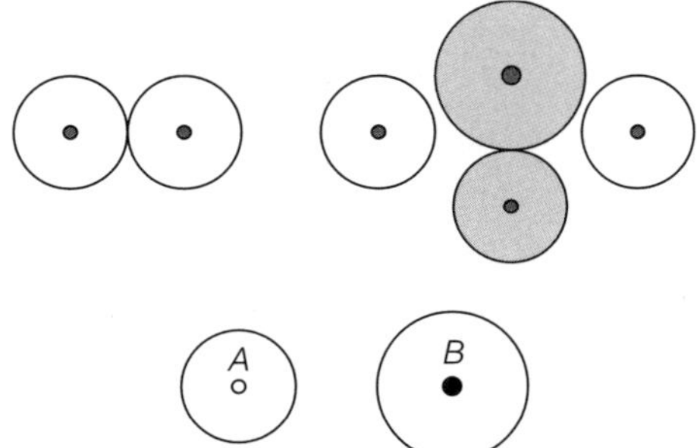

Figure 2.24. A schematic description of the competition between A and B to fill the spaces between two A's (the smaller circles).

peak reflects the inability of the system to "make a decision" as to which kind of particle should be filling the space between the two selected particles. We shall see in the next section an equivalent interpretation in terms of the force acting between the two particles. As in section 2.5, we stress again that all our considerations here are valid for spherical particles. For mixtures of more complicated molecules, the location of the various peaks is determined both by the abundance of molecules occupying the space between the tagged particles as well as by the strength of the intramolecular forces between the various species. For example, for the pair distribution function for two methane molecules in water, the second peak is determined by the structure of water and less by its molecular volume.

2.10 **Potential of mean force in mixtures**

In section 2.8, we defined the potential of mean force (PMF) between two tagged particles in a one-component system. This definition can be extended to any pair of species; for example, for the A–A pair, the potential of average force is defined by

$$g_{AA}\,(R) = \exp[-\beta W_{AA}(R)]. \tag{2.154}$$

Similar definitions apply to other pairs of species. Repeating exactly the same procedure as in section 2.8, we can show that the gradient of $W_{AA}\,(R)$ is related to the average force between the two tagged particles. The generalization of the expression (2.123) is quite straightforward. The force acting on the first A particle at R', given a second A particle at R'', can be written as

$$\begin{aligned} \boldsymbol{F}_1 = {} & -\boldsymbol{\nabla}' U_{AA}(\boldsymbol{R}', \boldsymbol{R}'') - \int d\boldsymbol{R}_A[\boldsymbol{\nabla}' U_{AA}\,(\boldsymbol{R}_A, \boldsymbol{R}')]\rho(\boldsymbol{R}_A/\boldsymbol{R}', \boldsymbol{R}'') \\ & - \int d\boldsymbol{R}_B[\boldsymbol{\nabla}' U_{BA}\,(\boldsymbol{R}_B, \boldsymbol{R}')]\rho(\boldsymbol{R}_B/\boldsymbol{R}', \boldsymbol{R}''). \end{aligned} \tag{2.155}$$

The first term on the rhs of (2.155) is simply the direct force exerted on the first A particle at $\boldsymbol{R}'$, by the second A particle at $\boldsymbol{R}''$. The average force exerted by the solvent now has two terms, instead of one in (2.123). The quantity $-\boldsymbol{\nabla}' U_{AA}(\boldsymbol{R}_A, \boldsymbol{R}')$ is the force exerted by any A particle (other than the two selected A's) located at $\boldsymbol{R}_A$ on the particle at the fixed position $\boldsymbol{R}'$ and $\rho(\boldsymbol{R}_A/\boldsymbol{R}', \boldsymbol{R}'')$ is the conditional density of A particles at $\boldsymbol{R}_A$, given two A's at $\boldsymbol{R}'$ and $\boldsymbol{R}''$. Integration over all locations of $\boldsymbol{R}_A$ gives the average force exerted by the A component on the A particle at $\boldsymbol{R}'$. Similarly, the third term on the rhs of (2.155) is the average force exerted by the B component on the A particle at $\boldsymbol{R}'$. The combination of the two last terms can be referred to as the "solvent" induced force (the term "solvent" is used here for all the particles in the system except the two selected or "tagged" particles).

Two extreme cases of equation (2.155) are the following. If $\rho_B \to 0$, then $\rho(\boldsymbol{R}_B/\boldsymbol{R}', \boldsymbol{R}'') \to 0$ also and the third term on the rhs of (2.155) vanishes. This is the case of a pure A. The "solvent" in this case will consist of all the A particles other than the two tagged particles at $\boldsymbol{R}'$ and $\boldsymbol{R}''$.

The second extreme case occurs when $\rho_A \to 0$. Note, however, that we still have two A's at fixed positions ($\boldsymbol{R}'$, $\boldsymbol{R}''$), but otherwise the solvent (here in the conventional sense) is pure B. We have the case of an extremely dilute solution of A in pure B. Note also that at the limit $\rho_A \to 0$, both the pair and the singlet distribution functions of A tend to zero, i.e.,

$$\rho_{AA}^{(2)}(\boldsymbol{R}', \boldsymbol{R}'') \to 0 \tag{2.156}$$

$$\rho_{A}^{(1)}(\boldsymbol{R}') \to 0. \tag{2.157}$$

However, the pair correlation function as well as the potential of average force are finite at this limit. We can think of $W_{AA}(R)$ in the limit of $\rho_A \to 0$ as the work required to bring two A's from infinite separation to the distance R in a *pure* solvent B at constant T and V (or T, P depending on the ensemble we use).

As in the case of pure liquids, the solvent-induced force can be attractive or repulsive even in regions where the direct force is negligible. An attractive force corresponds to a positive slope of $W(R)$, or equivalently, to a negative slope of $g(R)$. The locations of attractive and repulsive regions change when the composition of the system changes. Specifically, for $x_A \to 1$ we have the second peak of $g_{AA}(R)$ at about $\sigma_{AA} + \sigma_{AA} \approx 2$. On the other hand, for $x_A \to 0$, the second peak of $g_{AA}(R)$ is at $\sigma_{AA} + \sigma_{AB} \approx 2.5$ Clearly, there are regions that are attractive when $x_A \to 1$ (say $2 \leq R \leq 2.5$), but become repulsive when $x_A \to 0$. Therefore, when we change the composition of the system continuously, there are regions in which the two terms on the rhs of (2.155) produce forces in different directions. The result is a net diminishing of the overall solvent-induced force between the two tagged A particles. This corresponds to the flattening of $g(R)$ or of $W(R)$ which we have observed in figures 2.23a and 2.23b at $x_A \approx 0.64$.

Another useful way of examining the behavior of say, $g_{AA}(R)$ in a mixture of A and B is to look at the first-order expansion of $g_{AA}(R)$ in ρ_A and ρ_B. The generalization of (2.49) for two-component systems is

$$g_{AA}(R) = \exp[-\beta U_{AA}(R)] \left[1 + \rho_A \int f_{AA}(\mathbf{R}', \mathbf{R}_A) f_{AA}(\mathbf{R}_A, \mathbf{R}'') \, d\mathbf{R}_A \right.$$
$$\left. + \rho_B \int f_{AB}(\mathbf{R}', \mathbf{R}_B) f_{BA}(\mathbf{R}_B, \mathbf{R}'') \, d\mathbf{R}_B + \cdots \right] \tag{2.158}$$

where $f_{\alpha\beta}$ are defined as

$$f_{\alpha\beta}(\mathbf{R}', \mathbf{R}'') = \exp[-\beta U_{\alpha\beta}(\mathbf{R}', \mathbf{R}'')] - 1. \tag{2.159}$$

As we have discussed in section 2.5, we expect that the first integral, on the rhs of equation (2.158), will contribute an attractive force (even for hard-sphere particles) in the region $\sigma_{AA} \leq R \leq 2\sigma_{AA}$, whereas the second integral will have an attractive region at $\sigma_{AB} \leq R \leq 2\sigma_{AB}$.

THREE

Thermodynamic quantities expressed in terms of molecular distribution functions

In this chapter, we derive some of the most important relations between thermodynamic quantities and the molecular distribution function (MDF). As in the previous chapter, we shall first present the relation for a one-component system. This is done mainly for notational convenience. One can easily repeat exactly the same steps to derive the generalized relation for a multicomponent system. This is, in general, not necessary to do. As a rule, once we have the relation for a one-component system, we can almost straightforwardly write down the generalized relation without resorting to a full derivation. All that is needed is a clear understanding of the meaning of the various terms of the relations. An exception to this rule is the relation for the isothermal compressibility. Here, the one-component equation does not provide any clue for its generalization. We shall devote part of chapter 4 to derive the generalization of the compressibility equation, along with other relations between thermodynamic quantities and the MDF.

Most of the relations discussed in this chapter apply to systems obeying the assumption of pairwise additivity for the total potential energy. We shall indicate, however, how to modify the relations when higher order potentials are to be incorporated in the formal theory. In general, higher order potentials bring in higher order MDFs. Since very little is known about the analytical behavior of the latter, such relationships are rarely useful in applications.

Of particular importance to solution chemistry is the expression for the chemical potential, first derived by Kirkwood (1933). We shall devote a relatively large part of this chapter to discuss various expressions for the chemical potential.

The derivations carried out in this chapter apply to systems of simple spherical particles. We shall also point out the appropriate generalizations for non-spherical particles that do not possess internal rotations. For particles with internal rotations, one needs to take the appropriate average over all conformations. An example of such an average is discussed in chapter 7.

There are some steps common to most of the procedures leading to the relations between thermodynamic quantities and the pair distribution function. Therefore, in the next section we derive a general theorem connecting averages of pairwise quantities and the pair distribution function.

3.1 **Average values of pairwise quantities**

Consider an average of a general function of the configuration, $F(\mathbf{X}^N)$, in the T, V, N ensemble:

$$\langle F\rangle = \int\cdots\int d\mathbf{X}^N P(\mathbf{X}^N)F(\mathbf{X}^N), \tag{3.1}$$

with

$$P(\mathbf{X}^N) = \frac{\exp[-\beta U_N(\mathbf{X}^N)]}{Z_N}. \tag{3.2}$$

A pairwise quantity is defined as a function that is expressible as a sum of terms, each of which depends on the configuration of a pair of particles, namely

$$F(\mathbf{X}^N) = \sum_{i\neq j}\sum_{j} f(\mathbf{X}_i, \mathbf{X}_j) \tag{3.3}$$

where the sum is over all *different* pairs. In most of the applications, we shall have a factor of $\frac{1}{2}$ in (3.3) to account for the fact that this sum counts each pairwise function $f(\mathbf{X}_i, \mathbf{X}_j)$ twice, i.e., $f(\mathbf{X}_1, \mathbf{X}_2)$ appears when $i=1$ and $j=2$ and when $i=2$ and $j=1$. In the present treatment, all of the N particles are presumed to be equivalent, so that the function f is the same for each pair of indices. (The extension to mixtures will be discussed at the end of this section.)

Substituting (3.3) in (3.1) we get

$$\begin{aligned}\langle F\rangle &= \int\cdots\int d\mathbf{X}^N P(\mathbf{X}^N)\sum_{i\neq j} f(\mathbf{X}_i, \mathbf{X}_j)\\ &= \sum_{i\neq j}\int\cdots\int d\mathbf{X}^N\ P(\mathbf{X}^N) f(\mathbf{X}_i, \mathbf{X}_j)\end{aligned}$$

$$= N(N-1)\int\cdots\int d\mathbf{X}^N\ P(\mathbf{X}^N)\ f(\mathbf{X}_1,\mathbf{X}_2)$$
$$= \int d\mathbf{X}_1\int d\mathbf{X}_2\ f(\mathbf{X}_1,\mathbf{X}_2)\left[N(N-1)\int\cdots\int d\mathbf{X}_3\ldots d\mathbf{X}_N\ P(\mathbf{X}^N)\right]$$
$$= \int d\mathbf{X}_1\int d\mathbf{X}_2\ f(\mathbf{X}_1,\mathbf{X}_2)\ \rho^{(2)}(\mathbf{X}_1,\mathbf{X}_2). \tag{3.4}$$

It is instructive to go through the steps in (3.4) since these are standard steps in the theory of classical fluids. In the first step, we have merely interchanged the signs of summation and integration. In the second step, we exploit the fact that all particles are equivalent; thus each term in the sum has the same numerical value, independent of the indices i, j. Hence, we replace the sum over $N(N-1)$ terms by $N(N-1)$ times one integral. In the latter, we have chosen the (arbitrary) indices 1 and 2.

Clearly, due to the equivalence of the particles, we could have chosen any other two indices. The third and fourth steps make use of the definition of the pair distribution function defined in section 2.2.

We can rewrite the final result of (3.4) as

$$\langle F\rangle = \int d\mathbf{X}'\int d\mathbf{X}''f(\mathbf{X}',\mathbf{X}'')\rho^{(2)}(\mathbf{X}',\ \mathbf{X}'') \tag{3.5}$$

where we have changed to primed vectors to stress the fact that we do not refer to any specific pair of particles.

A simpler version of (3.5) may be obtained for spherical particles, for which each configuration $\mathbf{X}$ consists only of the locational vector $\mathbf{R}$. This is the most frequently used case in the theory of simple fluids. The corresponding expression for the average quantity in this case is

$$\langle F\rangle = \int d\mathbf{R}'\int d\mathbf{R}''f(\mathbf{R}',\mathbf{R}'')\rho^{(2)}(\mathbf{R}',\mathbf{R}''). \tag{3.6}$$

Normally the function $f(\mathbf{R}', \mathbf{R}'')$ depends only on the separation between the two points $R = |\mathbf{R}'' - \mathbf{R}'|$. In addition, for homogeneous and isotropic fluids, $\rho^{(2)}(\mathbf{R}', \mathbf{R}'')$ depends only on the scalar R. This permits the transformation of (3.6) into a one-dimensional integral. To do this, we first transform to relative coordinates

$$\overline{\mathbf{R}} = \mathbf{R}', \quad \mathbf{R} = \mathbf{R}'' - \mathbf{R}'. \tag{3.7}$$

Hence,

$$\langle F\rangle = \int d\overline{\mathbf{R}}\int d\mathbf{R}f(\mathbf{R})\rho^{(2)}(\mathbf{R}) = V\int d\mathbf{R}f(\mathbf{R})\rho^{(2)}(\mathbf{R}). \tag{3.8}$$

The integration over the entire volume yields the volume V.

Next, we transform to polar coordinates:

$$d\mathbf{R} = dx\,dy\,dz = R^2 \sin\theta\, d\theta\, d\phi\, dR. \tag{3.9}$$

Since the integrand in the last form of (3.1.8) depends only on the scalar R, we can integrate over all the orientations to get the final form

$$\begin{aligned}\langle F\rangle &= V\int_0^\infty f(R)\rho^{(2)}(R)4\pi R^2\,dR \\ &= \rho^2 V\int_0^\infty f(R)g(R)4\pi R^2\,dR.\end{aligned} \tag{3.10}$$

It is clear from (3.10) that a knowledge of the pairwise function $f(R)$ in (3.3), together with the radial distribution function $g(R)$, is sufficient to evaluate the average quantity $\langle \mathrm{F}\rangle$. Note that we have taken as infinity the upper limit of the integral in (3.10). This is not always permitted. In most practical cases, however, $f(R)$ will be of finite range, i.e., $f(R)\approx 0$ for $R > R_C$. Since $g(R)$ tends to unity at distances of a few molecular diameters (excluding the region near the critical point), the upper limit of the integral can be extended to infinity without affecting the value of the integral.

Now, we briefly mention two straightforward extensions of equation (3.5).

(1) For mixtures of, say, two components, A and B, a pairwise function is defined as

$$\begin{aligned}F(\mathbf{X}^{N_A+N_B}) &= \sum_{i\neq j} f_{AA}(\mathbf{X}_i, \mathbf{X}_j) + \sum_{i\neq j} f_{BB}(\mathbf{X}_i, \mathbf{X}_j) \\ &\quad + \sum_{i=1}^{N_A}\sum_{j=1}^{N_B} f_{AB}(\mathbf{X}_i, \mathbf{X}_j) + \sum_{i=1}^{N_B}\sum_{j=1}^{N_A} f_{BA}(\mathbf{X}_i, \mathbf{X}_j)\end{aligned} \tag{3.11}$$

where $\mathbf{X}^{N_A+N_B}$ stands for the configuration of the whole system of $N_A + N_B$ particles of species A and B. Here, $f_{\alpha\beta}$ is the pairwise function for the pair of species α and $\beta(\alpha = A,\ B$ and $\beta = A,\ B)$. Altogether, we have in (3.11) $N_A(N_A - 1) + N_B(N_B - 1) + 2N_AN_B$ terms which correspond to the total of $(N_A + N_B)\ (N_A + N_B - 1)$ pairs in the system. Note that here we count the pair i, i and j, i as different pairs†.

Note that in (3.11) we have assumed summation over $i\neq j$ for pairs of the same species. This is not required for pairs of different species. Using exactly

† We also note that, as in (3.3), for most quantities of interest we need only half of the sums in (3.11). See the example in the next section.

the same procedure as for the one-component system, we get for the average quantity the result

$$\begin{aligned}\langle F\rangle = &\int d\mathbf{X}' \int d\mathbf{X}''\, f_{AA}(\mathbf{X}', \mathbf{X}'')\rho_{AA}^{(2)}(\mathbf{X}', \mathbf{X}'')\\ &+ \int d\mathbf{X}' \int d\mathbf{X}'' f_{BB}(\mathbf{X}', \mathbf{X}'')\rho_{BB}^{(2)}(\mathbf{X}', \mathbf{X}'')\\ &+ \int d\mathbf{X}'' \int d\mathbf{X}'' f_{AB}(\mathbf{X}', \mathbf{X}'')\rho_{AB}^{(2)}(\mathbf{X}', \mathbf{X}'')\\ &+ \int d\mathbf{X}' \int d\mathbf{X}'' f_{BA}(\mathbf{X}', \mathbf{X}'')\rho_{BA}^{(2)}(\mathbf{X}', \mathbf{X}'')\end{aligned} \tag{3.12}$$

where $\rho_{\alpha\beta}^{(2)}$ are the pair distribution functions for species α and β.

(2) For functions F that depend on pairs as well as on triplets of particles of the form

$$F(\mathbf{X}^N) = \sum_{i\neq j} f(\mathbf{X}_i, \mathbf{X}_j) + \sum_{i\neq j\neq k} h(\mathbf{X}_i, \mathbf{X}_j, \mathbf{X}_k) \tag{3.13}$$

the corresponding average is

$$\begin{aligned}\langle F\rangle = &\int d\mathbf{X}' \int d\mathbf{X}'' f(\mathbf{X}', \mathbf{X}'')\rho^{(2)}(\mathbf{X}', \mathbf{X}'')\\ &+ \int d\mathbf{X}' \int d\mathbf{X}'' \int d\mathbf{X}''' h(\mathbf{X}', \mathbf{X}'', \mathbf{X}''')\rho^{(3)}(\mathbf{X}', \mathbf{X}'', \mathbf{X}''').\end{aligned} \tag{3.14}$$

The arguments leading to (3.14) are the same as those for (3.6). The new element which enters here is the triplet distribution function. Similarly, we can write formal relations for average quantities which depend on larger numbers of particles. The result would be integrals involving successively higher order molecular distribution functions. Unfortunately, even (3.14) is rarely useful since we do not have sufficient information on $\rho^{(3)}$.

3.2 Internal energy

We now derive an important expression for the internal energy of a liquid. Consider a system in the T, V, N ensemble and assume that the total potential energy of the interaction is pairwise additive, namely,

$$U_N(\mathbf{X}^N) = \tfrac{1}{2}\sum_{i\neq j} U(\mathbf{X}_i, \mathbf{X}_j). \tag{3.15}$$

The factor $\frac{1}{2}$ is included in (3.15) since the sum over $i \neq j$ counts each pair interaction twice.

The canonical partition function for the system is

$$Q(T, V, N) = \frac{q^N}{N!} Z_N = \frac{q^N}{N!} \int \cdots \int d\mathbf{X}^N \exp[-\beta U_N(\mathbf{X}^N)] \tag{3.16}$$

where the momentum partition function is included in q^N.

The internal energy of the system is given by†

$$E = -T^2 \frac{\partial(A/T)}{\partial T} = kT^2 \frac{\partial \ln Q}{\partial T} = NkT^2 \frac{\partial \ln q}{\partial T} + kT^2 \frac{\partial \ln Z_N}{\partial T}. \tag{3.17}$$

The first term on the rhs includes the internal and the kinetic energy of the individual molecules. For instance, for spherical and structureless molecules, we have $q = \Lambda^{-3}$ and hence

$$N\varepsilon^K = NkT^2 \left(\frac{\partial \ln q}{\partial T} \right) = \tfrac{3}{2} NkT \tag{3.18}$$

which is the average translational kinetic energy of the molecules. This consists of $\frac{1}{2}kT$ for the average kinetic energy per particle along the x, y, and z coordinates.

The second term on the rhs (3.17) is the average energy of interaction among the particles. This can be seen immediately by performing the derivative of the configurational partition function:

$$\begin{aligned} kT^2 \frac{\partial \ln Z_N}{\partial T} &= \frac{\int \cdots \int d\mathbf{X}^N \exp[\beta U_N(\mathbf{X}^N)] \; U_N(\mathbf{X}_N)}{Z_N} \\ &= \int \cdots \int d\mathbf{X}^N P(\mathbf{X}^N) U_N(\mathbf{X}^N) = \langle U_N \rangle. \end{aligned} \tag{3.19}$$

Hence, the total internal energy is

$$E = N\varepsilon^K + \langle U_N \rangle \tag{3.20}$$

where $N\varepsilon^K$ originates from the first term on the rhs of (3.17), which in general can include contributions from the translational, rotational, and vibrational energies.

The average potential energy in (3.20), with the assumption of pairwise additivity (3.15), fulfills the conditions of the previous section; hence, we can immediately apply theorem (3.5) to obtain

$$E = N\varepsilon^K + \tfrac{1}{2} \int d\mathbf{X}' \int d\mathbf{X}'' \; U(\mathbf{X}', \mathbf{X}'') \; \rho^{(2)}(\mathbf{X}', \mathbf{X}''). \tag{3.21}$$

† Note that E is referred to as the internal energy in the thermodynamic sense. ε^K designates the internal energy of a *single* molecule.

For spherical particles, we can transform relation (3.16) into a one-dimensional integral. Using the same arguments as were used to derive (3.10), we get from (3.21)

$$E = N\varepsilon^K + \tfrac{1}{2}N\rho \int_0^\infty U(R)g(R)4\pi R^2\, dR. \qquad (3.22)$$

Note again that integration in (3.22) extends to infinity. The reason is that $U(R)$ usually has a range of a few molecular diameters; hence, the main contribution to the integral on the rhs (3.22) comes from the finite region around the origin.

The interpretation of the second term on the rhs of (3.22) is quite simple. We select a particle and compute its total interaction with the rest of the system. Since the local density of particles at a distance R from the center of the selected particle is $\rho g(R)$, the average number of particles in the spherical element of volume $4\pi R^2\, dR$ is $\rho g(R)4\pi R^2\, dR$. Hence, the average interaction of a given particle with the rest of the system is

$$\int_0^\infty U(R)\rho g(R)4\pi R^2\, dR. \qquad (3.23)$$

We now repeat the same computation for each particle. Since the N particles are identical, the average interaction of each particle with the medium is the same. However, if we multiply (3.23) by N, we will be counting each pair interaction twice. Hence, we must multiply by N and divide by two to obtain the average interaction energy for the whole system, i.e.,

$$\tfrac{1}{2}N\rho \int_0^\infty U(R)g(R)4\pi R^2\, dR. \qquad (3.24)$$

Once we have an analytical form for $U(R)$ and acquired information (from either theoretical or experimental sources) on $g(R)$, we can compute the energy of the system by a one-dimensional integration.

The generalization of the result (3.21) or (3.22) is quite straightforward once we recognize the meaning of each term. The first term is the total kinetic and internal energy of all the particles. Instead of $N\varepsilon^K$ we simply have to write a sum over all species in the system, i.e., $\sum_{i=j}^{c} N_i\, \varepsilon_i^K$ where the sum extends over all the c species. Similarly, the second term on the rhs of (3.21) should be replaced by a double sum over all pairs of species. The final result for a c-component system is thus

$$E = \sum_{i=1}^{c} N_i\, \varepsilon_i^k + \tfrac{1}{2}\sum_{i=1}^{c}\sum_{j=1}^{c} \int d\mathbf{X}' \int d\mathbf{X}''\, U_{ij}(\mathbf{X}', \mathbf{X}'')\rho_{ij}^{(2)}(\mathbf{X}', \mathbf{X}''). \qquad (3.25)$$

3.3 The pressure equation

We first derive the pressure equation for a one-component system of spherical particles. This choice is made only because of notational convenience. We shall quote the equation for nonspherical particles at the end of this section, along with the generalization for multicomponent systems.

The pressure in the T, V, N ensemble is obtained from the Helmholtz energy by

$$P = -\left(\frac{\partial A}{\partial V}\right)_{T,N} \tag{3.26}$$

where

$$A = -kT \ln Q(T, V, N). \tag{3.27}$$

Note that the dependence of Q on the volume comes only through the configurational partition, hence

$$P = kT\left(\frac{\partial \ln Z_N}{\partial V}\right)_{T,N}. \tag{3.28}$$

To continue, we first express Z_N explicitly as a function of V. For macroscopic systems, we assume that the pressure is independent of the geometric form of the system. Hence, for convenience, we choose a cube of edge $L = V^{\frac{1}{3}}$ so that the configurational partition function is written as

$$Z_N = \int_0^L \cdots \int_0^L dx_1\, dy_1\, dz_1 \cdots dx_N\, dy_N\, dz_N \exp[-\beta U_N(\mathbf{R}^N)]. \tag{3.29}$$

Next we transform variables:

$$x_i' = V^{\frac{-1}{3}} x_i \quad y_i' = V^{\frac{-1}{3}} y_i, \quad z_i' = V^{\frac{-1}{3}} z_i \tag{3.30}$$

so that the limits of the integral in (3.29) become independent of V, hence we write

$$Z_N = V^N \int_0^1 \cdots \int_0^1 dx_1'\, dy_1'\, dz_1' \cdots dx_N'\, dy_N'\, dz_N' \exp(-\beta U_N). \tag{3.31}$$

After the transformation of variables, the total potential becomes a function of the volume, i.e.,

$$U_N = \tfrac{1}{2}\sum_{i\neq j} U(R_{ij}) = \tfrac{1}{2}\sum_{i\neq j} U\left(V^{\frac{1}{3}} R_{ij}'\right). \tag{3.32}$$

The relation between the distances expressed by the two sets of variables is

$$\begin{aligned} R_{ij} &= \left[(x_j - x_i)^2 + (y_j - y_i)^2 + (z_j - z_i)^2\right]^{\frac{1}{2}} \\ &= V^{\frac{1}{3}}\left[\left(x'_j - x'_i\right)^2 + \left(y'_j - y'_i\right)^2 + \left(z'_j - z'_i\right)^2\right]^{\frac{1}{2}} \\ &= V^{\frac{1}{3}} R'_{ij}. \end{aligned} \tag{3.33}$$

We now differentiate (3.31) with respect to the volume to obtain

$$\begin{aligned} \left(\frac{\partial Z_N}{\partial V}\right)_{T,N} &= NV^{N-1}\int_0^1 \cdots \int_0^1 dx'_1 \cdots dz'_N \exp(-\beta U_N) \\ &+ V^N \int_0^1 \cdots \int_0^1 dx'_1 \cdots dz'_N[\exp(-\beta U_N)]\left(-\beta \frac{\partial U_N}{\partial V}\right). \end{aligned} \tag{3.34}$$

From (3.32), we also have

$$\begin{aligned} \frac{\partial U_N}{\partial V} &= \tfrac{1}{2}\sum_{i \neq j} \frac{\partial U(R_{ij})}{\partial R_{ij}} \frac{\partial R_{ij}}{\partial V} \\ &= \tfrac{1}{2}\sum_{i \neq j} \frac{\partial U(R_{ij})}{\partial R_{ij}} \tfrac{1}{3} V^{\frac{-2}{3}} R'_{ij} \\ &= \frac{1}{6V}\sum_{i \neq j} \frac{\partial U(R_{ij})}{\partial R_{ij}} R_{ij}. \end{aligned} \tag{3.35}$$

Combining (3.34) and (3.35) and transforming back to the original variables, we obtain

$$\left(\frac{\partial \ln Z_N}{\partial V}\right)_{T,N} = \frac{N}{V} - \frac{\beta}{6V}\int \cdots \int d\mathbf{R}^N P(\mathbf{R}^N) \sum_{i \neq j} \frac{\partial U(\mathbf{R}_{ij})}{\partial R_{ij}} R_{ij}. \tag{3.36}$$

The second term on the rhs of (3.36) is an average of a pairwise quantity. Therefore, we can apply the general theorem of section 3.1 to obtain

$$P = kT\left(\frac{\partial \ln Z_N}{\partial V}\right)_{T,N} = kT\rho - \frac{\rho^2}{6}\int_0^\infty R \frac{\partial U(R)}{\partial R} g(R) 4\pi R^2 \, dR. \tag{3.37}$$

This is the pressure equation for a one-component system of spherical particles obeying the pairwise additivity for total potential energy. Note that the first term is the "ideal gas" pressure. The second is due to the effect of the intermolecular forces on the pressure. Note that in general, $g(R)$ is a function of the density; hence, the second term in (3.37) is not the second-order term in the density expansion of the pressure.

The pressure equation is very useful in computing the equation of state of a system based on the knowledge of the form of the function $g(R)$. Indeed, such computations have been performed to test theoretical methods of evaluating $g(R)$.

In a mixture of c components, the generalization of (3.37) is straightforward. Instead of the density ρ in the first term on the rhs of (3.37), we use the total density $\rho_T = \sum_i \rho_i$. Also, the second term is replaced by a double sum over all pair of species. The result is

$$P = \sum_{\alpha=1}^{c} kT\rho_\alpha - \frac{1}{6}\sum_{\alpha,\beta=1}^{c} \rho_\alpha\rho_\beta \int_0^\infty \frac{\partial U_{\alpha\beta}(R)}{\partial R} g_{\alpha\beta}(R) 4\pi R^3\, dR \tag{3.38}$$

where ρ_α is the density of the α species and $g_{\alpha\beta}(R)$ is the pair correlation function for the pair of species α and β.

For a system of rigid, nonspherical molecules, the derivation of the pressure equation is essentially the same as that for spherical molecules. The result is

$$P = kT\rho - \left(\frac{1}{6V}\right) \int d\mathbf{X}' \int d\mathbf{X}''[\mathbf{R} \cdot \nabla_{\mathbf{R}} U(\mathbf{X}', \mathbf{X}'')]\rho^{(2)}(\mathbf{X}', \mathbf{X}'') \tag{3.39}$$

where $\mathbf{R} = \mathbf{R}'' - \mathbf{R}'$ and $\nabla_{\mathbf{R}}$ is the gradient with respect to the vector $\mathbf{R}$.

3.4 The chemical potential

3.4.1 Introduction

The chemical potential is the most important quantity in chemical thermodynamics and, in particular, in solution chemistry. There are several routes for obtaining a relationship between the chemical potential and the pair correlation function. Again we start with the expression for the chemical potential in a one-component system, and then generalized to multicomponent systems simply by inspection and analyzing the significance of the various terms.

In this section, we discuss several different routes to "build up" the expression for the chemical potential. Note, however, that in actual applications only differences in chemical potentials can be measured.

The chemical potential is defined, in the T, V, N ensemble, by

$$\mu = \left(\frac{\partial A}{\partial N}\right)_{T,V}. \tag{3.40}$$

For reasons that will become clear in the following paragraphs, the chemical potential cannot be expressed as a simple integral involving the pair correlation function. Consider, for example, the pressure equation that we have derived in section 3.3 which we write symbolically as

$$P = P[g(R); \rho, T]. \tag{3.41}$$

By this notation, we simply mean that we have expressed the pressure as a function of ρ and T, and also in terms of $g(R)$, which is itself a function of ρ and T.

Since the pressure is also given by

$$P = -\left(\frac{\partial A}{\partial V}\right)_{T,N} = \left[\frac{\partial a}{\partial(\rho^{-1})}\right]_T \tag{3.42}$$

where $a = A/N$ and $\rho^{-1} = V/N$, we can integrate (3.42) to obtain

$$a = -\int P[g(R); \rho, T] d(\rho^{-1}). \tag{3.43}$$

Clearly, in order to express a in terms of $g(R)$, we must know the explicit dependence of $g(R)$ on the density. Thus, if we used the pressure equation in the integrand of (3.43) we need a second integration over the density to get the Helmholtz energy per particle.

The chemical potential can then be obtained as

$$\mu = a + Pv \tag{3.44}$$

with $v = V/N$.

A second method of computing the chemical potential is to use the energy equation derived in section 3.2, which we write symbolically as

$$E = E[g(R); \rho, T]. \tag{3.45}$$

The relation between the energy per particle and the Helmholtz energy is

$$e = \frac{E}{N} = -T^2\left\{\frac{\partial(a/T)}{\partial T}\right\}_\rho \tag{3.46}$$

which can be integrated to obtain

$$\frac{a}{T} = -N^{-1}\int E[g(R); \rho, T]/T^2 dT. \tag{3.47}$$

Again, we see that if we use the energy expression [in terms of $g(R)$] in the integrand of (3.47), we must also know the dependence of $g(R)$ on the temperature.

The two illustrations above show that in order to obtain a relation between μ and $g(R)$, it is not sufficient to know the function $g(R)$ at a given ρ and T; one needs the more detailed knowledge of $g(R)$ and its dependence on either ρ or T. This difficulty follows from the fact that the chemical potential is not an average of a pairwise quantity, and therefore the general theorem of section 3.1 is not applicable here. Nevertheless, the two procedures above are useful in the numerical computation of the chemical potential.

3.4.2 **Insertion of one particle into the system**

The chemical potential in the *T, V, N* ensemble may be written as

$$\mu = \left(\frac{\partial A}{\partial N}\right)_{T,V} = \lim_{dN \to 0}\left[\frac{A(N+dN) - A(N)}{dN}\right] = \frac{A(N+1) - A(N)}{1}. \quad (3.48)$$

In (3.48), we start with the definition of the chemical potential in the *T, V, N* system, then take the limit $dN \to 0$ as if *N* was a continuous variable. If *N* is very large, the addition of *one* particle may be viewed as an "infinitesimal" change in the variable *N*.[†]

The replacement of a derivative with respect to *N* by a difference is justified since the Helmholtz energy is an *extensive* function, i.e., it has the property $A(T, \alpha V, \alpha N) = \alpha A(T, V, N)$ for any $\alpha \geq 0$. Now define $\alpha = 1/dN$, $M = N/dN$, and $Y = V/dN$. Instead of taking the limit $dN \to 0$, we take the limits $M \to \infty$ and $Y \to \infty$, but M/Y is kept constant (this is the thermodynamic limit).

Thus, we rewrite (3.48) as

$$\begin{aligned}
\mu &= \lim_{dN \to 0}\left[\frac{A(T,V,N+dN) - A(T,V,N)}{dN}\right] \\
&= \lim_{dN \to 0}\left[A\left(T, \frac{V}{dN}, \frac{N}{dN}+1\right) - A\left(T, \frac{V}{dN}, \frac{N}{dN}\right)\right] \\
&= \lim_{\substack{Y\to\infty \\ M\to\infty \\ \rho = M/Y = \text{const.}}} A(T,Y,M+1) - A(T,Y,M). \qquad (3.49)
\end{aligned}$$

Relation (3.49) simply means that in order to compute the chemical potential, it is sufficient to compute the change of the Helmholtz energy upon the addition of *one* particle. We now use the connection between the Helmholtz energy and the canonical partition function to obtain

$$\begin{aligned}
\exp(-\beta\mu) &= \exp\{-\beta[A(T,V,N+1) - A(T,V,N)]\} = \frac{Q(T,V,N+1)}{Q(T,V,N)} \\
&= \frac{[q^{N+1}/\Lambda^{3(N+1)}(N+1)!]\int\cdots\int d\mathbf{R}_0\cdots d\mathbf{R}_N \exp(-\beta U_{N+1})}{(q^N/\Lambda^{3N}N!)\int\cdots\int d\mathbf{R}_1\cdots d\mathbf{R}_N \exp(-\beta U_N)}.
\end{aligned}$$

(3.50)

[†] Clearly, this "approximation" is not valid for any function. Take for instance $\sin(N)$; one cannot approximate the derivative $\lim_{dN\to 0}((\sin(N+dN) - \sin(N))/(dN))$, by taking $dN = 1$, no matter how large *N* is.

Note that the added particle has been given the index zero. Using the assumption of pairwise additivity of the total potential, we may split U_{N+1} into two terms:

$$\begin{aligned} U_{N+1}(\boldsymbol{R}_0, \ldots, \boldsymbol{R}_N) &= U_N(\boldsymbol{R}_1, \ldots, \boldsymbol{R}_N) + \sum_{j=1}^{N} U(\boldsymbol{R}_0, \boldsymbol{R}_j) \\ &= U_N(\boldsymbol{R}_1, \ldots, \boldsymbol{R}_N) + B(\boldsymbol{R}_0, \ldots \boldsymbol{R}_N). \end{aligned} \tag{3.51}$$

In (3.51), we have included all the interactions of the zeroth particle with the rest of the system into the quantity $B(\boldsymbol{R}_0, \ldots, \boldsymbol{R}_N)$. The quantity $B(\boldsymbol{R}_0, \ldots, \boldsymbol{R}_N)$ may be referred to as the *binding energy* of the particle at $\boldsymbol{R}_0$ to the rest of the particles at $\boldsymbol{R}_1, \ldots, \boldsymbol{R}_N$. Using (3.51) and the general expression for the basic probability density in the T, V, N ensemble, we rewrite (3.50) as

$$\begin{aligned} \exp(-\beta\mu) = \frac{q}{\Lambda^3(N+1)} \int \cdots \int d\boldsymbol{R}_0 d\boldsymbol{R}_1 \ldots d\boldsymbol{R}_N P(\boldsymbol{R}_1, \ldots, \boldsymbol{R}_N) \\ \times \exp[-\beta B(\boldsymbol{R}_0, \ldots, \boldsymbol{R}_N)]. \end{aligned} \tag{3.52}$$

Next, we transform to coordinates relative to $\boldsymbol{R}_0$, i.e.,

$$\boldsymbol{R}'_i = \boldsymbol{R}_i - \boldsymbol{R}_0, \quad i = 1, 2, \ldots, N.$$

Note that $B(\boldsymbol{R}_0, \ldots, \boldsymbol{R}_N)$ is actually a function only of the relative coordinates $\boldsymbol{R}'_1, \ldots, \boldsymbol{R}'_N$; for instance, $U(\boldsymbol{R}_0, \boldsymbol{R}_j)$ is a function of $\boldsymbol{R}'_j$ and not of both $\boldsymbol{R}_0$ and $\boldsymbol{R}_j$.

Hence, we rewrite the chemical potential as

$$\begin{aligned} \exp(-\beta\mu) = \frac{q}{\Lambda^3(N+1)} \int d\boldsymbol{R}_0 \int \cdots \int d\boldsymbol{R}'_1 \ldots d\boldsymbol{R}'_N P(\boldsymbol{R}'_1, \ldots, \boldsymbol{R}'_N) \\ \times \exp[-\beta B(\boldsymbol{R}'_1, \ldots, \boldsymbol{R}'_N)]. \end{aligned} \tag{3.53}$$

In this form, the integrand is independent of $\boldsymbol{R}_0$. Therefore, we may integrate over $\boldsymbol{R}_0$ to obtain the volume. The inner integral is simply the average in the T, V, N ensemble of the quantity $\exp(-\beta B)$, i.e.,

$$\exp(-\beta\mu) = \frac{qV}{(N+1)\Lambda^3} \langle \exp(-\beta B) \rangle. \tag{3.54}$$

Since $\rho = N/V \cong (N+1)/V$ (macroscopic system), we can rearrange (3.54) to obtain the final expression for the chemical potential:

$$\mu = kT \ln(\rho\Lambda^3 q^{-1}) - kT \ln\langle \exp(-\beta B) \rangle. \tag{3.55}$$

This is a very important and very useful expression for the chemical potential. As we shall soon see, this form is retained almost unchanged upon generalization to non- spherical particles, mixtures of species, or in different ensembles.

Note that when the newly added particle does not interact with the other particles in the system, i.e., $B \equiv 0$, the second term on the rhs of (3.55) is zero (of course, this is also true when there are no interactions among *all* the particles, in which case we have an ideal gas). The addition of a new particle at $\mathbf{R}_0$ (or equivalently at $\mathbf{R}'_0 = \mathbf{0}$) can be viewed as "turning on" of an "external field" acting on the system of N particles. This external field introduces the factor $\exp(-\beta B)$ in the expression for the chemical potential. More explicitly, if the potential energy is pairwise additive, then

$$\exp[-\beta B(\mathbf{R}_0, \ldots, \mathbf{R}_N)] = \prod_{j=1}^{N} \exp[-\beta U(\mathbf{R}_0, \mathbf{R}_j)]. \tag{3.56}$$

Clearly, this is *not a pairwise additive quantity* in the sense of (3.3), i.e., it is not a *sum*, but a *product* of pairwise functions. This is the reason why we cannot express the chemical potential as a simple integral involving only the pair distribution function.

The expression (3.54) follows directly from the definition of the chemical potential in (3.48). It was first derived in a slightly different notation by Widom (1963, 1982).

We now re-express the second term on the rhs of (3.55) in terms of the pair distribution function.

3.4.3 **Continuous coupling of the binding energy**

In section 3.4.1, we have seen that the chemical potential could be expressed in terms of $g(R)$ provided that we also know the dependence of $g(R)$ on either T or ρ. We now derive a third expression due to Kirkwood (1933), which employs the idea of a coupling parameter ξ.[†] The ultimate expression for the chemical potential would be an integral over both R and ξ involving the function $g(R, \xi)$.

We start by defining an auxillary potential function as follows:

$$U(\xi) = U_N(\mathbf{R}_1, \ldots, \mathbf{R}_N) + \xi \sum_{j=1}^{N} U(\mathbf{R}_0, \mathbf{R}_j) \tag{3.57}$$

which can be compared with (3.51). Clearly, we have the following two limiting cases:

$$U(\xi = 0) = U_N(\mathbf{R}_1, \ldots, \mathbf{R}_N) \tag{3.58}$$

$$U(\xi = 1) = U_{N+1}(\mathbf{R}_0, \ldots, \mathbf{R}_N). \tag{3.59}$$

[†] This idea is a generalization of the charging process employed in the theory of ionic solutions.

The idea is that by changing ξ from zero to unity, the function $U(\xi)$ changes continuously from U_N to U_{N+1}. Another way of saying the same thing is that by changing ξ from zero to unity the *binding energy* of the newly added particle at R_0 is "turned on" continuously. This is, of course, a thought experiment. We mentally "add" the new particle by "switching" on its interaction with the rest of the particles in the system.

Note, however, that within the assumption of pairwise additivity of the total potential energy, the quantity U_N is unaffected by this coupling of the binding energy of the newly added particle.

For each function $U(\xi)$, we also define the corresponding configurational partition function by

$$Z(\xi) = \int \cdots \int d\mathbf{R}_0\, d\mathbf{R}_1 \ldots d\mathbf{R}_N \exp[-\beta U(\xi)]. \tag{3.60}$$

Clearly, we have the following two limiting cases:

$$Z(\xi = 0) = \int \cdots \int d\mathbf{R}_0\, d\mathbf{R}_1 \ldots d\mathbf{R}_N \exp(-\beta U_N) = VZ_N \tag{3.61}$$

and

$$Z(\xi = 1) = Z_{N+1}. \tag{3.62}$$

The expression (3.50) for the chemical potential can be rewritten using the above notation as

$$\mu = kT \ln(\rho \Lambda^3 q^{-1}) - kT \ln Z(\xi = 1) + kT \ln Z(\xi = 0) \tag{3.63}$$

or, using the identity

$$kT \ln Z(\xi = 1) - kT \ln Z(\xi = 0) = kT \int_0^1 \frac{\partial \ln Z(\xi)}{\partial \xi}\, d\xi \tag{3.64}$$

we get

$$\mu = kT \ln(\rho \Lambda^3 q^{-1}) - kT \int_0^1 \frac{\partial \ln Z(\xi)}{\partial \xi}\, d\xi. \tag{3.65}$$

We can now differentiate $Z(\xi)$ in (3.60) with respect to ξ to obtain

$$\begin{aligned} kT \frac{\partial \ln Z(\xi)}{\partial \xi} &= \frac{kT}{Z(\xi)} \int \cdots \int d\mathbf{R}_0 \cdots d\mathbf{R}_N \{\exp[-\beta U(\xi)]\} \left[-\beta \sum_{j=1}^{N} U(\mathbf{R}_0, \mathbf{R}_j) \right] \\ &= -\int \cdots \int d\mathbf{R}_0 \cdots d\mathbf{R}_N P(\mathbf{R}^{N+1}, \xi) \sum_{j=1}^{N} U(\mathbf{R}_0, \mathbf{R}_j) \end{aligned}$$

$$
\begin{aligned}
&= -\sum_{j=1}^{N} \int \cdots \int d\boldsymbol{R}_0 \cdots d\boldsymbol{R}_N P(\boldsymbol{R}^{N+1}, \xi) U(\boldsymbol{R}_0, \boldsymbol{R}_j) \\
&= -N \int \int d\boldsymbol{R}_0 d\boldsymbol{R}_1 U(\boldsymbol{R}_0, \boldsymbol{R}_1) \int \cdots \int d\boldsymbol{R}_2 \cdots d\boldsymbol{R}_N P(\boldsymbol{R}^{N+1}, \xi) \\
&= -\frac{1}{N+1} \int \int d\boldsymbol{R}_0 d\boldsymbol{R}_1 U(\boldsymbol{R}_0, \boldsymbol{R}_1) \rho^{(2)}(\boldsymbol{R}_0, \boldsymbol{R}_1, \xi) \\
&= -\rho \int_0^\infty U(R) g(R, \xi) 4\pi R^2 dR. \qquad (3.66)
\end{aligned}
$$

It is instructive to go through the formal steps in (3.66). They are very similar to those in section 3.1. The only new feature in (3.66) is the appearance of the parameter ξ, in the pair distribution functions.

We now combine (3.66) with (3.65) to obtain the final expression for the chemical potential:

$$\mu = kT \ln(\rho \Lambda^3 q^{-1}) + \rho \int_0^1 d\xi \int_0^\infty U(R) g(R, \xi) 4\pi R^2 dR. \qquad (3.67)$$

We can also define the *standard chemical potential* in the ideal gas phase by

$$\mu^{0g} = kT \ln(\Lambda^3 q^{-1}) \qquad (3.68)$$

and the corresponding activity coefficient

$$kT \ln \gamma^{\text{ideal gas}} = \rho \int_0^1 d\xi \int_0^\infty U(R) g(R, \xi) 4\pi R^2 dR, \qquad (3.69)$$

to rewrite (3.67) in the form

$$\mu = \mu^{0g} + kT \ln(\rho \gamma^{\text{ideal gas}}). \qquad (3.70)$$

In (3.69) we have an explicit expression for the activity coefficient $\gamma^{\text{ideal gas}}$, which measures the extent of deviation of the chemical potential from the ideal-gas form. The quantity $\rho g(R, \xi)$, is the local density of particles around a given particle that is coupled to the extent of ξ, to the rest of the system. Note that (3.67) is not a simple integral involving $g(R)$. A more detailed knowledge of the function $g(R, \xi)$ is required to calculate the chemical potential.

The interpretation of the terms in (3.67) is as follows. Suppose that we have a system of N interacting particles at a given T and ρ, we now add a hypothetical particle which carries the same momentum and internal partition function as all other particles of the system. This particle is initially *uncoupled* in the sense of $\xi = 0$. The corresponding chemical potential of this particular particle at this stage is

$$\mu' = kT \ln(\Lambda^3 q^{-1} V^{-1}). \qquad (3.71)$$

Note that since we have added one particle which is initially different from all the other N particles, its density is $\rho' = V^{-1}$. The volume V enters here because the particle can reach any point within the system.

We now "turn on" the coupling parameter ξ until it reaches the value of unity. The chemical potential of the added particle changes in two ways. First, we have the work required to build up the interaction between the added particle and the rest of the system. This is the second term on the rhs of (3.67). Second, as long as the new particle is *distinguishable* from all the other particles (i.e., $\xi \neq 1$), its density remains fixed $\rho' = V^{-1}$. At the point $\xi = 1$, it abruptly becomes identical to the other particles. This involves an assimilation Helmholtz energy of amount (see Appendices H and I)

$$\Delta A = kT \ln(N/1) = kT \ln(\rho V). \tag{3.72}$$

This, together with the coupling work, converts (3.71) into (3.67). A second way of interpreting the two terms in (3.67), or equivalently in (3.55), will be discussed in the next section.

3.4.4 Insertion of a particle at a fixed position: the pseudo-chemical potential

The chemical potential is the work (here, at constant T, V) associated with the addition of one particle to a macroscopically large system:

$$\mu = A(T, V, N + 1) - A(T, V, N). \tag{3.73}$$

The *pseudo*-chemical potential refers to the work associated with the addition of one particle to a *fixed position* in the system, say at $\mathbf{R}_0$.†

$$\mu^* = A(T, V, N + 1;\ \mathbf{R}_0) - A(T, V, N). \tag{3.74}$$

The statistical mechanical expression for the pseudo-chemical potential can be obtained in a similar way as in (3.50), i.e., as a ratio between two partition functions corresponding to the difference in the Helmholtz energies in (3.74), i.e.,

$$\exp(-\beta\mu*) = \frac{(q^{N+1}/\Lambda^{3N}N!)\int\cdots\int d\mathbf{R}_1\ldots d\mathbf{R}_N \exp[-\beta U_{N+1}(\mathbf{R}_0,\ldots,\mathbf{R}_N)]}{(q^{N}/\Lambda^{3N}N!)\int\cdots\int d\mathbf{R}_1\ldots d\mathbf{R}_N \exp[-\beta U_{N}(\mathbf{R}_1,\ldots,\mathbf{R}_N)].} \tag{3.75}$$

It is instructive to note carefully the differences between (3.50) and (3.75). Since the added particle in (3.74) is *devoid* of the translational degree of

† This process is meaningful in classical statistical mechanics. The particle at $\mathbf{R}_0$ is assumed to have an exact location and exact velocity.

freedom, it will not bear a momentum partition function. Hence, we have Λ^{3N} in (3.75) instead of $\Lambda^{3(N+1)}$ in (3.50). For the same reason, the integration in the numerator of (3.75) is over the N *locations* $\boldsymbol{R}_1, \ldots, \boldsymbol{R}_N$ and not over $\boldsymbol{R}_0, \ldots, \boldsymbol{R}_N$ as in (3.50). Furthermore, since we have added a particle to a fixed position, it is *distinguishable* from the other particles; hence, we have $N!$ in (3.75) instead of $(N+1)!$ in (3.50).

Once we have set up the statistical mechanical expression (3.75), the following formal steps are nearly the same as in the previous section. The result is

$$\begin{aligned}\mu^* &= kT \ln q^{-1} - kT \ln\langle \exp(-\beta B)\rangle \\ &= kT \ln q^{-1} + \rho \int_0^1 d\xi \int_0^\infty U(R) g(R, \xi) 4\pi R^2 \, dR\end{aligned} \tag{3.76}$$

which should be compared with (3.55) and (3.67). Note that we have added the particle to a fixed position $\boldsymbol{R}_0$; therefore, from the formal point of view, μ^* depends on $\boldsymbol{R}_0$. However, in a homogeneous fluid, all the points of the system are presumed to be equivalent (except for a small region near the boundaries, which is negligible for our present purposes), and therefore μ^* is effectively independent of $\boldsymbol{R}_0$.

Combining (3.76) with either (3.55) or (3.67), we obtain the expression for the chemical potential

$$\mu = \mu^* + kT \ln(\rho\Lambda^3). \tag{3.77}$$

Here, the work required to add a particle to the system is split into two parts. This is shown schematically in figure 3.1. First, we add the particle to a *fixed position,* say $\boldsymbol{R}_0$, the corresponding work being μ^*. Next, we remove the constraint imposed by fixing the position of the particle; the corresponding work is the second term on the rhs of (3.77). The last quantity was referred to as the *liberation* Helmholtz energy[†]. Since we are dealing with classical statistics $\rho\Lambda^3 \ll 1$ and therefore the liberation Helmholtz energy is always negative. Thus, liberating the particle from its fixed position is always associated with a decrease in free energy. Note also that the term $kT \ln(\rho\Lambda^3)$ is in general not the ideal-gas chemical potential of the particles. The latter is $kT \ln(\rho\Lambda^3 q^{-1})$ where q is the internal partition function of the particles.

It is instructive to recognize the *three different* sources that contribute to the liberation free energy. First, the particle at a fixed position is devoid of momentum partition function (though it still has all other internal partition functions such as rotational and vibrational). Upon liberation, the particle

[†] In some articles, this term is referred to as the "mixing free energy." Clearly, since no *mixing* process occurs, we prefer the term "liberation of free energy."

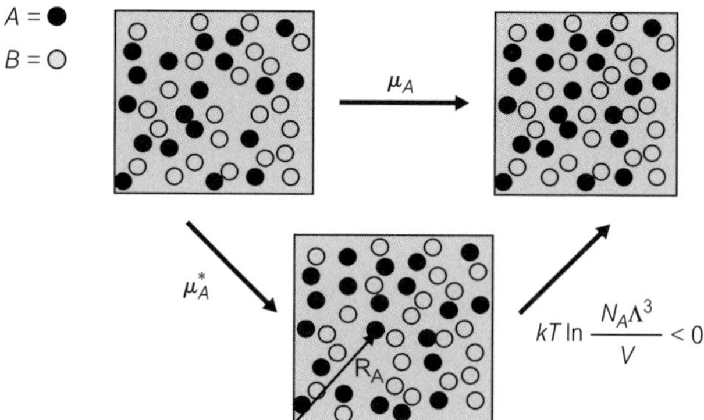

Figure 3.1 The process of adding one A particle to a solution is carried out in two steps. First, we insert the particle at a fixed position, then we release the particle to wander in the entire system. The corresponding free energy changes are μ_A^* and $kT \ln \rho_A\Lambda^3$, respectively.

acquires momentum, the distribution of which depends on the temperature. The corresponding contribution to the free energy is $kT \ln \Lambda^3$. Second, the released particle that was confined to a fixed position can now access the entire volume V. The corresponding contribution is $-kT \ln V$. Finally, and most importantly, the particle at $\boldsymbol{R}_0$ is *distinguishable* from all other particles in the system. Once it is released, it becomes *indistinguishable* from the other N members of the same particles. We call this process assimilation and the corresponding contribution to the change in free energy is $kT\ln N$. Together, the three contributions comprise the liberation free energy in which only the dimensionless quantity $\rho\Lambda^3$ features. It is important to realize that these three contributions are independent and conceptually arise from different sources. One can change one of these without changing the others (see also Appendices H and I).

3.4.5 **Building up the density of the system**

A third interpretation of the expression for the chemical potential in a one-component system may be obtained in terms of the Kirkwood–Buff integrals as discussed in section 3.5. We quote here only one result which we shall use for the purpose of this section [see equation (3.126) in section 3.5]:

$$\left(\frac{\partial \mu}{\partial \rho}\right)_T = kT\left(\frac{1}{\rho} - \frac{G}{1+\rho G}\right) \tag{3.78}$$

where G is defined by

$$G = \int_0^\infty [g(R) - 1]4\pi R^2\, dR \tag{3.79}$$

and $g(R)$ is the pair correlation function, defined in an open system (see section 3.5 for more details). Integrating (3.79) with respect to ρ (assuming that at $\rho = 0$ we have the ideal-gas behavior) we obtain

$$\mu = kT \ln \Lambda^3 q^{-1} + kT \ln \rho - kT \int_0^\rho \frac{G}{1 + \rho' G} d\rho'. \tag{3.80}$$

The third term on the rhs of (3.80) may be identified with the coupling work; i.e., comparing (3.80) with (3.55), we have

$$kT \ln \langle \exp(-\beta B) \rangle = kT \int_0^\rho \frac{G}{1 + \rho' G} d\rho'. \tag{3.81}$$

The coupling work is interpreted in (3.81) as the work required to increase the density from $\rho = 0$ to the final density ρ. A slightly different interpretation is obtained by rewriting (3.80) as

$$\begin{aligned} \mu &= (kT \ln \Lambda^3 q^{-1} + kT \ln \rho_0) + kT \int_{\rho_0}^\rho \left(\frac{1}{\rho'} - \frac{G}{1 + \rho' G} \right) d\rho' \\ &= (kT \ln \Lambda^3 q^{-1} + kT \ln \rho_0) + \left[kT \ln \frac{\rho}{\rho_0} - kT \int_{\rho_0}^\rho \frac{G}{1 + \rho' G} d\rho' \right]. \end{aligned} \tag{3.82}$$

The expression within the first set of parentheses corresponds to the work required to introduce one particle to an ideal-gas system (ρ_0 very low). The second term is the work involved in changing the density from ρ_0 to the final destiny ρ. This work is composed of two contributions; first, the change in the *assimilation* term $kT \ln \rho/\rho_0$ (note that V is constant in the process), and second, the coupling work (3.81).

Since (3.82) is valid for any $\rho_0 \approx 0$, we can put $\rho_0 = 0$ and get the expression (3.80). Note also that in order to express the chemical potential in terms of the pair correlation function, we need to take two integrations, one over R as in (3.79), and one over the density in (3.80).

3.4.6 **Some generalizations**

We now briefly summarize the modifications that must be introduced into the equation for the chemical potential for more complex systems.

(1) For systems that do not obey the assumption of pairwise additivity for the potential energy, equation (3.67) becomes invalid. In a formal way, one can derive an analogous relation involving higher order molecular distribution functions. This does not seem to be useful at present. However, in many applications for mixtures, one can retain the general expression (3.55) even

when the potential energy of the solvent does not obey any additivity assumption. We briefly discuss this case below.

(2) For rigid, nonspherical particles whose potential energy obeys the assumption of pairwise additivity, a relation similar to (3.67) holds. However, one must now integrate over the orientation as well as over the location of the particle. The generalized relation is

$$\mu = kT \ln(\rho\Lambda^3 q^{-1}) + \int_0^1 d\xi \int d\mathbf{X}'' U(\mathbf{X}', \mathbf{X}'')\rho(\mathbf{X}''/\mathbf{X}', \xi). \tag{3.83}$$

Here, q includes the rotational as well as the internal partition function of a single molecule. The quantity $\rho(\mathbf{X}'' \,/\, \mathbf{X}', \xi)$ is the local density of particles at $\mathbf{X}'$, given a particle at $\mathbf{X}'$, coupled to the extent of ξ. Clearly, the whole integral on the rhs of (3.83) does not depend on the choice of $\mathbf{X}'$ (for instance, we can take $\mathbf{R}' = \mathbf{0}$ and $\boldsymbol{\Omega}' = \mathbf{0}$ and measure $\mathbf{X}'$ relative to this configuration).

(3) For mixtures of c components, the expression for the chemical potential can be written upon inspection of the terms in the case of a one-component system. Consider first the expression (3.55), which is the more general one. Once we know the meaning of the two terms on the rhs (3.55), we can write down the chemical potential of any component i, immediately, i.e.,

$$\mu_i = kT \ln \rho_i \Lambda_i^3 q_i^{-1} - kT \ln\langle \exp[-\beta B_i] \rangle_0 \tag{3.84}$$

where the first term is the liberation term for particle of species i. This term does not depend on the presence of other species in the system, and it is the same as for pure i. The second term is the coupling work of i to the entire system. Note also that the significance of this term does not depend on any assumption of pairwise additivity i.e., B_i is defined simply as

$$B_i = U(N_1, N_2, \ldots, N_i + 1, \ldots, N_c) - U(N_1, N_2, \ldots, N_i, \ldots, N_c) \tag{3.85}$$

i.e., B_i is the change in the total potential energy of the system being at a specific configuration, upon the addition of one particle of type i at a fixed position, say $\mathbf{R}_0$.

Note that the average $\langle\ \rangle_0$ in (3.84) is over all the configurations of the "solvent" molecules, i.e., all the molecules of the system except the one placed at a fixed position.

If the total potential energy does fulfill the assumption of pairwise additivity, then we can obtain the generalization of equation (3.67) as

$$\mu_i = kT \ln \rho_i \Lambda_i^3 q_i^{-1} + \sum_{j=1}^{c} \rho_j \int_0^1 d\xi \int_0^\infty U_{ij}(R) g_{ij}(R, \xi) 4\pi R^2 dR. \tag{3.86}$$

(4) For molecules having internal rotational degrees of freedom (say polymers), the expression for the chemical potential should be modified to take into account all possible conformations of the molecules. In particular, the rotational partition function of the molecules (included in q) might be different for different conformations. We shall discuss a simple case of such molecules in chapter 7, section 7.8.

(5) The expression of the chemical potential in other ensembles. In all previous sections, we have used the definition of the chemical potential in the T, V, N ensemble. This was done mainly for convenience. In actual applications, and in particular when comparison with experimental results is required, it is necessary to use the T, P, N ensemble. In that case, the chemical potential is defined by

$$\mu = \left(\frac{\partial G}{\partial N}\right)_{T,P} \tag{3.87}$$

where G is the Gibbs energy of the system. It is easy to show that the formal split of μ into two terms as in (3.77) or (3.84) is maintained. In the T, P, N ensemble, $\rho = N/\langle V\rangle$ where $\langle V\rangle$ is the average volume, and $\langle\ \rangle$ should be interpreted as a T, P, N average. In the T, V, μ ensemble, μ is one of the independent variables used to describe the system. Yet it can also be written in the form (3.77), with the reinterpretation of the density $\rho = \langle N\rangle / V$, where $\langle N\rangle$ is the average in the T, V, μ ensemble; for more details see Ben-Naim (1987).

3.4.7 First-order expansion of the coupling work

We end this long section on the chemical potential with one simple and useful expression. We note first that in all of the expressions we had so far, the chemical potential was expressed as integrals over the pair correlation function. It is desirable to have at least one expression of the chemical potential in terms of *molecular interactions.* This can be obtained for very low densities, for which we know that the pair correlation function takes the form

$$g(R) = \exp[-\beta U(R)] \tag{3.88}$$

and hence for the added particle we write

$$g(R, \xi) = \exp[-\beta\xi U(R)]. \tag{3.89}$$

Substituting (3.89) into (3.67), we get an immediate integral over ξ, hence

$$\int_0^1 d\xi \int_0^\infty U(R) \exp[-\beta\xi U(R)] 4\pi R^2 dR$$
$$= -kT \int_0^\infty \{\exp[-\beta U(R)] - 1\} 4\pi R^2 dR. \tag{3.90}$$

Using the notation for the second virial coefficient (see section 1.5)

$$B_2(T) = -\tfrac{1}{2}\int_0^{\infty} \{\exp[-\beta U(R) - 1]\}4\pi R^2 dR \tag{3.91}$$

we can write (3.67) for this case as

$$\mu = \mu^{0g} + kT \ln \rho + 2kTB_2(T)\rho. \tag{3.92}$$

The last term on the rhs of (3.92) is the first-order term in the expansion of the coupling work in the density.

The virial expansion for the pressure may be recovered from (3.92) by using the thermodynamic relation

$$dP = \rho d\mu \quad (T \text{ constant}). \tag{3.93}$$

From (3.92) we have

$$d\mu = \frac{kT}{\rho} d\rho + 2kTB_2(T)d\rho. \tag{3.94}$$

Combining (3.93) and (3.94) yields

$$dP = [kT + 2kTB_2(T)\rho]d\rho. \tag{3.95}$$

This may be integrated between $\rho = 0$ and the final destiny ρ to yield

$$P = kT\rho + kTB_2(T)\rho^2 \tag{3.96}$$

which is the leading form of the virial expansion of the pressure.

It should be noted that for $\rho \to 0$, we obtain the ideal-gas expression for the chemical potential. In (3.92), we have the first-order term in the expansion of the nondivergent part of the chemical potential in the density.

The same result can be obtained by expanding the third term on the rhs of (3.80) to first order in the density, i.e.,

$$-kT \int_0^{\rho} \frac{G}{1 + \rho' G} d\rho' = -kT\rho G^0 \tag{3.97}$$

where we have denoted by

$$G^0 = \lim_{\rho \to 0} G. \tag{3.98}$$

From (3.97) and (3.92), we can identify G^0 as

$$G^0 = -2B_2(T) = \int_0^{\infty} \{\exp[-\beta U(R)] - 1\}4\pi R^2\, dR \tag{3.99}$$

We shall discuss the generalization of equation (3.92) for mixtures in the next chapter.

3.5 **The compressibility equation**

The compressibility relation is one of the simplest and most useful relations between a thermodynamic quantity and the pair correlation function. In this section, we derive this relation and point out some of its outstanding features.

We consider here a one-component system of rigid, nonspherical particles in the T, V, μ ensemble. We stress from the outset that no assumption of additivity of the potential energy is invoked at any stage of the derivation. As we shall soon see, the generalization of this equation for a multicomponent system is not straightforward.

We recall the normalization conditions for $\overline{\rho^{(1)}(\boldsymbol{X}_1)}$ and for $\overline{\rho^{(2)}(\boldsymbol{X}_1,\boldsymbol{X}_2)}$ in the T, V, μ ensemble:

$$\int d\boldsymbol{X}_1\overline{\rho^{(1)}(\boldsymbol{X}_1)} = \langle N\rangle \tag{3.100}$$

$$\int d\boldsymbol{X}_1 d\boldsymbol{X}_2\overline{\rho^{(2)}(\boldsymbol{X}_1,\boldsymbol{X}_2)} = \langle N^2\rangle - \langle N\rangle. \tag{3.101}$$

Either bars or the bracket $\langle\ \rangle$ stand for the average in the T, V, μ ensemble. We used bars for MDFs defined in the T, V, μ ensemble, whereas the symbol $\langle\ \rangle$ is used for averages computed with these MDFs.

By squaring equation (3.100) and subtracting from (3.101), we get

$$\int d\boldsymbol{X}_1 d\boldsymbol{X}_2[\overline{\rho^{(2)}(\boldsymbol{X}_1,\boldsymbol{X}_2)} - \overline{\rho^{(1)}(\boldsymbol{X}_1)}\ \ \overline{\rho^{(1)}(\boldsymbol{X}_2)}] = \langle N^2\rangle - \langle N\rangle^2 - \langle N\rangle. \tag{3.102}$$

For a homogeneous and isotropic fluid we also have

$$\overline{\rho^{(1)}(\boldsymbol{X}_1)} = \frac{\rho}{8\pi^2}. \tag{3.103}$$

The definition of the pair correlation function is

$$\overline{g(\boldsymbol{X}_1,\boldsymbol{X}_2)} = \frac{\overline{\rho^{(2)}(\boldsymbol{X}_1,\boldsymbol{X}_2)}}{\overline{\rho^{(1)}(\boldsymbol{X}_1)}\ \ \overline{\rho^{(1)}(\boldsymbol{X}_2)}} \tag{3.104}$$

and the corresponding spatial pair correlation function is defined by

$$\overline{g(\boldsymbol{R}_1,\boldsymbol{R}_2)} = \frac{1}{(8\pi^2)^2}\int d\Omega_1, d\Omega_2\overline{g(\boldsymbol{X}_1,\boldsymbol{X}_2)}. \tag{3.105}$$

We can rewrite (3.102) as

$$\rho^2 \int d\boldsymbol{R}_1 d\boldsymbol{R}_2 \overline{[g(\boldsymbol{R}_1, \boldsymbol{R}_2)} - 1] = \langle N^2 \rangle - \langle N \rangle^2 - \langle N \rangle. \tag{3.106}$$

Since $\overline{g(\boldsymbol{R}_1, \boldsymbol{R}_2)}$ depends only on the scalar distance $R = |\boldsymbol{R}_2 - \boldsymbol{R}_1|$, we can rewrite (3.106) as

$$\begin{aligned} 1 + \rho \int d\boldsymbol{R} \overline{[g(R)} - 1] &= \frac{\langle N^2 \rangle - \langle N \rangle^2}{\langle N \rangle} \\ &= 1 + \rho \int_0^\infty \overline{[g(R)} - 1] 4\pi R^2 dR. \end{aligned} \tag{3.107}$$

Relation (3.107) is an important connection between the radial distribution function and fluctuations in the number of particles. The fluctuations in the number of particles can be obtained directly from the grand partition function. The relation is (see section 1.3)

$$\langle N^2 \rangle - \langle N \rangle^2 = kTV\rho^2 \kappa_T \tag{3.108}$$

where κ_T is the isothermal compressibility of the system. Combining (3.108) with (3.107), we get the final result

$$\begin{aligned} \kappa_T &= \frac{1}{kT\rho} + \frac{1}{kT} \int_V d\boldsymbol{R} \overline{[g(\boldsymbol{R})} - 1] \\ &= \frac{1}{kT\rho} + \frac{1}{kT} \int_0^\infty \overline{[g(R)} - 1] 4\pi R^2 \, dR. \end{aligned} \tag{3.109}$$

This is known as the compressibility equation. We define the quantity†

$$G = \int_0^\infty \overline{[g(R)} - 1] 4\pi R^2 \, dR. \tag{3.110}$$

In terms of G, the compressibility equation is written as

$$kT\rho\kappa_T = 1 + \rho G. \tag{3.111}$$

Note that the first term on the rhs of (3.109) is the compressibility of an ideal gas. That is, for a system obeying the equation of state $P = \rho kT$, we have

$$\kappa_T = -\frac{1}{V}\left(\frac{\partial V}{\partial P}\right)_{T,N} = \left(\frac{\partial \ln \rho}{\partial P}\right)_T = \frac{1}{kT\rho}. \tag{3.112}$$

Hence, the second term on the rhs of (3.109) conveys the contribution to the compressibility due to the existence of interactions (and therefore correlation)

† We use the letter G for both the Gibbs energy and the Kirkwood–Buff integral, as defined in (3.110).

among the particles. Note also that the last expression for the isothermal compressibility holds either for an *ideal* gas in the sense of $U(R) \equiv 0$, for *any* density ρ, or for a *real* gas at very low density, for which the equation of state $P = \rho\kappa T$ holds. Clearly, in the limit $\rho \to 0$, $\kappa_T \to \infty$. Originally, the compressibility equation was used by Ornstein and Zernike (1914) in their theory of the well-known phenomenon of critical opalescence. Since κ_T diverges to infinity at the critical point, it follows also that G diverges at the critical point. Since $\rho g(R)$ has probabilistic meaning, the integrand in (3.110) must be bounded from above. Therefore, the divergence of G should be a result of *long-range correlations* near the critical point.

The compressibility equation has some outstanding features which we now highlight.

(1) We recall that no assumption of additivity on the total potential energy has been introduced to obtain (3.109). In the previous sections, we found relations between some thermodynamic quantities and pair correlation functions which were based explicitly on the assumption of the pairwise additivity of the total potential energy. We also recall that higher order molecular distribution functions must be introduced if higher order potentials are not negligible. Relation (3.109) does not depend on the additivity assumption; hence, it does not undergo any modification should high-order potentials be of importance. In this respect, the compressibility equation is far more general than the previously obtained relations (e.g., the energy or the pressure relation).

(2) The compressibility equation involves the *radial* distribution function even when the system consists of nonspherical particles. We recall that previously obtained relations between, say, the energy or the pressure, and the pair correlation function were dependent on the type of particle under consideration. The compressibility depends only on the *spatial* pair *correlation* function. If nonspherical particles are considered, it is understood that $\overline{g(R)}$ in (3.109) is the average over all orientations (3.105). In the following, we shall remove the bar over $g(R)$. We shall assume that the angle average has been taken before using the compressibility equation.

(3) The compressibility equation is a simple integral over $\overline{g(R)}$. It does not require explicit knowledge of $U(R)$ (or higher order potentials). It is true that $g(R)$ is a functional of $U(R)$. However, once we have obtained $g(R)$, we can use it directly to compute the compressibility by means of (3.109). This is not possible for the computation of, say, the energy.

One of the most important applications of the compressibility equation is to test the accuracy of various methods of computing $g(R)$. We recall that the pressure equation (3.37) has been found useful for computing the equation of state of a substance, and hence can be used as a test of the theory that has furnished $g(R)$. Similarly, by integrating the compressibility equation, we obtain the equation of state of the system, which may serve as a different test of the theory. Clearly, if we use the exact function $g(R)$ in either the pressure or in the compressibility equations, we must end up with the same equation of state. However, since we usually have only an approximation for $g(R)$, the results of the two equations may be different. Therefore, the discrepancy between the two results obtained with the same $g(R)$ using the pressure and the compressibility equations, can serve as a sensitive test of the accuracy of the method of computing $g(R)$.

In applying the compressibility equation (3.109), care must be exercised to use the pair correlation function $g(R)$ as obtained in the *grand canonical* ensemble, rather than the corresponding function $g(R)$ obtained in a closed system. Whenever this distinction is important, we use the notation $g_O(R)$ and $g_C(R)$ for open and closed systems, respectively. Although the difference between the two is in a term of the order of N^{-1} this small difference becomes important when integration over the entire volume is performed as in the definition of the quantity G (equation 3.110).

Let us first demonstrate the source of difficulty by a simple example. Consider an ideal gas in the T, V, N ensemble. In section 2.5, we saw that $g_C(R)$ in this case has the form (see also Appendix G)

$$g_C(R) = 1 - 1/N \quad \text{(ideal gas: } T, V, N \text{ ensemble).} \tag{3.113}$$

On the other hand, $g_O(R)$ in the T, V, μ ensemble is

$$g_O(R) = 1 \quad \text{(ideal gas: } T, V, \mu \text{ ensemble).} \tag{3.114}$$

The difference between the two results (3.113) and (3.114) arises from the *finite number* of particles in the T,V,N system. Even when there are no interactions, $U(\mathbf{R}^N) \equiv 0$, there is still correlation between the particles. The density at any point in the system is $\rho(\mathbf{R}) = N/V$. The conditional density at $\mathbf{R}$ given a particle at any other point $\mathbf{R}'$ is not $\rho(\mathbf{R}) = N/V$ but $(N-1)/V$. Fixing one particle at some point has an effect on the density at any other point merely because the number of particles was reduced from N to $N-1$. Such an effect does not exist if we open the system, in which case the pair correlation function $g_O(R)$ is unity everywhere for an ideal gas.

Clearly, we can always take the infinite-system size limit of (3.113) to obtain

$$\lim_{N\to\infty} g_C(R) = 1 \tag{3.115}$$

which can be used in the compressibility equation.

Thus, although the difference between $g_C(R)$ and $g_O(R)$ is extremely small for macroscopic systems ($N \approx 10^{23}$), the results obtained upon integration over a macroscopic volume are not negligible. The different results obtained using $g_C(R)$ and $g_O(R)$ in equation (3.109) for an ideal gas are

$$\kappa_T = \frac{1}{kT\rho} + \frac{1}{kT}\int_V d\mathbf{R}\left(\frac{-1}{N}\right) = \frac{1}{kT\rho} - \frac{1}{kT\rho} = 0 \quad [\text{using } g_C(R) \text{ from (3.113)}] \tag{3.116}$$

$$\kappa_T = \frac{1}{kT\rho} \quad [\text{using } g_O(R) \text{ from (3.114)}]. \tag{3.117}$$

Clearly, only the second result gives the correct compressibility of the ideal gas.

Relations (3.113) and (3.114) hold for an ideal gas. In the general case, the limiting behavior of $g_C(R)$ as $R \to \infty$ is (see also Appendix G)

$$g_C(R) \to 1 - \frac{\rho kT\kappa_T}{N} \quad (T, V, N \text{ ensemble}) \tag{3.118}$$

$$g_O(R) \to 1 \quad (T, V, \mu \text{ ensemble}). \tag{3.119}$$

Clearly, (3.119) can be obtained from (3.118) by taking the infinite-system-size limit ($N \to \infty$). Another way of demonstrating the discrepancy between the two results in the T,V,N and T,V,μ ensembles is in the difference in the normalization conditions for the molecular distribution functions. In particular, in the T, V, N ensemble, we have

$$\langle N^2 \rangle = \langle N \rangle^2 = N^2. \tag{3.120}$$

Hence, the normalization condition is

$$\int d\mathbf{X}_1 d\mathbf{X}_2 \left[\rho^{(2)}(\mathbf{X}_1, \mathbf{X}_2) - \rho^{(1)}(\mathbf{X}_1)\rho^{(1)}(\mathbf{X}_2)\right] = -N \tag{3.121}$$

which is equivalent to the normalization condition

$$\rho \int_0^\infty [g_C(R) - 1] 4\pi R^2 \, dR = -1, \quad (T, V, N \text{ ensemble}). \tag{3.122}$$

The last result simply means that the total number of particles in the system, N, is equal to the total number of particles around a given particle at the origin, plus that particle at the origin. This simple calculation does not hold for an open system where N is not a fixed number.

The corresponding normalization condition in the T, V, μ ensemble is (3.106) in which $\langle N^2 \rangle \neq \langle N \rangle^2$. Here, instead of (3.122), we have the compressibility equation (3.109) which we write again as

$$\rho \int_0^\infty \overline{[g_O(R) - 1]} 4\pi R^2 dR = -1 + kT\rho\kappa_T. \tag{3.123}$$

Clearly, the difference, $kT\rho\kappa_T$, between (3.122) and (3.123) is finite and arises from the difference in the long-range behavior of $g_C(R)$ and $g_O(R)$. For more details, see Appendix G.

The reader may wonder why we have dealt only now with the question of the limiting behavior of $g(R)$ as $R \to \infty$. The reason is quite simple. In all of our previous integrals, $g(R)$ appeared with another function in the integrand. For instance, in the equation for the energy, we have an integral of the form

$$\int_0^\infty U(R)g(R)4\pi R^2\, dR. \tag{3.124}$$

Clearly, since $U(R)$ is presumed to tend to zero, as, say, R^{-6} as $R \to \infty$, it is of no importance whether the limiting behavior of $g(R)$ is given by (3.118) or (3.119); in both cases the integrand will become practically zero as R becomes large enough so that $U(R) \approx 0$. The unique feature of the compressibility relation is that *only* $g(R)$ appears under the integral sign. Therefore, different results may be anticipated according to the different limiting behavior of $g(R)$ as $R \to \infty$.

As a corollary to this section, we derive a relation between the density derivative of the chemical potential and an integral involving $g(R)$. Recall the thermodynamic identity

$$\left(\frac{\partial \mu}{\partial \rho}\right)_T = \frac{1}{\kappa_T \rho^2}. \tag{3.125}$$

Combining (3.125) and (3.111) yields

$$\left(\frac{\partial \mu}{\partial \rho}\right)_T = \frac{kT}{\rho + \rho^2 G} = kT\left(\frac{1}{\rho} - \frac{G}{1 + \rho G}\right). \tag{3.126}$$

Relation (3.126) will be generalized in the next chapter for mixtures. Here, we note that by integrating (3.126) with respect to the density, we get the chemical potential, i.e.,

$$\mu = \int \frac{kT\, d\rho}{\rho + \rho^2 G} + \text{const.} \tag{3.127}$$

Thus, once we have G and its density dependence, we can determine μ from (3.127) up to a constant. The constant of integration is evaluated as follows. We choose a very low density ($\rho_0 \to 0$) in such a way that the chemical potential has the ideal-gas form, i.e.,

$$\mu(\rho_0) = kT \ln(\rho_0 \Lambda^3 q^{-1}) = \mu^{0g} + kT \ln \rho_0. \tag{3.128}$$

The chemical potential may be obtained by integrating (3.126) from ρ_0 to the final density ρ, i.e.,

$$\begin{aligned}\mu(\rho) &= \mu(\rho_0) + kT\int_{\rho_0}^{\rho}\left(\frac{1}{\rho'} - \frac{G}{1+\rho' G}\right)d\rho' \\ &= \mu^{0g} + kT\ln\rho_0 + kT\int_{\rho_0}^{\rho}\left(\frac{1}{\rho'} - \frac{G}{1+\rho' G}\right)d\rho' \\ &= \mu^{0g} + kT\ln\rho - kT\int_{0}^{\rho}\frac{G}{1+\rho' G}d\rho'. \end{aligned} \tag{3.129}$$

Note that in the last form on the rhs of (3.129), we have replaced the lower limit ρ_0 by $\rho_0 = 0$. This could not have been done when the divergent part $(\rho')^{-1}$ was in the integrand.

Finally, we note that unlike the procedure we have used to generalize previous expression to mixtures, here there is no straightforward generalization procedure. In all of the previous examples we have generalized for mixtures simply by inspection of the expression for the one-component system. Looking at the compressibility equation (3.109) or (3.123), we see no hint or clue that suggests a generalization for mixtures. We shall indeed see that the analog of the compressibility equation for mixtures is far more complicated than what we would have expected from our experience so far with the equation for the energy, the pressure, and the chemical potential. We shall devote the next chapter to obtain this generalization. In doing so, we shall also reach for new, interesting and very important relations between thermodynamic quantities and integrals over the pair correlation functions.

3.6 **Relations between thermodynamic quantities and generalized molecular distribution functions**

In section 2.7, we introduced the generalized molecular distribution functions GMDFs. Of particular importance are the singlet GMDF, which may be re-interpreted as the quasi-component distribution function (QCDF). These functions were deemed very useful in the study of liquid water. They provided a firm basis for the so-called mixture model approach to liquids in general, and for liquid water in particular (see Ben-Naim 1972a, 1973a, 1974).

In this section we shall derive some new relationships between thermodynamic quantities and GMDF. In previous sections we have derived a few relationships between thermodynamic quantities and pair distribution functions. It is well known and easy to see that if we try to express quantities such as heat capacity, compressibility, thermal expansion coefficients, etc., we shall need higher order MDFs. Since these are largely unknown, such relationships were not found to be useful. However, by using GMDFs, we can express these thermodynamic quantities in terms of singlet and pair distribution functions. It is hoped that once we gain information on the singlet and pair distribution function, these relations would be more useful.[†] However, even without knowing any details of these GMDFs, some of these relationships were found useful in interpreting some anomalous properties of water and aqueous solutions (Ben-Naim 1974).

In this section we shall be working in the T, P, N ensemble, and all the distribution functions are presumed to be defined in this ensemble. We denote by x either a vector or a function which serves as a QCDF. An appropriate subscript will be used to indicate the property employed in the classification procedure. For instance, using the coordination number (CN) as a property, the components of $\boldsymbol{x}_C$ are the quantities $x_C(K)$. Similarly, using the BE as a property, the components of $\boldsymbol{x}_B$ are the quantities $x_B(\nu)$. When reference is made to a general QCDF, we simply write $\boldsymbol{x}$ without a subscript. Once a QCDF is given, we can obtain the average number of each quasi-component directly from the components of the vector $\mathbf{N} = N\boldsymbol{x}$.[‡]

Let E be any extensive thermodynamic quantity expressed as a function of the variables T, P, and N (where N is the total number of molecules in the system). Viewing the same system as a *mixture* of quasi-components, we can express E as a function of the new set of variables T, P, and $\mathbf{N}$. For correctness, consider a QCDF based on the concept of CN. The two possible functions mentioned above are then

$$E(T, P, N) = E(T, P, N_C^{(1)}(0), N_C^{(1)}(1), \ldots). \tag{3.130}$$

For the sake of simplicity, we henceforth use $N(K)$ in place of $N_C^{(1)}(K)$, so that the treatment will be valid for any discrete QCDF. Since E is an extensive quantity, it has the property

$$E(T, P, \alpha N(0), \alpha N(1), \ldots) = \alpha E(T, P, N(0), N(1), \ldots) \tag{3.131}$$

for any real $\alpha \geq 0$; i.e., E is a homogeneous function of order one with respect to the variables $N(0)$, $N(1), \ldots$, keeping T, P constant. For such a function, the Euler theorem states that

$$E(T, P, \mathbf{N}) = \sum_{i=0}^{\infty} \overline{E}_i(T, P, \mathbf{N}) N(i) \tag{3.132}$$

[†] Matubayasi and Nakahara (2000, 2002, 2003) have recently investigated a related topic, specifically for dilute solutions.

[‡] Note that $\mathbf{N}$ is the vector $\mathbf{N} = (N_1, N_2, \ldots, N_c)$, but N is the sum of all N_i ($i = 1, \ldots, c$).

where $\overline{E}(T, P, \mathbf{N})$ is the partial molar (or molecular) quantity defined by

$$\overline{E}(T, P, \mathbf{N}) = \left[\frac{\partial E}{\partial N(i)}\right]_{T,P,N(j),j\neq i}. \tag{3.133}$$

In (3.132) and (3.133), we have stressed the fact that the partial molar quantities depend on the whole vector $\mathbf{N}$.

At this point, we digress to discuss the meaning of the partial derivatives introduced in (3.133). We recall that the variables $N(i)$ are *not* independent; therefore, it is impossible to take the derivatives of (3.133) experimentally. One cannot, in general, add $dN(i)$ of the i-species while keeping all the $N(j)$, $j \neq i$, constant, a process which can certainly be achieved in a mixture of *independent* components. However, if we assume that in principle E can be expressed in terms of the variables T, P and N, then $\overline{E}_i$ is the component of the gradient of E along its ith axis. Here, we must assume that in the neighborhood of the equilibrium vector $\mathbf{N}$, there is a sufficiently dense set of vectors (which describe various frozen-in systems) so that the gradient of E exists along each axis.

The generalization of (3.132) and (3.133) for the case of a continuous QCDF requires the application of the technique of functional differentiation. We introduce the generalized Euler theorem by way of analogy with (3.133). More details can be found in Appendix B.

The generalization can be easily visualized if we rewrite (3.132) in the form

$$E(T, P, \mathbf{N}) = \sum_{i=0}^{\infty} \overline{E}(T, P, \mathbf{N}; i) N(i) \tag{3.134}$$

where we have introduced the (discrete) variable i as one of the arguments of the function $\overline{E}$. If $\mathbf{N}$ is a vector derived from a QCDF based on a continuous variable, say v, then the generalization of (3.134) is simply

$$E(T, P, \mathbf{N}) = \int_{-\infty}^{\infty} \overline{E}(T, P, \mathbf{N}; v) N(v)\, dv \tag{3.135}$$

where $\overline{E}(T, P, \mathbf{N}; v)$ is the functional derivative of $E(T, P, \mathbf{N})$ with respect to $N(v)$, i.e.,

$$\overline{E}(T, P, \mathbf{N}; v) = \frac{\delta E(T, P, \mathbf{N})}{\delta N(v)}. \tag{3.136}$$

By analogy with the discrete case, we may assign to $\overline{E}(T, P, \mathbf{N}; v)$ the meaning of a partial molar quantity of the appropriate v -species. The functional derivative in (3.136) is viewed here as a limiting case of (3.133) when the index i becomes a continuous variable. As an example of (3.135), the volume of the system can be written as

$$V(T, P, \mathbf{N}_{\psi}^{(1)}) = \int_0^{\infty} \phi N_{\psi}^{(1)}(\phi)\, d\phi. \tag{3.137}$$

Note that this relation is based on the fact that the volumes of the Voronoi polyhedron (*VP*) of all the particles add up to build the total volume of the system. Here we have an example of an explicit dependence between V and $\mathbf{N}_{\psi}^{(1)}$ which could have been guessed. Therefore, the partial molar volume of the ϕ-species can be obtained by taking the functional derivative of V with respect to $\mathbf{N}_{\psi}^{(1)}(\phi)$, i.e.,

$$\overline{V}(T, P, \mathbf{N}_{\psi}^{(1)}; \phi') = \frac{\delta V(T, P, \mathbf{N}_{\psi}^{(1)})}{\delta N_{\psi}^{(1)}(\phi')} = \phi'. \tag{3.138}$$

This is a remarkable result. It states that the *partial molar volume* of the ϕ'-species is exactly equal to the *volume* of its VP. We note that, in general, the partial molar volume of a species is not related, in a simple manner, to the actual volume contributed by that species to the total volume of the system. We also note that in this particular example, the partial volume $\overline{V}(T, P, N_{\psi}^{(1)}; \phi')$ is independent of $T, P, \mathbf{N}_{\psi}^{(1)}$.

A second example is the average internal energy *E*, which in the *T*, *P*, *N* ensemble is given by

$$E(T, P, \mathbf{N}_{B}^{(1)}) = N\varepsilon^{K} + \langle U_N \rangle = N\varepsilon^{K} + \tfrac{1}{2}\int_{-\infty}^{\infty} v N_{B}^{(1)}(v)\, dv \tag{3.139}$$

where ε^{K} is the average kinetic energy of a single particle, and $N_{B}^{(1)}(v)$ is the singlet distribution function for the binding energy.

Note that in (3.139), *E* stands for the energy, whereas in previous expressions in this section, we have used *E* for any extensive thermodynamic quantity.

Since the normalization condition for $\mathbf{N}_{B}^{(1)}$ is

$$\int_{-\infty}^{\infty} N_{B}^{(1)}(v)\, dv = N \tag{3.140}$$

we can rewrite (3.139) as

$$E(T, P, \mathbf{N}_{B}^{(1)}) = \int_{-\infty}^{\infty} (\varepsilon^{K} + \tfrac{1}{2}v) N_{B}^{(1)}(v)\, dv. \tag{3.141}$$

This again is an explicit relation between the energy and the singlet generalized MDF, $\mathbf{N}_{B}^{(1)}$. By direct functional differentiation, we obtain

$$\overline{E}(T, P, \mathbf{N}_{B}^{(1)}; v') = \frac{\delta E(T, P, \mathbf{N}_{B}^{(1)})}{\delta N_{B}^{(1)}(v')} = \varepsilon^{K} + \tfrac{1}{2}v'. \tag{3.142}$$

Thus, the partial molar energy of the v'-species is equal to its average kinetic energy and half of its BE . Here again, the partial molar energy does not depend on the composition, although it still depends on T through ε^K. We recall, from section 3.2, that the energy of the system is expressed in terms of the *pair* distribution function. Here, the energy of the system is expressed in terms of the *singlet* GMDF.

Consider next the temperature derivatives of (3.137) and (3.141):

$$\left(\frac{\partial V}{\partial T}\right)_{P,N} = \int_0^\infty \phi \frac{\partial N_\psi^{(1)}(\phi)}{\partial T} d\phi \tag{3.143}$$

$$\left(\frac{\partial E}{\partial T}\right)_{P,N} = Nc^K + \tfrac{1}{2}\int_{-\infty}^\infty v \frac{\partial N_B^{(1)}(v)}{\partial T} dv \tag{3.144}$$

where Nc^K is the contribution of the kinetic energy to the heat capacity.

The first derivative (3.143) is related to the thermal expansivity; the second is part of the heat capacity at constant pressure (see below).

Similarly, the pressure derivatives of V and E are

$$\left(\frac{\partial V}{\partial P}\right)_{T,N} = \int_0^\infty \phi \frac{\partial N_\psi^{(1)}(\phi)}{\partial P} d\phi \tag{3.145}$$

$$\left(\frac{\partial E}{\partial P}\right)_{T,N} = \tfrac{1}{2}\int_{-\infty}^\infty v \frac{\partial N_B^{(1)}(v)}{\partial P} dv. \tag{3.146}$$

By taking the temperature and pressure derivatives of the singlet GMDF, $N_\psi^{(1)}(\phi)$ and $N_B^{(1)}(v)$, we can express all of the four derivatives as average quantities using the *singlet* and *pair* distributions only.

The simplest expression is for the isothermal compressibility defined by

$$\kappa_T = \frac{-1}{V}\left(\frac{\partial V}{\partial P}\right)_{T,N} \tag{3.147}$$

Here, V is the average volume in the T, P, N ensemble. This volume is given in equation (3.137) and can also be rewritten as

$$\langle V \rangle = N\langle \psi_1 \rangle \tag{3.148}$$

where $\langle \psi_1 \rangle$ is the average Voronoi polyhedron of a specific particle, say particle 1. The pressure derivative of the volume can now be written as

$$\left(\frac{\partial \langle V \rangle}{\partial P}\right)_{T,N} = \frac{-1}{kT}[N(\langle \psi_1^2 \rangle - \langle \psi_1 \rangle^2) + N(N-1)(\langle \psi_1 \psi_2 \rangle - \langle \psi_1 \rangle \langle \psi_2 \rangle)] \tag{3.149}$$

where $\langle\psi_1^2\rangle$ and $\langle\psi_1\rangle$ are averages taken with the *singlet* GMDF and $\langle\psi_1\psi_2\rangle$ is taken with respect to the *pair* GMDF. The pressure derivative of the volume, hence the isothermal compressibility, can be expressed explicitly in terms of the singlet and pair GMDFs, $N_\psi^{(1)}$ and $N_\psi^{(2)}$ (see Ben-Naim 1973b, 1974). Here we have expressed this quantity as fluctuations and cross fluctuations of the Voronoi polyhedra of one and two particles.

The second quantity is the thermal expansivity at constant pressure

$$\alpha_p = \frac{1}{V}\left(\frac{\partial V}{\partial T}\right)_{P,N}. \tag{3.150}$$

Again we note that V in (3.150) is the average volume in the T, P, N ensemble. Since we already have an expression for $\langle V\rangle$, we need to express only its derivative with respect to temperature. Using the definition of $N_\psi^{(1)}(\phi)$, see section 2.7.4, we get for (3.143)

$$\left(\frac{\partial V}{\partial T}\right)_{P,N} = \frac{1}{2kT^2}[N(\langle\psi_1 B_1\rangle - \langle\psi_1\rangle\langle B_1\rangle) + N(N-1)(\langle\psi_1 B_2\rangle - \langle\psi_1\rangle\langle B_2\rangle)] + \frac{P}{kT^2}[N(\langle\psi_1^2\rangle - \langle\psi_1\rangle^2) + N(N-1)(\langle\psi_1\psi_2\rangle - \langle\psi_1\rangle\langle\psi_2\rangle)]. \tag{3.151}$$

We note again that all the averages in (3.151) are taken with respect to the singlet and the pair GMDF. Explicit relations are given in Ben-Naim (1974). Similarly, the heat capacity at constant volume and constant pressure are

$$C_V = \left(\frac{\partial E}{\partial T}\right)_{V,N} = Nc^K + \frac{1}{4kT^2}[N(\langle B_1^2\rangle - \langle B_1\rangle^2) + N(N-1)(\langle B_1 B_2\rangle - \langle B_1\rangle\langle B_2\rangle)] \tag{3.152}$$

$$C_P = \left(\frac{\partial H}{\partial T}\right)_{P,N} = \left(\frac{\partial E}{\partial T}\right)_{P,N} + \left(\frac{\partial V}{\partial T}\right)_{P,N}. \tag{3.153}$$

Since we have already obtained the second term on the rhs of (3.153), we only need the first term:

$$\left(\frac{\partial E}{\partial T}\right)_{P,N} = Nc^K + \frac{P}{2kT^2}[N(\langle\psi_1 B_1\rangle - \langle\psi_1\rangle\langle B_1\rangle) + N(N-1)(\langle\psi_1 B_2\rangle - \langle\psi_1\rangle\langle B_2\rangle)] + \frac{1}{4kT^2}[N(\langle B_1^2\rangle - \langle B_1\rangle^2) + N(N-1)(\langle B_1 B_2\rangle - \langle B_1\rangle\langle B_2\rangle)]. \tag{3.154}$$

Note that the averages in (3.152) are in the T, V, N ensemble, whereas in (3.154) and (3.151), the averages are in the T, P, N ensemble.

The generalization of the relationships derived in this section to mixtures is quite straightforward, the only difficulty is notational. We therefore discuss only the relations which are of interest in the study of solvation in an ideal dilute system. We consider a system of N solvent molecules for which the total energy is given by (3.141). We now add *one* solute s at a fixed position $\boldsymbol{R}_s$ in the solvent. The solvation energy is (see chapter 7)

$$\Delta E_s^* = \langle B_s \rangle + \tfrac{1}{2} \int v[N_B^{(1)}(v/\boldsymbol{R}_s) - N_B^{(1)}(v)]\, dv. \tag{3.155}$$

Thus, the solvation energy here in the T, V, N ensemble has two contributions; an average binding energy to the solvent and the change in the average potential energy of the solvent molecules caused by placing the solute at a fixed position $\boldsymbol{R}_s$. Similarly, the solvation volume, here in the T, P, N ensemble, may be written as

$$\Delta V_s^* = \langle \psi_s \rangle + \int \phi[N_\psi^{(1)}(\phi/\boldsymbol{R}_s) - N_\psi^{(1)}(\phi)]\, d\phi. \tag{3.156}$$

Again, there are two contributions to ΔV_s^*: one is the average Voronoi polyhedra of the solute; the second is the change in the average Voronoi polyhedra of the solvent molecules caused by placing s at $\boldsymbol{R}_s$. The last term may also be interpreted as structural changes induced by the solute on the solvent. Similar interpretations hold for the second term on the rhs of 3.155. For more details see Ben-Naim (1992).

FOUR

The Kirkwood–Buff theory of solutions

The Kirkwood–Buff (KB) theory is the most important theory of solutions. This chapter is therefore central to the entire book. We devote this chapter to derive the main results of this theory. We start with some general historical comments. Then we derive the main results, almost exactly as Kirkwood and Buff did, only more slowly and in more detail, adding occasionally a comment of clarification that was missing in the original publication. We first derive the results for any multicomponent system, and thereafter specialize to the case of two-components system. In section 4, we present the *inversion* of the KB theory, which has turned a *potentially* useful theory into an *actually* useful, general and powerful tool for investigating solutions on a molecular level. Three-component systems and some comments on the application of the KB theory to electrolyte solutions are discussed in the last sections.

4.1 **Introduction**

The Kirkwood–Buff (KB) theory of solutions was published in 1951. In the original paper, Kirkwood and Buff derived some new relationships between thermodynamic quantities and molecular distribution functions for multicomponent systems in the T, V, μ ensemble. One of these is a generalization of the compressibility equation for the one-component system (section 3.5) to multicomponent systems. As we have noted in section 3.5, there is no obvious or straightforward way to generalize the compressibility equation even though we fully grasp the meaning and the origin of each of the terms in the equation. The same is true for the equation for the derivative of the chemical potential with respect to the density.

We have also noted in section 3.5 that the compressibility equation is outstanding in comparison with other relationships between thermodynamic quantities and MDFs. The same is true for the Kirkwood–Buff theory. It is also more general in its applicability than the McMillan–Mayer theory published in 1945 (see section 6.5). As such, the Kirkwood–Buff theory is the most general and most powerful theory of solutions. In essence, it provides a direct relationship between thermodynamic properties such as compressibility, partial molar volumes and derivatives of the chemical potentials, in terms of the so-called KB integrals (KBI), defined by

$$G_{ij} = \int_0^\infty [g_{ij}(R) - 1] 4\pi R^2 \, dR \tag{4.1}$$

where $g_{ij}(R)$ is the pair correlation function defined in the *open*, or the T, V, μ, system for the two species i and j. Thus, the theory may be used to compute the thermodynamic quantities based on our knowledge of the pair correlation function. Symbolically

$$\{g_{ij}\} \rightarrow \{\text{Thermodynamic quantities}\}. \tag{4.2}$$

Unfortunately, almost nothing was known at that time on the pair correlation functions in any mixture. Even today, most of the known MDFs for mixtures are obtained either from solving integral equations or from simulations. It is not surprising therefore that the Kirkwood–Buff theory, though general and potentially powerful, was practically dormant for many years. For almost 20 years, there were merely a handful of publications where the Kirkwood–Buff theory had been used[†]. Moreover, the KB theory was almost ignored by many authors of books on the theory of mixtures and solutions.

The first turning point occurred in the beginning of 1972 when the Kirkwood–Buff theory was found useful in interpreting some properties of water and aqueous solutions. The main idea was to apply the Kirkwood–Buff theory of *solutions*, to pure one-component systems viewed as a *mixture* of various quasi-component systems. The KB theory was also applied in the analysis of various ideal solutions on a molecular level (Ben-Naim 1973b, 1974).

A more dramatic turning point for the Kirkwood–Buff theory occurred in 1978 after the publication of the *inversion* of the Kirkwood–Buff theory (Ben-Naim 1978). Symbolically, the inversion theory may be written as

$$\{\text{Thermodynamic quantities}\} \rightarrow \{G_{ij}\} \tag{4.3}$$

where the quantities G_{ij} could be extracted from measurable thermodynamic quantities. In a strict sense, G_{ij} are not molecular properties. However, they do

[†] Soon after its publication, the KB theory was followed up and extended by Buff and Brout (1955) and by Mazo (1958), Buff and Schindler (1958), Münster and Sagel (1959). Much later, Debenedetti (1987) has generalized the KB theory. It seems however, that the KB theory as well as the followed-up articles never took off from the formal theoretical grounds into the realm of application.

convey information on the *local* mode of packing of the various species. As such, the theory provides a powerful tool to probe local properties of the mixtures. Ever since the publication of the inversion of the KB theory, the number of papers published grew steadily and dramatically.

4.2 General derivation of the Kirkwood–Buff theory

The derivation of the relationship between the thermodynamic quantities and KBIs consists of two parts. First, we use the normalization conditions for the singlet and the pair distribution functions in the T, V, μ ensemble. This provides relationships between the KBIs and the fluctuations in the number of the particles in the open system. Next, by differentiation of the grand partition function, we obtain relationships between thermodynamic quantities and fluctuations in the number of particles. Finally, by eliminating the fluctuations in the number of particles, we obtain the required relations between thermodynamic quantities and the KBIs.

We start by considering the grand canonical ensemble characterized by the variables T, V, and $\boldsymbol{\mu}$ where $\boldsymbol{\mu} = (\mu_1, \mu_2, \ldots, \mu_c)$ is the vector comprising the chemical potentials of all the c components of the system. The normalization conditions for the singlet and the pair distribution functions follow directly from their definitions. Here, we use the indices α and β to denote the species α, $\beta = 1, 2, \ldots, c$. The two normalization conditions are (for particles not necessarily spherical)

$$\int \overline{\rho_\alpha^{(1)}(\mathbf{X}')}\, d\mathbf{X}' = \langle N_\alpha \rangle \tag{4.4}$$

$$\begin{aligned}\int \overline{\rho_{\alpha\beta}^{(2)}(\mathbf{X}', \mathbf{X}'')} d\mathbf{X}'\, d\mathbf{X}'' &= \begin{cases} \langle N_\alpha N_\beta \rangle & \text{if } \alpha \neq \beta \\ \langle N_\alpha (N_\alpha - 1) \rangle & \text{if } \alpha = \beta \end{cases} \\ &= \langle N_\alpha N_\beta \rangle - \langle N_\alpha \rangle \delta_{\alpha\beta} \end{aligned} \tag{4.5}$$

where the symbol $\langle\ \rangle$ stands for an average in the grand canonical ensemble. In (4.5), we make a distinction between two cases: $\alpha \neq \beta$ and $\alpha = \beta$. The two cases can be combined into a single equation by using the Kronecker delta function: $\delta_{\alpha\beta} = 1$ for $\alpha = \beta$ and $\delta_{\alpha\beta} = 0$, for $\alpha \neq \beta$. For homogeneous and isotropic fluids, we also have the following relations (see chapter 2)

$$\overline{\rho_\alpha^{(1)}(\mathbf{X}')} = \frac{\rho_\alpha}{8\pi^2} \tag{4.6}$$

$$\overline{\rho_{\alpha\beta}^{(2)}(\boldsymbol{X}', \boldsymbol{X}'')} = \frac{\rho_\alpha \rho_\beta \, \overline{g_{\alpha\beta}(\boldsymbol{X}', \boldsymbol{X}'')}}{(8\pi^2)^2}. \tag{4.7}$$

Here, ρ_α is the average number density of molecules of species α, i.e., $\rho_\alpha = \langle N_\alpha \rangle / V$, with V the volume of the system. We also recall the definition of the spatial pair correlation function

$$\overline{g_{\alpha\beta}(\boldsymbol{R}', \boldsymbol{R}'')} = (8\pi^2)^{-2} \int d\boldsymbol{\Omega}' \, d\boldsymbol{\Omega}'' \, \overline{g_{\alpha\beta}(\boldsymbol{X}', \boldsymbol{X}'')} \tag{4.8}$$

which is a function of the scalar distance $R = |\boldsymbol{R}'' - \boldsymbol{R}'|$. The angular dependence of the pair correlation function has been averaged out in (4.8).

From (4.4) and (4.5) we obtain

$$\begin{aligned}
&\int \overline{\rho_{\alpha\beta}^{(2)}(\boldsymbol{X}', \boldsymbol{X}'')} \, d\boldsymbol{X}' \, d\boldsymbol{X}'' - \int \overline{\rho_\alpha^{(1)}(\boldsymbol{X}')} \, d\boldsymbol{X}' \int \overline{\rho_\beta^{(1)}(\boldsymbol{X}'')} \, d\boldsymbol{X}'' \\
&= \int [\overline{\rho_{\alpha\beta}^{(2)}(\boldsymbol{X}', \boldsymbol{X}'')} - \overline{\rho_\alpha^{(1)}(\boldsymbol{X}')} \; \overline{\rho_\beta^{(1)}(\boldsymbol{X}'')}] \, d\boldsymbol{X}' \, d\boldsymbol{X}'' \\
&= \langle N_\alpha N_\beta \rangle - \langle N_\alpha \rangle \delta_{\alpha\beta} - \langle N_\alpha \rangle \langle N_\beta \rangle.
\end{aligned} \tag{4.9}$$

Using relations (4.6) to (4.8), we can simplify (4.9) as

$$\rho_\alpha \rho_\beta \int [\overline{g_{\alpha\beta}(\boldsymbol{R}', \boldsymbol{R}'')} - 1] \, d\boldsymbol{R}' \, d\boldsymbol{R}'' = \langle N_\alpha N_\beta \rangle - \langle N_\alpha \rangle \delta_{\alpha\beta} - \langle N_\alpha \rangle \langle N_\beta \rangle \tag{4.10}$$

Next, we define the quantity, referred to as the KB integral (KBI), by

$$\overline{G}_{\alpha\beta} = \int_0^\infty [\overline{g_{\alpha\beta}(R)} - 1] 4\pi R^2 \, dR. \tag{4.11}$$

Combining (4.10) and (4.11) we get

$$\overline{G}_{\alpha\beta} = V \left(\frac{\langle N_\alpha N_\beta \rangle - \langle N_\alpha \rangle \langle N_\beta \rangle}{\langle N_\alpha \rangle \langle N_\beta \rangle} - \frac{\delta_{\alpha\beta}}{\langle N_\alpha \rangle} \right). \tag{4.12}$$

This concludes the first part of the derivation of the KB theory.

Equation (4.12) is a connection between the cross fluctuations in the number of particles of various species, and integrals involving only the *spatial* pair correlation functions for the corresponding pairs of species α and β.

Before we derive the second part, i.e., the connection between the KBIs, $\overline{G}_{\alpha\beta}$, and thermodynamics, it should be stressed that all the distribution functions used in this section are defined in the *open system*. This has been indicated by a bar over the various distribution functions. If we were in a closed system, the normalization conditions would have been

$$\int \rho_\alpha^{(1)}(\mathbf{X}')d\mathbf{X}' = N_\alpha \tag{4.13}$$

and

$$\int \rho_{\alpha\beta}^{(2)}(\mathbf{X}', \mathbf{X}'')\, d\mathbf{X}'\, d\mathbf{X}'' = N_\alpha N_\beta - N_\alpha \delta_{\alpha\beta}. \tag{4.14}$$

Thus, instead of relation (4.12), we would have the result

$$G_{\alpha\beta}^{(\text{closed})} = \frac{-V}{N_\alpha}\delta_{\alpha\beta} \tag{4.15}$$

where N_α and V are the exact number of α particles and the volume of the closed system, respectively.

The reason for this fundamentally different behavior of $G_{\alpha\beta}$ in the closed and open systems is the same as in the one-component system discussed in section 3.5. See also Appendix G.

Relation (4.15) can be written as

$$\rho_A G_{AA}^{(\text{closed})} = -1 \tag{4.16}$$

$$\rho_A G_{AB}^{(\text{closed})} = 0 \tag{4.17}$$

The difference in the values of $G_{\alpha\beta}$ in open and a closed systems should be noted carefully. In a *closed* system, placing an A at a fixed position say, $\boldsymbol{R}_0$, changes the average number of A particles in the entire surroundings of A at $\boldsymbol{R}_0$ by exactly -1. Placing an A at a fixed position does not change the total number of B's in its entire surroundings. This is a direct consequence of the *closure* of the system with respect to the number of particles. Since we shall use only the $G_{\alpha\beta}$ defined in the *open* system, we remove the bar over $\overline{G}_{\alpha\beta}$ in the following derivation of the KB theory*.

The next part of the theory involves a connection between the fluctuations in the number of molecules and thermodynamic quantities. We start with the grand canonical partition function for a c-component system:

$$\Xi(T, V, \boldsymbol{\mu}) = \sum_{\mathbf{N}} Q(T, V, \mathbf{N}) \exp(\beta \boldsymbol{\mu} \cdot \mathbf{N}) \tag{4.18}$$

where $\mathbf{N} = (N_1, N_2, \ldots, N_c)$ and the summation is over each of the N_i, from zero to infinity. The exponential function includes the scalar product

$$\boldsymbol{\mu} \cdot \mathbf{N} = \sum_{i=1}^{c} \mu_i N_i. \tag{4.19}$$

* Note that the KB theory does not impose any restrictions on the *signs* of $\overline{G}_{\alpha\beta}$. Unfortunately the literature is replete with erroneous claims regarding the possible signs of $\overline{G}_{\alpha\beta}$.

The average number of, say, α molecules in the system is*

$$\langle N_\alpha \rangle = \Xi^{-1} \sum_{N} N_\alpha Q(T, V, \boldsymbol{N}) \exp(\beta \boldsymbol{\mu} \cdot \boldsymbol{N}) = kT \left[\frac{\partial \ln \Xi(T, V, \mu)}{\partial \mu_\alpha} \right]_{T, V, \mu'_\alpha} \tag{4.20}$$

where μ'_α stands for the vector $(\mu_1, \mu_2, \ldots, \mu_c)$, excluding μ_α, i.e., $\mu'_\alpha = \mu_1, \ldots, \mu_{\alpha-1}, \mu_{\alpha+1}, \ldots, \mu_c$

Differentiating $\langle N_\alpha \rangle$ in (4.20) with respect to μ_β, we get

$$kT \left(\frac{\partial \langle N_\alpha \rangle}{\partial \mu_\beta} \right)_{T,V,\mu'_\beta} = \Xi^{-1} \sum_{N} N_\alpha N_\beta Q(T, V, N) \exp(\beta \boldsymbol{\mu} \cdot \boldsymbol{N}) - \langle N_\alpha \rangle \langle N_\beta \rangle = \langle N_\alpha N_\beta \rangle - \langle N_\alpha \rangle \langle N_\beta \rangle. \tag{4.21}$$

Since all the equations are symmetrical with respect to interchanging the indices α and β, we have

$$kT \left(\frac{\partial \langle N_\alpha \rangle}{\partial \mu_\beta} \right)_{T,V,\mu'_\beta} = kT \left(\frac{\partial \langle N_\beta \rangle}{\partial \mu_\alpha} \right)_{T,V,\mu'_\alpha} = \langle N_\alpha N_\beta \rangle - \langle N_\alpha \rangle \langle N_\beta \rangle. \tag{4.22}$$

We now combine the results of the two parts, relations (4.22) with (4.12), to eliminate the fluctuations in the number of particles. The result it

$$B_{\alpha\beta} \equiv \frac{kT}{V} \left(\frac{\partial \langle N_\alpha \rangle}{\partial \mu_\beta} \right)_{T,V,\mu'_\beta} = kT \left(\frac{\partial \rho_\alpha}{\partial \mu_\beta} \right)_{T,\mu'_\beta} = \rho_\alpha \rho_\beta G_{\alpha\beta} + \rho_\alpha \delta_{\alpha\beta}. \tag{4.23}$$

Note that $G_{\alpha\beta} = G_{\beta\alpha}$ by virtue of the symmetry with respect to interchanging the α and β indices.

The result (4.23) is already a relation between thermodynamic quantities and molecular distribution functions. However, since the derivatives in (4.23) are taken at constant chemical potentials, these relations are of importance mainly in osmotic systems. Here, we are interested in derivatives at constant *temperature* and *pressure.* Obtaining these require some simple transformations of the partial derivatives. We first define the elements of the matrix $\boldsymbol{A}$ by

$$A_{\alpha\beta} \equiv \frac{V}{kT} \left(\frac{\partial \mu_\alpha}{\partial \langle N_\beta \rangle} \right)_{T,V,N'_\beta} = \frac{1}{kT} \left(\frac{\partial \mu_\alpha}{\partial \rho_\beta} \right)_{T,\rho'_\beta}. \tag{4.24}$$

Note again that we use N'_β and ρ'_β to denote vectors from which we have excluded the components N_β and ρ_β, respectively. Using the chain rule of

* Note that μ'_α, N'_α and ρ'_α are vectors. However we shall not use bold-face better for these quantities.

differentiation, we get the identity

$$\delta_{\alpha\gamma} = \left(\frac{\partial \mu_\alpha}{\partial \mu_\gamma}\right)_{T,\mu'_\gamma} = \sum_{\beta=1}^{c} \left(\frac{\partial \mu_\alpha}{\partial \rho_\beta}\right)_{T,\rho'_\beta} \left(\frac{\partial \rho_\beta}{\partial \mu_\gamma}\right)_{T,\mu'_\gamma} = \sum_{\beta=1}^{c} A_{\alpha\beta} B_{\beta\gamma}. \tag{4.25}$$

The elements $B_{\alpha\beta}$ were defined in (4.23). In equation (4.25) we have a product of two matrices. It can be rewritten in matrix notation as

$$\boldsymbol{A} \cdot \boldsymbol{B} = \boldsymbol{I} \tag{4.26}$$

where $\boldsymbol{I}$ is the unit matrix of order $c \times c$. From (4.26), we can solve for $\boldsymbol{A}$ if we know $\boldsymbol{B}$. Taking the inverse[†] of the matrix $\boldsymbol{B}$, we get for the elements of the matrix $\boldsymbol{A}$,

$$A_{\alpha\beta} = B^{\alpha\beta}/|\boldsymbol{B}| \tag{4.27}$$

where $B^{\alpha\beta}$ stands for the cofactor of the element $B_{\alpha\beta}$ in the determinant $|\boldsymbol{B}|$. The cofactor of $B_{\alpha\beta}$ is obtained by eliminating the row and the column containing $B_{\alpha\beta}$ in the determinant $|\boldsymbol{B}|$, and multiplying[‡] the result by $(-1)^{\alpha+\beta}$. The existence of the inverse of the matrix $\boldsymbol{B}$ is equivalent to a stability condition of the system. Since the $B_{\alpha\beta}$ are already expressible in terms of the $G_{\alpha\beta}$ through (4.23), relation (4.27) also connects $A_{\alpha\beta}$ with the molecular quantities $G_{\alpha\beta}$.

Next, we transform from the volume as an independent variable, into the pressure. This can be achieved by using the thermodynamic identity[¶] (see also Appendix A)

$$\left(\frac{\partial \mu_\alpha}{\partial N_\beta}\right)_{T,V,N'_\beta} = \left(\frac{\partial \mu_\alpha}{\partial N_\beta}\right)_{T,P,N'_\beta} + \left(\frac{\partial \mu_\alpha}{\partial P}\right)_{T,N} \left(\frac{\partial P}{\partial N_\beta}\right)_{T,V,N'_\beta}. \tag{4.28}$$

We also use the identity (see Appendix A)

$$\left(\frac{\partial P}{\partial N_\beta}\right)_{T,V,N'_\beta} \left(\frac{\partial N_\beta}{\partial V}\right)_{T,P,N'_\beta} \left(\frac{\partial V}{\partial P}\right)_{T,N} = -1 \tag{4.29}$$

together with the definitions of the partial molar volumes

$$\overline{V}_\alpha = \left(\frac{\partial V}{\partial N_\alpha}\right)_{T,P,N'_\alpha} = \left(\frac{\partial \mu_\alpha}{\partial P}\right)_{T,N} \tag{4.30}$$

to get from (4.28) the relation

$$\mu_{\alpha\beta} \equiv \left(\frac{\partial \mu_\alpha}{\partial N_\beta}\right)_{T,P,N'_\beta} = \left(\frac{\partial \mu_\alpha}{\partial N_\beta}\right)_{T,V,N'_\beta} - \frac{\overline{V}_\alpha \overline{V}_\beta}{V\kappa_T} \tag{4.31}$$

[†] The existence of the inverse of $\boldsymbol{B}$ is guaranteed by the stability condition of the system. See also section 4.3 below. Equation (4.27) is known as Cramer's rule for solving a set of linear equations.

[‡] Here, α and β must take numerical values, otherwise $(-1)^{\alpha+\beta}$ is meaningless. In the following applications, we shall take α and β to stand, for say, components A and B, respectively. In this case, we may assign the number 1, say, to A, and the number 2 to B.

[¶] From hereon, for convenience of notation, we use N_α instead of $\langle N_\alpha \rangle$. It should be clear from the context whether we refer to an exact or an average quantity.

where κ_T is the isothermal compressibility of the system, defined as

$$\kappa_T = -\frac{1}{V}\left(\frac{\partial V}{\partial P}\right)_{T,N}. \tag{4.32}$$

We now have all the necessary relations to express the thermodynamic quantities $\mu_{\alpha\beta}$, $\overline{V}_\alpha$, and κ_T in terms of the $G_{\alpha\beta}$.

In order to obtain the explicit expressions for the c^2+c+1 quantities [c^2 derivatives $\mu_{\alpha\beta}$ (α, $\beta = 1, 2, \ldots, c$), c partial molar volumes $\overline{V}_i$ ($i = 1, \ldots, c$), and the isothermal compressibility κ_T], we need to solve the following c^2+c+1 equations:

c^2 equations for $\mu_{\alpha\beta}$

$$\mu_{\alpha\beta} = \frac{kT}{V}\frac{B^{\alpha\beta}}{|\mathbf{B}|} - \frac{\overline{V}_\alpha \overline{V}_\beta}{V\kappa_T} \qquad \text{(for each } \alpha \text{ and } \beta\text{)} \tag{4.33}$$

c Gibbs–Duhem equations

$$\sum_\alpha \rho_\alpha \mu_{\alpha\beta} = 0 \qquad \text{(for each } \beta\text{)} \tag{4.34}$$

and the identity

$$\sum_{i=1}^{C} \rho_i \overline{V}_i = 1. \tag{4.35}$$

Solving the c^2+c+1 equations for $\mu_{\alpha\beta}$ (α, $\beta = 1, 2, \ldots, c$), $\overline{V}_i$ ($i = 1, \ldots, c$), and κ_T, we obtain the final result[†]

$$\kappa_T = \frac{|\mathbf{B}|}{kT\sum_{i,j}\rho_i\rho_j B^{ij}} \tag{4.36}$$

$$\overline{V}_\alpha = \frac{\sum_i \rho_i B^{i\alpha}}{\sum_{i,j}\rho_i\rho_j B^{ij}} \tag{4.37}$$

$$\mu_{\alpha\beta} = \left(\frac{\partial \mu_\alpha}{\partial N_\beta}\right)_{T,P,N'_\beta} = \frac{kT}{V|\mathbf{B}|}\frac{\sum_{i,j}\rho_i\rho_j[B^{\alpha\beta}B^{ij} - B^{i\alpha}B^{j\beta}]}{\sum_{i,j}\rho_i\rho_j B^{ij}}. \tag{4.38}$$

In equations (4.36)–(4.38), we have expressed the quantities $\mu_{\alpha\beta}$, $\overline{V}_\alpha$, and κ_T in terms of the G_{ij} (included in the matrix $\boldsymbol{B}$ and its various cofactors). Clearly, these are quite involved expressions in the general case of c components.

[†] In their original paper, Kirkwood and Buff (1951) derived an equation which is useful whenever one component is the solvent, say, water. A detailed derivation of this equation [equation (12) in the original article] may be found in Münster (1969), p. 341. The results in this chapter are more general and apply to mixtures of arbitrary composition. The author is grateful to Dr. R.M. Mazo for his comment on this specific result of Kirkwood and Buff.

Therefore, we shall discuss some special cases of two- and three-component systems in the following sections.

Before turning to the specific cases, we note that for one-component system, the above equations reduce to

$$\kappa_T = \frac{\rho^2 G + \rho}{kT\rho^2} = \frac{\rho G + 1}{kT\rho} \tag{4.39}$$

$$\overline{V} = \frac{\rho}{\rho^2} = \frac{1}{\rho} \tag{4.40}$$

$$\left(\frac{\partial \mu}{\partial N}\right)_{T,V} = \frac{kT}{V(\rho + \rho^2 G)} \quad \text{or} \quad \left(\frac{\partial \mu}{\partial \rho}\right)_T = \frac{kT}{\rho + \rho^2 G} \tag{4.41}$$

$$\left(\frac{\partial \mu}{\partial N}\right)_{T,P} = 0. \tag{4.42}$$

Equation (4.39) is simply the compressibility equation for a one-component system. Equation (4.40) is simply the molar volume of a one-component system and (4.41) is the derivative of the chemical potential at constant volume. Note that since μ is an intensive quantity, its derivative with respect to N, at constant P, T, is zero.

As we have noted in section 3.5, the generalization from a one-component system to a multicomponent system is not straightforward. There are no clues in equations (4.39)–(4.42) to indicate how to generalize to multicomponent systems. That is probably the reason why Kirkwood and Buff had to go through the lengthy derivation of equations (4.36)–(4.38).

Before closing this long section, we recap the main features of the KB theory, which make it so general and powerful.

First, the theory is valid for any kind of particles, not necessarily spherical particles. Only the *spatial* pair correlation function features in $G_{\alpha\beta}$, even when the particles are not spherical. Second, no assumption on pairwise additivity of the total potential energy is invoked in the theory. Finally, we note that in this book, we discuss only classical systems; the Kirkwood–Buff results, however, hold for quantum systems as well.

In the following sections, we shall discuss in more detail some aspects of the KB theory for two- and three-component mixtures.

4.3 Two-component systems

The KB theory, as well as its inversion, has been used mainly for two-component systems. The KB results for a two-component system may be obtained from the general equations (4.36)–(4.38), simply by taking the summation over only two species, say A and B. For these systems, the determinant $|\boldsymbol{B}|$ reduces to

$$|\boldsymbol{B}| = \begin{vmatrix} \rho_A + \rho_A^2 G_{AA} & \rho_A\rho_B G_{AB} \\ \rho_A\rho_B G_{AB} & \rho_B + \rho_B^2 G_{BB} \end{vmatrix} = \rho_A\rho_B\left[1 + \rho_A G_{AA} + \rho_B G_{BB} + \rho_A\rho_B\left(G_{AA}G_{BB} - G_{AB}^2\right)\right]. \tag{4.43}$$

The four cofactors of $|\boldsymbol{B}|$ are[†]

$$B^{AA} = \rho_B + \rho_B^2 G_{BB},\ B^{AB} = B^{BA} = -\rho_A\rho_B G_{AB},\ B^{BB} = \rho_A + \rho_A^2 G_{AA}. \tag{4.44}$$

Also, we have

$$\sum_i \sum_j \rho_i\rho_j B^{ij} = \rho_A\rho_B[\rho_A + \rho_B + \rho_A\rho_B(G_{AA} + G_{BB} - 2G_{AB})]. \tag{4.45}$$

It is convenient to define the two auxiliary quantities:

$$\eta = \rho_A + \rho_B + \rho_A\rho_B(G_{AA} + G_{BB} - 2G_{AB}) \tag{4.46}$$

$$\zeta = 1 + \rho_A G_{AA} + \rho_B G_{BB} + \rho_A\rho_B\left(G_{AA}G_{BB} - G_{AB}^2\right). \tag{4.47}$$

With this notation we can express all the thermodynamic quantities $\mu_{\alpha\beta}$, $\overline{V}_\alpha$, and κ_T in terms of the Kirkwood–Buff integrals, $G_{\alpha\beta}$:

$$\kappa_T = \frac{\zeta}{kT\eta} \tag{4.48}$$

$$\overline{V}_A = \frac{1 + \rho_B(G_{BB} - G_{AB})}{\eta} \tag{4.49}$$

$$\overline{V}_B = \frac{1 + \rho_A(G_{AA} - G_{AB})}{\eta} \tag{4.50}$$

$$\mu_{AA} = \frac{\rho_B kT}{\rho_A V\eta}, \qquad \mu_{BB} = \frac{\rho_A kT}{\rho_B V\eta}, \qquad \mu_{AB} = \mu_{BA} = -\frac{kT}{V\eta}. \tag{4.51}$$

In equations (4.48)–(4.51), we have completed the process of expressing the thermodynamic quantities in terms of the molecular quantities.[‡] We now examine a few limiting cases. In the limit $\rho_B \to 0$, we have

$$\lim_{\rho_B\to 0} \eta = \rho_A \qquad \text{and} \qquad \lim_{\rho_B\to 0} \zeta = 1 + \rho_A^0 G_{AA}^0. \tag{4.52}$$

[†] Note that the letter B is used for both the matrix $\boldsymbol{B}$ and as one of the species.

[‡] From the stability conditions of the system, it can be proven that $\eta > 0$ and $\zeta > 0$ always. The first follows from the stability condition applied to the chemical potential. We must have $\mu_{AB} < 0$ and $\mu_{AA} > 0$, $\mu_{BB} > 0$, hence $\eta > 0$. Furthermore, since $\kappa_T > 0$, it follows that $\zeta > 0$ also. These conditions ensure that the inverse of the matrix $\boldsymbol{B}$ exists (here for the two-component system).

In this limit, the compressibility in (4.3.6) reduces to

$$\lim_{\rho_B \to 0} \kappa_T = \frac{1 + \rho_A^0 G_{AA}^0}{kT\rho_A^0}. \tag{4.53}$$

This is just the compressibility equation for a one-component system (G_{AA}^0 and ρ_A^0 are the limiting values of G_{AA} and ρ_A as $\rho_B \to 0$, respectively). Similarly, from (4.49) and (4.50) we obtain, in this limit,

$$\lim_{\rho_B \to 0} \overline{V}_A = \frac{1}{\rho_A^0} \qquad \text{and} \qquad \lim_{\rho_B \to 0} \overline{V}_B = \frac{1 + \rho_A^0 (G_{AA}^0 - G_{AB}^0)}{\rho_A^0}. \tag{4.54}$$

Thus, for the component *A*, we simply get the molar (or molecular) volume of *pure A*, whereas for component *B*, we get the partial molar volume at infinite dilution.

Also, in this limit, we have

$$\mu_{AA} = 0, \quad \mu_{AB} = \mu_{BA} = \frac{-kT}{\rho_A^0 V} = \frac{-kT}{N_A}, \quad \mu_{BB} \approx \frac{kT}{\rho_B V} = \frac{kT}{N_B}. \tag{4.55}$$

Next, we derive some relations which will prove useful in later applications of the theory. All of the following relations are obtainable by the application of simple identities between partial derivatives, such as (see also Appendix A)

$$\rho_A \left(\frac{\partial \mu_A}{\partial \rho_B}\right)_{T,P} + \rho_B \left(\frac{\partial \mu_B}{\partial \rho_B}\right)_{T,P} = 0 \tag{4.56}$$

$$\left(\frac{\partial \mu_A}{\partial \rho_B}\right)_{T,\mu_B} \left(\frac{\partial \rho_B}{\partial \mu_B}\right)_{T,\mu_A} \left(\frac{\partial \mu_B}{\partial \mu_A}\right)_{T,\rho_B} = -1 \tag{4.57}$$

$$\left(\frac{\partial \mu_B}{\partial \rho_B}\right)_{T,P} = \left(\frac{\partial \mu_B}{\partial \rho_B}\right)_{T,\mu_A} + \left(\frac{\partial \mu_B}{\partial \mu_A}\right)_{T,\rho_B} \left(\frac{\partial \mu_A}{\partial \rho_B}\right)_{T,P}. \tag{4.58}$$

From (4.56)–(4.58), we can eliminate the required derivative at constant *P* and *T*, to obtain

$$\left(\frac{\partial \mu_B}{\partial \rho_B}\right)_{T,P} = \frac{\rho_A (\partial \mu_B/\partial \rho_B)_{T,\mu_A} (\partial \mu_A/\partial \rho_B)_{T,\mu_B}}{\rho_A (\partial \mu_A/\partial \rho_B)_{T,\mu_B} - \rho_B (\partial \mu_B/\partial \rho_B)_{T,\mu_A}}. \tag{4.59}$$

On the rhs of (4.59), we only have quantities that are expressible in terms of the $G_{\alpha\beta}$. Explicitly:

$$\begin{aligned}\left(\frac{\partial \mu_B}{\partial \rho_B}\right)_{T,P} &= \frac{kT}{\rho_B (1 + \rho_B G_{BB} - \rho_B G_{AB})} \\ &= kT \left(\frac{1}{\rho_B} - \frac{G_{BB} - G_{AB}}{1 + \rho_B G_{BB} - \rho_B G_{AB}}\right)\end{aligned} \tag{4.60}$$

The second form on the rhs of (4.60) will be found useful for the study of very dilute solutions of B in A.

From equations (4.60) and (4.56), we also get

$$\left(\frac{\partial\mu_A}{\partial\rho_B}\right)_{T,P} = -\frac{\rho_B}{\rho_A}\left(\frac{\partial\mu_B}{\partial\rho_B}\right)_{T,P} = \frac{-kT}{\rho_A(1+\rho_B G_{BB} - \rho_B G_{AB})}. \tag{4.61}$$

Similarly, if we interchange the roles of A and B, we obtain

$$\left(\frac{\partial\mu_A}{\partial\rho_A}\right)_{T,P} = \frac{kT}{\rho_A(1+\rho_A G_{AA} - \rho_A G_{AB})} \tag{4.62}$$

$$\left(\frac{\partial\mu_B}{\partial\rho_A}\right)_{T,P} = \frac{-kT}{\rho_B(1+\rho_A G_{AA} - \rho_A G_{AB})}. \tag{4.63}$$

Note that unlike the derivatives in (4.22), here the two derivatives are not equal, i.e.,

$$\left(\frac{\partial\mu_B}{\partial\rho_A}\right)_{T,P} \neq \left(\frac{\partial\mu_A}{\partial\rho_B}\right)_{T,P}. \tag{4.64}$$

The relation between these two derivatives can be obtained by taking the ratio of (4.61) and (4.63), i.e.,

$$\left(\frac{\partial\mu_A}{\partial\rho_B}\right)_{T,P} = \left(\frac{\partial\mu_B}{\partial\rho_A}\right)_{T,P}\frac{\rho_B(1+\rho_A G_{AA}-\rho_A G_{AB})}{\rho_A(1+\rho_B G_{BB}-\rho_B G_{AB})} = \left(\frac{\partial\mu_B}{\partial\rho_A}\right)_{T,P}\frac{\rho_B \overline{V}_B}{\rho_A \overline{V}_A}. \tag{4.65}$$

Another useful relation is

$$\left(\frac{\partial\rho_A}{\partial\rho_B}\right)_{T,P} = \frac{(\partial\rho_A/\partial\mu_A)_{T,P}}{(\partial\rho_B/\partial\mu_A)_{T,P}} = -\frac{1+\rho_A G_{AA} - \rho_A G_{AB}}{1+\rho_B G_{BB} - \rho_B G_{AB}} = -\frac{\overline{V}_B}{\overline{V}_A}. \tag{4.66}$$

Similarly, we can get the following derivatives of the chemical potentials

$$\left(\frac{\partial\mu_B}{\partial x_A}\right)_{T,P} = \frac{-kT(\rho_A+\rho_B)^2}{\rho_B\eta} \tag{4.67}$$

$$\left(\frac{\partial\mu_A}{\partial x_B}\right)_{T,P} = \frac{-kT(\rho_A+\rho_B)^2}{\rho_A\eta} \tag{4.68}$$

where on the rhs of (4.67) and (4.68), we have expressions in terms of $G_{\alpha\beta}$.

Again, we note that, in general, the two derivatives in (4.67) and (4.68) are not equal, i.e.,

$$\left(\frac{\partial \mu_A}{\partial x_B}\right)_{T,P} \neq \left(\frac{\partial \mu_B}{\partial x_A}\right)_{T,P}. \tag{4.69}$$

The relationship between the two can be obtained either from the Gibbs–Duhem relation, or from (4.67) and (4.68), namely

$$\left(\frac{\partial \mu_A}{\partial x_B}\right)_{T,P} = \left(\frac{\partial \mu_B}{\partial x_A}\right)_{T,P} \frac{N_B}{N_A}. \tag{4.70}$$

Another useful derivative of the chemical potential with respect to the mole fraction is obtained from

$$\left(\frac{\partial \mu_A}{\partial x_A}\right)_{T,P} = \left(\frac{\partial \mu_A}{\partial \rho_A}\right)_{T,P} \left(\frac{\partial \rho_A}{\partial x_A}\right)_{T,P} = \left(\frac{\partial \mu_A}{\partial \rho_A}\right)_{T,P} (\rho_A + \rho_B)^2 \overline{V}_B. \tag{4.71}$$

In the last form on the rhs of (4.71), we have used the derivative of x_A with respect to ρ_A, i.e.,

$$\left(\frac{\partial x_A}{\partial \rho_A}\right)_{T,P} = \frac{1}{(\rho_A + \rho_B)^2 \overline{V}_B} = \frac{1}{\rho^2 \overline{V}_B} \tag{4.72}$$

where $\rho = \rho_A + \rho_B$ is the total density

From (4.71), (4.50), (4.62), we get the final important result

$$\left(\frac{\partial \mu_A}{\partial x_A}\right)_{T,P} = \frac{kT\rho^2}{\rho_A \eta} = kT\left(\frac{1}{x_A} - \frac{\rho_B \Delta_{AB}}{1 + \rho_B x_A \Delta_{AB}}\right) \tag{4.73}$$

where we have defined the quantity Δ_{AB} as

$$\Delta_{AB} = G_{AA} + G_{BB} - 2G_{AB}. \tag{4.74}$$

Relation (4.73) will be most useful for the study of various concepts of ideality carried out in the next chapter.

4.4 Inversion of the Kirkwood–Buff theory

The Kirkwood–Buff theory of solutions was originally formulated to obtain thermodynamic quantities from molecular distribution functions. This formulation is useful whenever distribution functions are available either from analytical calculations or from computer simulations. The inversion procedure of the same theory reverses the role of the thermodynamic and molecular quantities, i.e., it allows the evaluation of integrals over the pair correlation functions from thermodynamic quantities. These integrals G_{ij}, referred to as the Kirkwood–Buff integrals (KBIs), were found useful in the study of mixtures on a molecular level. They are also used in the theory of preferential solvation, discussed in chapter 8.

The main result of the KB theory can be symbolically written as:

$$\{G_{ij}\} \rightarrow \{\overline{V}_i, \kappa_T, \partial\mu_i/\partial\rho_j\}. \tag{4.75}$$

Having information on the G_{ij}, one can compute the thermodynamic quantities[†]. However, the original KB theory could have been used only in rare cases where G_{ij} could be obtained from theoretical work. In principle, having an approximate theory for computing the various pair correlation functions $g_{ij}(R)$, it is possible to evaluate the integrals G_{ij} and then compute the thermodynamic quantities through the KB theory. Comparison between the thermodynamic quantities thus obtained, and the corresponding experimental data, could serve as a test of the theory that provides the pair correlation functions.

The inversion procedure may be symbolically written as

$$\{\overline{V}_i, \kappa_T, \partial\mu_i/\partial\rho_j\} \rightarrow \{G_{ij}\}. \tag{4.76}$$

In this form, the thermodynamic quantities are used as input to compute the molecular quantities G_{ij}. Since it is relatively easier to measure the required thermodynamic quantities, the inversion procedure provides a new and powerful tool to investigate the characteristics of the local environments of each species in a multicomponent system.

It should be noted that there are some difficulties in obtaining accurate values of the KBI from the available thermodynamic data (Kato 1984; Zaitsev et al. 1985, 1989). Matteoli and Lepori (1984) have made an extensive comparison between the values of G_{ij} calculated by different authors (e.g., Ben-Naim 1977; Donkersloot 1979a, b; Patil 1981) and found large discrepancies between the reported results. Another method of obtaining the KBI is the small-angle x-ray or neutron scattering intensities from mixtures; see, for example, Nishikawa (1986), Nishikawa et al. (1989), Hayashi et al. (1990), Misawa and Yoshida (2000), Almasy et al. (2002), and Dixit et al. (2002).

In the following we shall discuss only the mathematical aspects of the inversion procedure and not delve into the problem of the accuracies of the results.

The inversion procedure can be carried out in principle for any mixture of c components. In (4.36)–(4.38), we have c^2+c+1 expressions for the thermodynamic quantities $\mu_{\alpha\beta}$, $\overline{V}_\alpha$, and κ_T. These are not independent equations, because of the $c+1$ relationships (4.34) and (4.35). Hence, only c^2 independent relationships exist between the thermodynamic quantities and the c^2 KBIs G_{ij}.

[†] There is an equivalent set of relationships between integrals over the direct correlation functions (see Appendix C) and the thermodynamic quantities. (O'Connell [1971, 1975, 1981, 1990], Hamad et al. [1989]).

Since the inversion procedure becomes increasingly complicated for larger values of c, we shall outline the procedure for two-component mixtures here and further discuss the three-component case in section 4.5.

For two-component systems, we have already written the KB results in (4.48)–(4.51). Altogether, these are seven equations. However, because of the following three equations

$$\rho_A\mu_{AA} + \rho_B\mu_{AB} = 0 \tag{4.77}$$

$$\rho_B\mu_{BB} + \rho_A\mu_{AB} = 0 \tag{4.78}$$

$$\rho_A\overline{V}_A + \rho_B\overline{V}_B = 1 \tag{4.79}$$

we are left with only *four* independent relationships between the thermodynamic quantities and the *four* KBIs $G_{\alpha\beta}$. In fact, since $\mu_{AB} = \mu_{BA}$ we have only three equations for the three quantities G_{AA}, G_{BB}, and $G_{AB} = G_{BA}$.

To solve for G_{ij}, we first eliminate η from any of the relationships in (4.51) to obtain:

$$\eta = \frac{kT\rho_B}{\rho_A V\mu_{AA}} = \frac{kT\rho_A}{\rho_B V\mu_{BB}} = \frac{-kT}{V\mu_{AB}}. \tag{4.80}$$

Next, we eliminate ζ from (4.48)

$$\zeta = kT\eta\kappa_T. \tag{4.81}$$

Now, we can use (4.49) and (4.50) together with (4.79) to express all of the $G_{\alpha\beta}$ in terms of experimental quantities. The results are

$$G_{AB} = kT\kappa_T - \rho\overline{V}_A\overline{V}_B/D \tag{4.82}$$

$$G_{AA} = kT\kappa_T - \frac{1}{\rho_A} + \frac{\rho_B\overline{V}_B^2\rho}{\rho_A D} \tag{4.83}$$

$$G_{BB} = kT\kappa_T - \frac{1}{\rho_B} + \frac{\rho_A\overline{V}_A^2\rho}{\rho_B D} \tag{4.84}$$

where $\rho = \rho_A + \rho_B$ and D denotes

$$D = \frac{x_A}{kT}\left(\frac{\partial\mu_A}{\partial x_A}\right)_{T,P}. \tag{4.85}$$

The three equations (4.82)–(4.84) can be cast in a more condensed form as

$$G_{\alpha\beta} = kT\kappa_T - \frac{\delta_{\alpha\beta}}{\rho_\alpha} + \rho kT\frac{\left(1-\rho_\alpha\overline{V}_\alpha\right)\left(1-\rho_\beta\overline{V}_\beta\right)}{\rho_\alpha\rho_\beta\mu_{\alpha\beta}} \tag{4.86}$$

where $\rho = \rho_A + \rho_B$ and $\mu_{\alpha\beta}$ is $(\partial\mu_\alpha/\partial N_\beta)_{P,T,N'_\beta}$.

In (4.86), we have expressed all of the $G_{\alpha\beta}$ in terms of the thermodynamic quantities κ_T, $\overline{V}_\alpha$, and $\mu_{\alpha\beta}$. In practice, one uses the derivative (4.85) obtained either from the second derivative of the excess Gibbs energy of the system, or from data on the vapor pressure of one of the components.

The excess Gibbs energy (per mole of the mixture) of the two-component system is defined by

$$g^{EX} = \frac{G^{EX}}{N_A + N_B} = x_A\mu_A + x_B\mu_B - x_A\left(\mu_A^P + kT\ln x_A\right) - x_B\left(\mu_B^P + kT\ln x_B\right) \quad (4.87)$$

where μ_A^P and μ_B^P are the chemical potentials of pure A and B, respectively.

Taking the second derivative with respect to x_A and using the Gibbs–Duhem relationship, we obtain

$$D = \frac{x_A}{kT}\left(\frac{\partial\mu_A}{\partial x_A}\right)_{P,T} = 1 + \frac{x_A x_B}{kT}\left(\frac{\partial^2 g^{EX}}{\partial x_A^2}\right)_{P,T} \quad (4.88)$$

which can be used in equations (4.82)–(4.84).

Another source of experimental information can be used if the vapor above the mixture may be considered to be an ideal-gas mixture, in which case the chemical potential of each component in the gaseous phase has the form

$$\mu_\alpha^l = \mu_\alpha^g = \mu_\alpha^{0g} + kT\ln p_\alpha \quad (4.89)$$

where p_α is the partial vapor pressure of the component α.

Hence, in this case

$$D = \frac{x_A}{kT}\left(\frac{\partial\mu_A}{\partial x_A}\right)_{P,T} = x_A\frac{\partial\ln p_A}{\partial x_A} = x_B\frac{\partial\ln p_B}{\partial x_B}. \quad (4.90)$$

Clearly, only one of the derivatives on the rhs of (4.90) is needed.

4.5 Three-component systems

The general equations for κ_T, $\overline{V}_\alpha$, and $\mu_{\alpha\beta}$ are given in (4.36)–(4.38). As we have seen in the case of $c=2$ (two-component systems), it is easy to write the explicit expressions for the thermodynamic equations in terms of G_{ij} For three-component systems, $c=3$, these expressions become very long and complicated, especially the expression for $\mu_{\alpha\beta}$ which contains a sum over nine determinants, each of which when fully expanded consists of a large number of terms. Fortunately, there exists a simplification of equation (4.38) which reads (see Appendix K)

$$\mu_{\alpha\beta} = \frac{kT}{\rho_\alpha \rho_\beta V} \frac{|\boldsymbol{E}(\alpha, \beta)|}{|\boldsymbol{D}|} \tag{4.91}$$

where $\boldsymbol{E}(\alpha, \beta)$ and $\boldsymbol{D}$ are two matrixes derived from the matrix

$$\boldsymbol{G} = \begin{pmatrix} G_{11} + \rho_1^{-1} & G_{12} & G_{13} \dots \\ G_{21} & G_{22} + \rho_2^{-1} & G_{23} \\ \vdots & \vdots & \vdots \end{pmatrix}. \tag{4.92}$$

The general element of the matrix $\boldsymbol{G}$ is

$$(\boldsymbol{G})_{ij} = G_{ij} + \delta_{ij}\rho_i^{-1} \tag{4.93}$$

where G_{ij} are the KBIs. Details of the derivation of equation (4.91) are provided in Appendix K.

The expressions for the thermodynamic quantities can be written in somewhat shorter forms by defining the quantities:

$$\Delta_{\alpha\beta} = G_{\alpha\alpha} + G_{\beta\beta} - 2G_{\alpha\beta} \tag{4.94}$$

$$\delta_{\alpha\beta} = G_{\alpha\alpha} G_{\beta\beta} - G_{\alpha\beta}^2 \tag{4.95}$$

$$\begin{aligned} \eta = \rho_A + \rho_B + \rho_C + \rho_A\rho_B\Delta_{AB} + \rho_B\rho_C\Delta_{BC} + \rho_A\rho_C\Delta_{AC} \\ - \tfrac{1}{4}\rho_A\rho_B\rho_C(\Delta_{AB}^2 + \Delta_{BC}^2 + \Delta_{AC}^2 - 2\Delta_{AC}\Delta_{BC} \\ - 2\Delta_{AB}\Delta_{AC} - 2\Delta_{AB}\Delta_{BC}) \end{aligned} \tag{4.96}$$

which is the generalization of η defined for a two-component system in section 4.3.

Similarly, we define

$$\begin{aligned} \zeta = 1 + \rho_A G_{AA} + \rho_B G_{BB} + \rho_C G_{CC} + \rho_A\rho_B\delta_{AB} + \rho_A\rho_C\delta_{AC} + \rho_B\rho_C\delta_{BC} \\ + \rho_A\rho_B\rho_C(G_{AA}\delta_{BC} + G_{BB}\delta_{AC} + G_{CC}\delta_{AB} - 2G_{AA}G_{BB}G_{CC} + 2G_{AB}G_{AC}G_{BC}) \end{aligned} \tag{4.97}$$

which is the generalization of ζ defined for a two-component system in section 4.3.

In terms of these quantities, the thermodynamic quantities for a three-component system are:

$$\mu_{AA} = \frac{kT(\rho_B + \rho_C + \rho_B\rho_C\Delta_{BC})}{V\eta\rho_A} \tag{4.98}$$

$$\mu_{BB} = \frac{kT(\rho_A + \rho_C + \rho_A\rho_C\Delta_{AC})}{V\eta\rho_B} \tag{4.99}$$

$$\mu_{CC} = \frac{kT(\rho_A + \rho_B + \rho_A\rho_B\Delta_{AB})}{V\eta\rho_C} \tag{4.100}$$

$$\mu_{AB} = \frac{kT[1 + \rho_C(G_{AB} + G_{CC} - G_{AC} - G_{BC})]}{V\eta} \tag{4.101}$$

$$\mu_{AC} = \frac{-kT[1 + \rho_B(G_{AC} + G_{BB} - G_{AB} - G_{BC})]}{V\eta} \tag{4.102}$$

$$\mu_{BC} = \frac{-kT[1 + \rho_A(G_{BC} + G_{AA} - G_{AC} - G_{AB})]}{V\eta} \tag{4.103}$$

$$\begin{aligned} V_A =& (1/\eta)[1 + \rho_B(G_{BB} - G_{AB}) + \rho_C(G_{CC} - G_{AC}) \\ &+ \rho_B\rho_C(G_{AB}G_{BC} + G_{AC}G_{BC} + G_{BB}G_{CC} \\ &- G_{AC}G_{BB} - G_{AB}G_{CC} - G_{BC}^2)] \end{aligned} \tag{4.104}$$

$$\begin{aligned} V_B =& (1/\eta)[1 + \rho_A(G_{AA} - G_{AB}) + \rho_C(G_{CC} - G_{BC}) \\ &+ \rho_A\rho_C(G_{AB}G_{AC} + G_{AC}G_{BC} + G_{AA}G_{CC} \\ &- G_{AA}G_{BC} - G_{AB}G_{CC} - G_{AC}^2)] \end{aligned} \tag{4.105}$$

$$\begin{aligned} V_C =& (1/\eta)[1 + \rho_B(G_{BB} - G_{BC}) + \rho_A(G_{AA} - G_{AC}) \\ &+ \rho_A\rho_B(G_{AB}G_{AC} + G_{AB}G_{BC} + G_{AA}G_{BB} \\ &- G_{AA}G_{BC} - G_{AC}G_{BB} - G_{AB}^2)] \end{aligned} \tag{4.106}$$

$$\kappa_T = \frac{\zeta}{kT\eta}. \tag{4.107}$$

We note again that as in the extension from one-component to two-component systems, the generalization to a three-component system is not straightforward and cannot be done only by inspection of the expressions for the two-component case.

The inversion of the KB theory for a three-component system is quite complicated. The present author tried unsuccessfully to find a simple expression such as (4.86) for the three-component system. Ruckenstein and Shulgin (2001a, b, c) spelled out these long and complicated expressions explicitly. Matteoli and Lepori (1995), however, suggested that it is simpler to use the relations (4.23) to express all $G_{\alpha\beta}$ in terms of $B_{\alpha\beta}$. This is already a relation between $G_{\alpha\beta}$ and thermodynamic quantities. However, one can go further and express all the $B_{\alpha\beta}$ in terms of $A_{\alpha\beta}$, by solving equation (4.6), and then express $A_{\alpha\beta}$ in terms of the required thermodynamic quantities $\mu_{\alpha\beta}$, V_α, and κ_T via (4.31). This procedure is straightforward for practical calculations of $G_{\alpha\beta}$, though it does not provide simple, explicit expressions in terms of the thermodynamic quantities.

4.6 Dilute system of *S* in *A* and *B*

In the previous section, we have seen that the application of the KB theory for three or more components become very complicated. Therefore, the main application of the KB theory has been for dilute solution[†].

In this section, we shall derive only the equations for the system of three components, *A*, *B* and *S*, where *S* is very dilute in the mixture of *A* and *B*, i.e., we examine the limit of $\rho_S \to 0$. This case is important in the study of solvation phenomena (chapter 7).

From the general expressions from the previous section, we rename *C* as *S* and take the limit $\rho_S \to 0$. We get

$$\kappa_T = \frac{\zeta}{kT\eta} \tag{4.108}$$

$$\mu_{AA} = \frac{kT\rho_B}{\rho_A V\eta}, \quad \mu_{BB} = \frac{kT\rho_A}{\rho_B V\eta}, \quad \mu_{AB} = \mu_{BA} = -\frac{kT}{V\eta} \tag{4.109}$$

$$\overline{V}_A = \frac{1 + \rho_B(G_{BB} - G_{AB})}{\eta}, \quad \overline{V}_B = \frac{1 + \rho_A(G_{AA} - G_{AB})}{\eta} \tag{4.110}$$

where η is the same as in (4.46). As expected, the quantities (4.108)–(4.110) are the same as in a two-component system, see equations (4.48)–(4.51). The new expressions for the three-component system as $\rho_S \to 0$ are:

$$\mu_{AS} = \mu_{SA} = \frac{-kT[1 + \rho_B(G_{BB} - G_{BS}) + \rho_B(G_{AS} - G_{AB})]}{V\eta} \tag{4.111}$$

$$\mu_{BS} = \mu_{SB} = \frac{-kT[1 + \rho_A(G_{AA} - G_{AS}) + \rho_B(G_{BS} - G_{AB})]}{V\eta} \tag{4.112}$$

and

$$\begin{aligned}\overline{V}_S &= (1 + \rho_A(G_{AA} - G_{AS}) + \rho_B(G_{BB} - G_{BS}) + \rho_A\rho_B[G_{AA}G_{BB} - G_{AB}^2 \\ &\quad + G_{AS}(G_{AB} - G_{BB}) + G_{BS}(G_{AB} - G_{AA})])/\eta \\ &= kT\kappa_T - \rho_A\overline{V}_A G_{AS} - \rho_B\overline{V}_B G_{BS}.\end{aligned} \tag{4.113}$$

[†] As an example, Shulgin and Ruckenstein (2002) have derived an expression for Henry's constant in binary solvent mixtures.

The only term which diverges as $\rho_S \to 0$ is μ_{SS}; in this limit $\mu_{SS} \approx kT/N_s$. Hence, we look for the limiting behavior of the nondivergent part

$$\begin{aligned}
&\lim_{\rho_s \to 0}\left[\mu_{SS} - \frac{kT}{N_S}\right] \\
&\quad = (-kT[4(1 + \rho_A \Delta_{AS} + \rho_B \Delta_{BS}) + \rho_\alpha \rho_\beta \\
&\qquad \times (\Delta_{AB}^2 + \Delta_{AS}^2 + \Delta_{BS}^2 - 2\Delta_{AB}\Delta_{BS} - 2\Delta_{AS}\Delta_{BS})])/(4\eta V) \qquad (4.114)
\end{aligned}$$

where

$$\Delta_{\alpha\beta} = G_{\alpha\alpha} + G_{\beta\beta} - 2G_{\alpha\beta}. \qquad (4.115)$$

A more useful derivative of the chemical potential of the solute, in this limit, is

$$\begin{aligned}
\left(\frac{\partial \mu_S}{\partial x_S}\right)_{T,P,N_A,N_B} &= \left(\frac{\partial \mu_S}{\partial N_S}\right)_{T,P,N_A,N_B} \left(\frac{\partial N_S}{\partial x_S}\right)_{T,P,N_A,N_B} \\
&= \mu_{SS}\left(\frac{\rho_T^2 V}{\rho_A + \rho_B}\right) \qquad (4.116)
\end{aligned}$$

where $\rho_T = \rho_A + \rho_B + \rho_S$. Again, this is a complicated function of G_{ij}. However, one can express this derivative as a function of $\Delta_{\alpha\beta}$ and then show that when all $\Delta_{\alpha\beta} = 0$, we get

$$\left(\frac{\partial \mu_S}{\partial x_S}\right)_{T,P,N_A,N_B} = \frac{kT}{x_S}. \qquad (4.117)$$

This is the case of SI solutions discussed in the next chapter.

4.7 **Application of the KB theory to electrolyte solutions**

The simplest system of electrolyte solution is one solvent, say water (W) and one, completely dissociable solute, say KCl, which we denote by *S*. It is assumed that *S* dissociates completely into two fragments

$$S \to A + B. \qquad (4.118)$$

We stress from the outset that the fragments *A* and *B* could be either neutral or ions.

The above system could be viewed in two equivalent ways. One is to ignore the dissociation (4.118) and view the system as a *two-component* system of *W* and *S*. For such a system, the KB theory as derived in section 4.3 applies; simply change notation from *A* and *B* in section 4.3 into *W* and *S*, and we can use all the results from the KB theory. Specifically, the matrix **B**, written explicitly for this case, is

$$\boldsymbol{B} = \begin{pmatrix} \rho_W + \rho_W^2 G_{WW} & \rho_S \rho_W G_{SW} \\ \rho_S \rho_W G_{SW} & \rho_S + \rho_S^2 G_{SS} \end{pmatrix}. \tag{4.119}$$

The second view is to admit the occurrence of dissociation into fragments (4.118) and to treat the system as a *three-component* system. Again, all the results of section 4.3 apply here. Specifically the matrix **B**, for the system of three components, *W*, *A* and *B*, is

$$\boldsymbol{B} = \begin{pmatrix} \rho_W + \rho_W^2 G_{WW} & \rho_W \rho_A G_{WA} & \rho_W \rho_B G_{WB} \\ \rho_W \rho_A G_{WA} & \rho_A + \rho_A^2 G_{AA} & \rho_A \rho_B G_{AB} \\ \rho_W \rho_B G_{WB} & \rho_A \rho_B G_{AB} & \rho_B + \rho_B^2 G_{BB} \end{pmatrix}. \tag{4.120}$$

All the derivations of the KB results can be followed as in section 4.5.

However, when the dissociation of *S* into fragments (4.118) produces ionic species, say

$$\mathrm{KCl} \rightarrow \mathrm{K}^+ + \mathrm{Cl}^- \tag{4.121}$$

one invokes the so-called electro-neutrality (electro-neutrality) conditions. These are statements on the electro-neutrality of the entire system, i.e., the total charge around a solvent molecule must be zero, hence

$$\rho G_{WA} = \rho G_{WB} \tag{4.122}$$

where $\rho = \rho_S = \rho_A = \rho_B$ is the number density of the solute *S*. Equivalently

$$G_{WA} = G_{WB} \tag{4.123}$$

Similarly, we must have the conservation of the total charge around *A* and around *B*, hence

$$1 + \rho G_{AA} = \rho G_{AB} \tag{4.124}$$

$$1 + \rho G_{BB} = \rho G_{AB}. \tag{4.125}$$

The three conditions (4.123)–(4.125) are referred to as the electro-neutrality conditions. Note that from (4.124) and (4.125), one may also obtain

$$G_{AA} = G_{BB}. \tag{4.126}$$

Substituting these conditions in the matrix **B**, we obtain

$$B = \begin{pmatrix} \rho_W + \rho_W^2 G_{WW} & \rho_W \rho G_{WA} & \rho_W \rho G_{WA} \\ \rho_W \rho G_{WA} & \rho^2 G_{AB} & \rho^2 G_{AB} \\ \rho_W \rho G_{WA} & \rho^2 G_{AB} & \rho^2 G_{AB} \end{pmatrix}. \tag{4.127}$$

Clearly, since two rows and two columns in this matrix are identical **B** is a singular matrix and the corresponding determinant $|\boldsymbol{B}|$ is zero. This renders the solution of the matrix equation (4.26) impossible, i.e., **B** has no inverse.

Therefore, one cannot proceed with the KB theory.[†] Sometimes, this rendering of the KB theory impossible, is attributed to the long-range interactions between the ionic solutes. However, the KB theory does not require any specific behavior of the intermolecular interactions. The KB theory can be applied for ionic solutions without any restriction on the type of interactions. The apparent impossibility of obtaining an inverse of the matrix B is not due to any special electro-neutrality condition, but is a result of mismatch of the KBIs, $G_{\alpha\beta}$, defined in different ensembles. To clarify the situation, we consider the following two examples

(1) **One-component system**

Here, we have the compressibility equation

$$\rho G_O = -1 + \rho kT\kappa_T. \tag{4.128}$$

The corresponding $\boldsymbol{B}$ matrix for this case is simply

$$\boldsymbol{B} = \rho^2 G_O + \rho \tag{4.129}$$

where we used the subscript O for an *open* system.

If, however, we use the KBI, as defined in the closed system (C), i.e., the one with the normalization

$$\rho G_C = -1 \tag{4.130}$$

we obtain the matrix

$$\boldsymbol{B} = -\rho + \rho = 0 \tag{4.131}$$

which is singular. Furthermore, if we use (4.130) in the compressibility equation, we get the absurd result

$$-1 = -1 + \rho kT\kappa_T$$

or equivalently

$$\rho kT\kappa_T = 0. \tag{4.132}$$

Clearly, the singularity of the matrix $\boldsymbol{B}$ results from using the *wrong* G, i.e., the KBI for the *closed* system in (4.129), where the KBI for an *open* system, G_O, should be used.

(2) **Two-component system**

Suppose we have a two-component system of A and B. The matrix $\boldsymbol{B}$ in this case is

$$\boldsymbol{B} = \begin{pmatrix} \rho_A + \rho_A^2 G_{AA} & \rho_A\rho_B G_{AB} \\ \rho_A\rho_B G_{AB} & \rho_B + \rho_B^2 G_{BB} \end{pmatrix} \tag{4.133}$$

[†] The problem of the occurrence of the singularity of the matrix $\boldsymbol{B}$, hence noninvertability, was pointed out by many authors Friedman and Ramanathan 1970: Kusalik and Patey 1987; Behera 1998; Newman 1988, 1989a, b, 1990, 1994, Beeby 1973. In particular Kusalik and Patey (1987) wrote: "The condition renders indeterminate all the thermodynamic quantities obtained by direct substitution into the KB equations."

where again, one should use the KBI in an *open* system. If, on the other hand, we use $G_{\alpha\beta}$ from the *closed* system, i.e., with the normalization conditions

$$\rho_\alpha G_{\alpha\beta,C} = -\delta_{\alpha\beta} \tag{4.134}$$

in (4.133), we get a singular matrix

$$\boldsymbol{B} = \begin{pmatrix} 0 & 0 \\ 0 & 0 \end{pmatrix}. \tag{4.135}$$

Again, we note that the singularity of the matrix $\boldsymbol{B}$ in (4.135) arises from using the *wrong* $G_{\alpha\beta,C}$ (i.e., the KBIs for the *closed* system) in a theory derived in an open system.

It is now clear that the singularity of the matrix $\boldsymbol{B}$ as written in (4.127) is *not* the result of some special features of the interactions between the ionic species (the KB theory applies for any type of intermolecular interactions), but from using the wrong $G_{\alpha\beta}$ in the KB theory.

To see this, we first note that though it is true that for ionic species, equations (4.123)–(4.126) *can* result from the electro-neutrality conditions, the conditions themselves are *not necessarily* a result of the electric charge neutrality. They arise from the *closure* condition with respect to the fragments A and B. Thus, for a solute S dissociating into two *neutral* fragments A and B, as in (4.118) not necessarily ionic species as in (4.121), we still have the following conservation relations:

(a) The total number of A and B particles must be the same, viewed from a solvent molecule at the center, i.e.,

$$\rho_A \int_V g_{AW}(\boldsymbol{R})\, d\boldsymbol{R} = \rho_B \int_V g_{BW}(\boldsymbol{R})\, d\boldsymbol{R}. \tag{4.136}$$

From this condition the relation (4.123) follows

(b) The total number of A's must be equal to the total number of B's, when viewed from either an A or from a B particle at the center, i.e.,

$$1 + \rho_A \int_V g_{AA}(\boldsymbol{R})\, dR = \rho_B \int_V g_{AB}(\boldsymbol{R})\, d\boldsymbol{R} \tag{4.137}$$

$$1 + \rho_B \int_V g_{BB}(\boldsymbol{R})\, d\boldsymbol{R} = \rho_A \int_V g_{AB}(\boldsymbol{R})\, d\boldsymbol{R} \tag{4.138}$$

from which relations (4.124) and (4.125) follow.

Thus, we see that the requirement that the solute fully dissociates as stated in (4.118) or (4.121) imposes closure conditions on G_{AA}, G_{AB}, G_{BB}, G_{AW}, and G_{BW}. Therefore, care must be exercised to label these KBIs properly, e.g., we should write instead of (4.123)–(4.125) the relations

$$G_{WA,C} = G_{WB,C} \tag{4.139}$$

$$1 + \rho G_{AA,C} = \rho G_{AB,C} \tag{4.140}$$

$$1 + \rho G_{BB,C} = \rho G_{AB,C}. \tag{4.141}$$

The subscript C stands for closure with respect to A and B *individually*. The system can still be opened with respect to W and S, but we do not allow the concentrations of A and B to change independently. With this labeling, it is clear that we *cannot* use relations (4.139)–(4.141) in the KB theory which was derived in the grand ensemble and hence all the KBIs in the matrix $\boldsymbol{B}$ must be *defined* in a system open with respect to W, A, B. Failing to take this precaution leads to absurd results, exactly as in the example given in (4.131) and (4.134). However, if we *open* the system for all the species A, B, and W, then we can use the KB theory as it is without any modification. It is true though that if we open the system with respect to A, B, and W, and if A and B are charged particles, then there will be fluctuations in the total charge of the system. In this case the problem is not the applicability of the KB theory nor in the application of its inversion. The problem is that the thermodynamic quantities on the rhs of (4.75) are simply not available.

It should be noted that Kirkwood and Buff in their original publication commented that equation (20) in their publication (4.60 in this chapter) is "completely general" and provides an alternative to the usual "charging process." They also added that "in the absence of long-range intermolecular forces, the integrals $G_{\alpha\beta}$ may be developed in power series in the solute concentration." It is clear that Kirkwood and Buff did not see any difficulty in applying their results for ionic solutions as long as one uses the open-system $G_{\alpha\beta}$.

To summarize, for ionic solutions, one can either adopt the view that the system is a two-component mixture and use the KB theory for a two non-ionic species, say water and salt [as has been done by Friedman and Ramanathan (1970) and by Chitra and Smith (2002)], where the system is open with respect to the water and the salt (as a single entity). In this view, there is no place for ion-ion correlations hence the electro-neutrality condition is irrelevant. In this view only correlations between neutral molecules are meaningful. Or we can view the system as water, cation, and anion – but in this case one must *open* the system with respect to each of the species individually (hence allow also charged fluctuations in the system). In this case, the KBI are meaningful, but the thermodynamic quantities such as $\overline{V}_i$ or $\partial\mu_i/\partial N_j$ are not available. In the second view, the ion-ion correlations do enter into the ***KB*** theory, but again the electro-neutrality condition is irrelevant.

FIVE

Ideal Solutions

There exist several reference states of solutions referred to as "ideal state," for which we can say something on the behavior of the thermodynamic functions of the system. The most important "ideal states" are the ideal-gas mixtures, the symmetric ideal solutions and the dilute ideal solution. The first arises from either the total lack of interactions between the particles (the theoretical ideal gas), or because of a very low total number density (the practical ideal gas). The second arises when the two (or more) components are "similar." We shall discuss various degrees of similarities in sections 5.2. The last arises when one component is very dilute in the system (the system can consist of one or more components). Clearly, these are quite different ideal states and caution must be exercised both in the usage of notation and in the interpretations of the various thermodynamic quantities. Failure to exercise caution is a major reason for confusion, something which has plagued the field of solution chemistry.

5.1 Ideal-gas mixtures

As in the case of a one-component system, ideal-gas (IG) mixtures also enjoy having a simple and solvable *molecular* theory, in the sense that one can calculate all the thermodynamic properties of the system from molecular properties of *single* molecules. We also have a truly *molecular* theory of mixtures of slightly nonideal gases, in which case one needs in addition to molecular properties of single molecules, also interactions between two or more molecules.

As in the case of a one-component system, we should also make a distinction between the *theoretical* ideal gas, and the *practical* ideal gas. The former is a system of noninteracting particles; the latter applies to any *real* system at very low densities. Occasionally, the former serves as a model for the latter. For instance, to obtain the equation of state of an ideal gas

$$\beta P = \rho \tag{5.1}$$

we can either take a *real* gas and let $\rho \to 0$, or we can envisage a model system of noninteracting particles at any finite density. Both come up with the same equation of state (5.1). However, if we are to study deviation from the ideal-gas state, we write a density expansion of the form

$$\beta P = \rho + B_2\rho^2 + B_3\rho^3 + \cdots \tag{5.2}$$

where $B_i(T)$ are the virial coefficients which depend on temperature and on the interactions among the i particles. Clearly, for a theoretical ideal gas, all the $B_i(T)$ are zero, hence (5.2) reduces to (5.1). However, for a *real* gas, the coefficient are non-zero. For any finite value of $B_i(T)$, (5.2) reduces to (5.1) in the limit of $\rho \to 0$. Thus, the ideal-gas equation of state is the same for the *theoretical* and the *practical* ideal gas, but the reasons for reaching (5.1) are different: in one, because all $B_i(T) = 0$; in the second, because $\rho \to 0$.

The theoretical ideal-gas partition function for a system of c components of composition $\mathbf{N} = N_1, N_2, \ldots, N_c$ contained in a volume V at temperature T is

$$Q(T, V, N) = V^N \prod_{i=1}^{C} \frac{q_i^{N_i}}{N_i!\,\Lambda_i^{3N_i}} \tag{5.3}$$

where $V^N = V^{\sum N_i}$ is obtained from the configurational partition function of the system. Hence, we have $\sum N_i$ integrations over the volume V. q_i is the internal partition function of a molecule of species i, excluding the momentum partition function Λ_i^3.

The Helmholtz energy of the system is obtained from

$$A(T, V, N) = -kT \ln Q(T, V, N) \tag{5.4}$$

The chemical potential of the species i can be obtained by direct differentiation with respect to N_i, i.e.,

$$\mu_i = \left(\frac{\partial A}{\partial N_i}\right)_{T,V,N_i'} = kT \ln \frac{N_i}{V} \Lambda_i^3 q_i^{-1}. \tag{5.5}$$

From (5.3) and (5.4), we can derive all the thermodynamic quantities for our system. However, in this chapter, we shall be interested in the T, P, N system; these are the most common variables that are controlled in the experiments. Hence, we use the T, P, $\mathbf{N}$ ensemble to derive the quantities of interest. Here the chemical potential is defined as[†]

$$\mu_i = \left(\frac{\partial G}{\partial N_i}\right)_{T,P,N_i'} = G(T, P, \mathbf{N} + 1_i) - G(T, P, \mathbf{N}) \tag{5.6}$$

[†] Note that we add one particle of species i (denoted 1_i) to obtain the derivative defining the chemical potential.

where G is the Gibbs energy and N_i' is the vector obtained from $\mathbf{N}=N_1, N_2, \ldots, N_c$, by eliminating the ith component, N_i namely $N_i' = (N_1, N_2, \ldots, N_{i-1}\ N_{i+1}, \ldots, N_c)$. Using the T, P, $\mathbf{N}$ partition function, we obtain for the chemical potential

$$\exp[-\beta\mu_i] = \frac{\Delta(T,P,\mathbf{N}+1_i)}{\Delta(T,P,\mathbf{N})} = \frac{q_i}{\Lambda_i^3(N_i+1)}\int dV\, P(V)V = \frac{q_i\langle V\rangle}{\Lambda_i^3 N_i} \quad (5.7)$$

where $P(V)$ is the probability density of finding the system with volume V.

This is the same as (5.5) except for the replacement of the *exact* volume V in the T, V, $\mathbf{N}$ ensemble by the *average* volume $\langle V\rangle$ in the T, P, $\mathbf{N}$ ensemble. We have also replaced N_i+1 by N_i for the macroscopic system.

We define the average density $\rho_i = N_i/\langle V\rangle$ in the T, P, $\mathbf{N}$ ensemble and rewrite (5.7) as

$$\begin{aligned}\mu_i &= kT\ \ln \rho_i\Lambda_i^3 q_i^{-1}\\ &= kT\ \ln \Lambda_i^3 q_i^{-1} + kT\ \ln \rho_i\\ &= kT\ \ln \Lambda_i^3 q_i^{-1}\rho_T + kT\ \ln x_i\\ &= \mu_i^P + kT\ \ln x_i \end{aligned} \quad (5.8)$$

where $\rho_T = \sum \rho_i$ is the total (average) density of the system, and x_i is the mole fraction of the species i in the system. $\mu_i^P(T,P)$ is the chemical potential of the *pure* component i, at the same T and P. Its density is ρ_T, i.e., one can obtain the *pure* i at the same T, P, $\mathbf{N}$ by replacing each of the particles of the system by an i particle keeping the total density ρ_T fixed. It is clear from (5.8) that knowing the molecular properties of the system like the mass, vibrational energy, etc., we can calculate μ_i and hence all the properties of the system.

The Gibbs energy of this system is

$$G = \sum N_i\mu_i = \sum N_i\mu_i^P + \sum kTN_i \ln x_i. \quad (5.9)$$

The first term on the rhs of (5.9) is the Gibbs energy of c systems each containing the pure component i at the *same* T, P. Sometimes, this is referred to as the system *before* mixing (figure 5.1). Since each of these systems contains N_i particles at the same P, T, the volumes of each system *before mixing* are $V_i^P = N_i kT/P$. Removing the partitions separating the pure systems, we obtain the mixture at the rhs of figure 5.1. The system's volume is now $V = \sum N_i kT/P$ (same T, P). The quantity $\sum kTN_i \ln x_i$ (or the corresponding quantity per mole of the mixture $\sum kTx_i \ln x_i$) is referred to *erroneously* as the free energy of mixing. However, tracing the origin of the change in the Gibbs energy in the process depicted in figure 5.1 shows that this change in due only to the

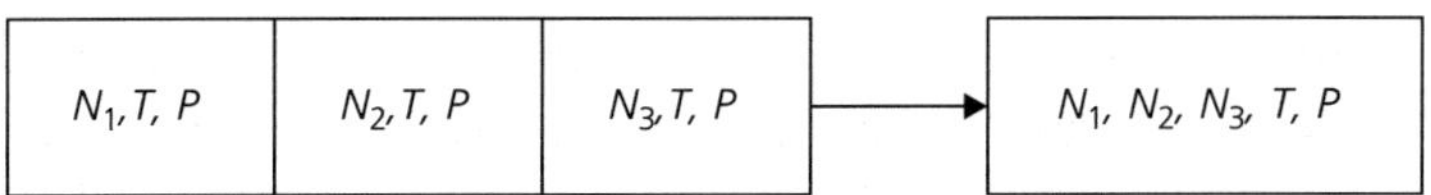

Figure 5.1 A process of mixing ideal gases. Initially we have three systems, each compartment containing N_i particles of species i, at the same T and P. The final state is obtained by removing the partitions between the systems. The temperature and the total pressure remains the same T and P.

expansion of each component from the initial volume V_i^P into the final volume V. More precisely, for the ith component we have

$$\begin{aligned} N_i(\mu_i - \mu_i^P) &= N_i(kT \ln \rho_i - kT \ln \rho_i^P) \\ &= N_i\left(kT \ln \frac{N_i}{V} - kT \ln \frac{N_i}{V_i^P}\right) \\ &= kTN_i \ln \frac{V_i^P}{V} = kT\, N_i \ln x_i \end{aligned} \tag{5.10}$$

and the total change in the Gibbs energy is

$$\sum N_i(\mu_i - \mu_i^P) = \sum N_i kT \ln x_i \tag{5.11}$$

which clearly shows that the *decrease* of the total Gibbs energy is due to the *expansion* of each component, from the initial volume V_i^P at P, T, to the final volume $V\ (= \sum V_i^P)$, at the same P, T.

By taking the temperature derivative of (5.9), one can obtain the corresponding entropy and enthalpy of the system

$$S = -\left(\frac{\partial G}{\partial T}\right)_{P,N} = \sum_i N_i S_i^P - \sum kN_i \ln x_i \tag{5.12}$$

$$H = G - TS = \sum N_i H_i^P \tag{5.13}$$

where S_i^P and H_i^P are the entropy and the enthalpy (per molecule) of the pure ith component at the same P, T, respectively. Note again that the quantity $-\sum kN_i \ln x_i$ is erroneously referred to as the *entropy of mixing*. Clearly, by the same argument given above, one can easily show that this change in entropy arises from the *expansion* of each component from the initial volume V_i^P to the final volume V (at the same P, T). The *mixing* of ideal gases in itself has no effect on any thermodynamic quantity of the system. More on that in Appendices H, I and J.

The (average) volume of the system may be obtained from the pressure derivative of the Gibbs energy, i.e.,

$$V = \left(\frac{\partial G}{\partial P}\right)_{T,N} = \sum N_i V_i^P. \tag{5.14}$$

Note that since each of the pure components is an ideal gas at P, T, we have $V_i^P = N_i kT/P$, hence

$$V = \frac{kT}{P} \sum N_i \tag{5.15}$$

or equivalently

$$\rho_T = \sum \rho_i = \frac{P}{kT} \tag{5.16}$$

which is the equation of state for the ideal-gas mixture.

We note again that the equation of state (5.16) is obtained for either the theoretical model of noninteracting particles at *any* density ρ_T, or for a *real* system of interacting particles but at very low density, where encounters between particles, hence interactions, are rare events. In *both* cases, the pressure of the system is a result of the interactions of the particles with the walls of the system. (By noninteracting particles, one assumes that the *intramolecular* interactions among the $\sum N_i$ particles in the system are switched off. There must still be an interaction between the particles and the wall, otherwise the particles will not be confined to the volume V.)

Finally, we note that once we have the molecular properties of the molecules, we can calculate all the thermodynamic quantities of the system, such as the Gibbs energy, entropy, enthalpy, etc., Note also that the equation of state does not depend on the specific properties of the system, only on the total number of the particles in the system, at a given P, T. The same is true for the derivatives of the volume with respect to pressure and temperature.

5.2 **Symmetrical ideal solutions**

In the previous section, we have derived the equations for the thermodynamic quantities of ideal-gas mixtures. We could also compute all of these quantities from the knowledge of the molecular properties of the single molecules. Once we get into the realm of liquid densities, we cannot expect to obtain that amount of detailed information on the thermodynamics of the system.

We have seen that in an ideal-gas mixture, the chemical potential of each species can be written in one of the following forms:

$$\mu_i = kT \ln \rho_i \Lambda_i^3 q_i^{-1} \tag{5.17}$$

$$= \mu_i^* + kT \ln \rho_i \Lambda_i^3 \tag{5.18}$$

$$= \mu_i^P + kT \ln x_i. \tag{5.19}$$

Clearly knowing Λ_i^3 and q_i allows us to compute the chemical potential μ_i. This is not possible for real mixtures at high densities. Nevertheless, both theory and experiments show that under certain conditions, the chemical potential of species i depends on the density ρ_i or on the mole fraction x_i in the same way as in (5.18) and (5.19) with some constants μ_i^* or μ_i^P which are independent of ρ_i or x_i, respectively, but whose actual dependence on the molecular properties of the system is not known.

In this section, we shall discuss a class of mixture for which the chemical potential of each species i depends on x_i as in (5.19). In the next section, we shall study the condition under which the chemical potential depends on ρ_i as in (5.18) with μ_i^* independent of ρ_i. In both cases, we shall be satisfied in obtaining conditions under which the chemical potential has this particular dependence on ρ_i or on x_i, even though the constants μ_i^* or μ_i^P cannot be calculated from the theory.

We start by defining a symmetrical ideal (SI) solution as a system for which the chemical potential of each species has the form

$$\mu_i = \mu_i^P + kT \ln x_i \tag{5.20}$$

where μ_i^P is the chemical potential of pure i at the same P, T as in the mixture, and we require that this form is valid in the entire range of compositions $0 \leq x_i \leq 1$.

Clearly, an ideal-gas mixture is a particular example of a SI solution, as we have seen in the previous section (particularly equation 5.18). Here we discuss a *real* mixture at normal liquid densities, consisting of interacting molecules. We shall examine first the conditions on the molecular properties (specifically on the intermolecular interactions) that lead to this particular form of the chemical potential. In section 5.2.2, we shall examine the *local* conditions under which relation (5.20) is achieved.

5.2.1 **Very similar components: A sufficient condition for SI solutions**

For notational convenience, we shall discuss a two-component mixture of A and B. The generalization for multicomponent system is quite straightforward. We consider a system of two components in the T, P, N_A, N_B ensemble. We have chosen the T, P, N_A, N_B ensemble because the isothermal-isobaric systems are the most common ones in actual experiments. By *very similar* components, we mean, in the present context, that the potential energy of interaction among a group of n molecules in a configuration X^n is *independent* of the species we assign to each configuration X_i[†]. For example, the pair potential $U_{AA}(X', X'')$ is nearly the same as the pair potential $U_{AB}(X', X'')$ or $U_{BB}(X', X'')$, provided that the configuration of the pair is the same in each case. Clearly, we do not expect that this property will be fulfilled exactly for any pair of *different* real molecules. However, for molecules differing in, say, isotopic constitution, it may hold to a good approximation.[‡]

The chemical potential of component A is defined by

$$\mu_A = \left(\frac{\partial G}{\partial N_A}\right)_{T,P,N_B} = G(T, P, N_A + 1, N_B) - G(T, P, N_A, N_B) \qquad (5.21)$$

where G is the Gibbs energy and the last equality is valid by virtue of the same reasoning as given in section 3.4.

The connection between the chemical potential and statistical mechanics follows directly from the definition of the chemical potential in (5.21), i.e.,

$$\begin{aligned} \exp(-\beta\mu_A) &= \frac{\Delta(T,P,N_A+1,N_B)}{\Delta(T,P,N_A,N_B)} \\ &= \frac{q_A \int dV \int d\mathbf{X}^{N_A+1}\, d\mathbf{X}^{N_B} \exp[-\beta PV - \beta U_{N_A+1,N_B}(\mathbf{X}^{N_A+1},\mathbf{X}^{N_B})]}{\Lambda_A^3(N_A+1) \int dV \int d\mathbf{X}^{N_A}\, d\mathbf{X}^{N_B} \exp[-\beta PV - \beta U_{N_A,N_B}(\mathbf{X}^{N_A},\mathbf{X}^{N_B})]} \end{aligned} \qquad (5.22)$$

where Λ_A^3 and q_A are the momentum and the internal partition function of an A molecule, respectively. An obvious shorthand notation has been used for the total potential energy of the system. The configuration $(\mathbf{X}^{N_A}, \mathbf{X}^{N_B})$ denotes the total configuration of N_A molecules of type A and N_B molecules of type B.

Next, consider a system of N particles of type A only. The chemical potential for such a system (at the same P and T as before) is

$$\exp(-\beta\mu_A^P) = \frac{q_A \int dV \int d\mathbf{X}^{N+1} \exp[-\beta PV - \beta U_{N+1}(\mathbf{X}^{N+1})]}{\Lambda_A^3(N+1) \int dV \int d\mathbf{X}^{N} \exp[-\beta PV - \beta U_{N}(\mathbf{X}^{N})]} \qquad (5.23)$$

where we have denoted by μ_A^P the chemical potential of pure A at the same P and T as for the mixture,

Now we choose N in (5.23) to be equal to $N_A + N_B$ in (5.22). The assumption of *very similar components* implies, according to its definition, the

[†] If the particles are spherical, we need to specify only the *locations* of the particles. In case of non-spherical particles, we specify the configuration of each particle by the same set of locational and orientational coordinates.

[‡] It is interesting to note that even mixtures of isotopes sometimes show measurable deviations from SI solutions. Examples are mixtures of ^{36}Ar and ^{40}Ar (Calado et al. 2000) and mixtures of CH_4 and CD_4 (Calado et al. 1994). However, mixtures of H_2O and D_2O do not show any significant deviations from SI solutions (Jancso and Jakli 1980).

two equalities[†]

$$U_N(\boldsymbol{X}^N) = U_{N_A,N_B}(\boldsymbol{X}^{N_A}, \boldsymbol{X}^{N_B}) \tag{5.24}$$

$$U_{N+1}(\boldsymbol{X}^{N+1}) = U_{N_A+1,N_B}(\boldsymbol{X}^{N_A+1}, \boldsymbol{X}^{N_B}). \tag{5.25}$$

Using (5.24) and (5.25) in (5.22), we get

$$\exp(-\beta\mu_A + \beta\mu_A^P) \approx \frac{(N+1)}{(N_A+1)}. \tag{5.26}$$

Rearranging (5.26) and noting that for macroscopic systems

$$x_A = \frac{N_A}{N} \approx \frac{(N_A+1)}{(N+1)}$$

we get the final result

$$\mu_A(T, P, x_A) = \mu_A^P(T, P) + kT \ln x_A. \tag{5.27}$$

Thus, the chemical potential, when expressed in terms of the intensive variables T, P and x_A, has this explicit dependence on the mole fraction x_A. A system for which relation of the form (5.27) is obeyed by each component, in the entire range of composition, is called a *symmetrical ideal* solution[‡]. It is symmetrical in the sense that from the assumptions (5.24) and (5.25), it follows that relation (5.27) holds true for *any* component in the system. In a two-component system, it is sufficient to define SI behavior for one component only. The same behavior of the second component, follows from the Gibbs–Duhem relation

$$x_A \frac{\partial\mu_A}{\partial x_A} + x_B \frac{\partial\mu_B}{\partial x_A} = 0. \tag{5.28}$$

Thus, whenever (5.27) is true for all $0 \leq x_A \leq 1$, it follows that

$$\mu_B = \mu_B^P + kT \ln x_B \qquad \text{for } 0 \leq x_B \leq 1. \tag{5.29}$$

The generalization for multicomponent systems is quite straightforward. The condition on the equality of all interaction potentials is sufficient for the SI behavior of all the components in the system. In the multicomponent system, we have to require that the SI behavior of the type (5.27) holds for $c-1$ components; the Gibbs–Duhem relation ensures the validity of the SI behavior for the cth component.

[†] This is the same as requiring that replacing any A at X by a B molecule at the same configuration $\boldsymbol{X}$ will have no effect on the total potential energy.

[‡] It should be noted that this nomenclature is not universally accepted. Sometimes the term "perfect" solution is used instead (Guggenhein 1967; Prigogine 1957). See also the end of section 5.4.

Relation (5.27) is important since it gives an explicit dependence of the chemical potential on the composition, the fruitfulness of which was recognized long ago. This relation has been obtained at the expense of the very strong requirement that the two components be *very similar*. We know from experiment that a relation such as (5.27) holds also under much weaker conditions. We shall see in the next section that relations of the form (5.27) can be obtained under much weaker assumptions on the extent of the "similarity" of the two components. In fact, relation (5.27) could not have been so useful had it been restricted to the extreme case of very similar components, such as two isotopes.

An alternative derivation of (5.20), which employs a somewhat weaker assumption, is the following; we write the general expression for the chemical potential of A in pure A denoted by μ_A^P as

$$\mu_A^P = \mu_A^* + kT \ln \rho_A \Lambda_A^3 = W(A|A) + kT \ln \rho_A^P \Lambda_A^3 q_A^{-1}. \tag{5.30}$$

We use the notation $W(A \mid A)$ to designate the coupling work of A against an environment which is pure A, ρ_A^P being the density of pure A at the same P, T.

A straightforward generalization of (5.30) for a two-component mixture is

$$\mu_A = W(A|A+B) + kT \ln \rho_A \Lambda_A^3 q_A^{-1} \tag{5.31}$$

where $W(A \mid A+B)$ is a shorthand notation for the coupling work of A against an environment composed of a mixture of A and B at the same P and T.

We now replace each B molecule in the environment of the A molecule (for which we have written the chemical potential) by an A' molecule. By A' molecules, we mean molecules that interact with A in exactly the same manner as B interacts with A. However, A' is still distinguishable from A. If we do that, then the particle A that is being coupled to its environment would not notice the difference in its environment. Hence, the coupling work in (5.31), $W(A \mid A+B)$, will be the same as $W(A \mid A)$ in (5.30). But note that since the A' molecules are *distinguishable* from the A molecules, the density ρ_A in (5.31) *does not* change. Thus, substituting $W(A \mid A+A') = W(A \mid A)$ from (5.31) into (5.30), we obtain

$$\begin{aligned}\mu_A &= \mu_A^P - kT \ln \rho_A^P \Lambda_A^3 q_A^{-1} + kT \ln \rho_A \Lambda_A^3 q_A^{-1} \\ &= \mu_A^P + kT \ln x_A\end{aligned} \tag{5.32}$$

where

$$x_A = \frac{\rho_A}{\rho_A^P} = \frac{\rho_A}{\rho_A + \rho_B}. \tag{5.33}$$

This is the same result as (5.27). Although we have used the assumption that A and B are "very similar," it is clear that the requirements in the second derivation are somewhat weaker. We only need that $W(A \mid A+B)$ be the same for any replacement of A and B in the environment of A. This condition is weaker since it involves only an *average* quantity and not the bare pair potentials themselves. We shall make this statement more precise in the next section.

The symmetrical ideal behavior is equivalent to the well-known Raoult law. Suppose that a mixture of A and B is in equilibrium with an ideal-gas phase; let P_A be the partial pressure of A. The chemical potential of A in the gas phase is

$$\mu_A^g = kT \ln\left(\frac{\Lambda_A^3 q_A^{-1} P_A}{kT}\right). \tag{5.34}$$

From the equilibrium condition $\mu_A^g = \mu_A(T, P, x_A)$, we obtain from (5.27) and (5.34)

$$P_A = P_A^0 x_A \qquad 0 \leq x_A \leq 1.$$

The proportionality constant P_A^0 can be identified as the vapor pressure of *pure* A at the same temperature and total pressure P.

5.2.2 Similar components: A necessary and sufficient condition for SI solutions

In the previous section, we showed that for a mixture of "very similar" (in the sense defined there) components, the chemical potential of each component i has the form†

$$\mu_i = \mu_i^p + kT \ln x_i \qquad (0 \leq x_i \leq 1) \tag{5.35}$$

in the entire range of compositions. We have referred to a mixture for which (5.35) is valid for all its components as an SI solution.

Experimentally, it had been known long ago that many mixtures are SI although the two (or more) components are far from being "very similar" in the sense of section 5.2.1.

A classical example is a mixture of ethylene bromide (EB) and propylene bromide (PB). Figure 5.2a shows the partial and the total pressures of these mixtures as a function of the mole fraction of (PB) at 85 °C, Based on the work of von Zawidzki (1900) quoted by Guggenheim (1952). These two components clearly cannot be considered as being identical, or "very similar." Yet, the fact that they form an SI solution in the entire range of compositions is equivalent

† The superscript p on μ_α^p stands for the *pure* species. We reserve the symbol μ_α^0 to denote the standard chemical potential (see chapter 7).

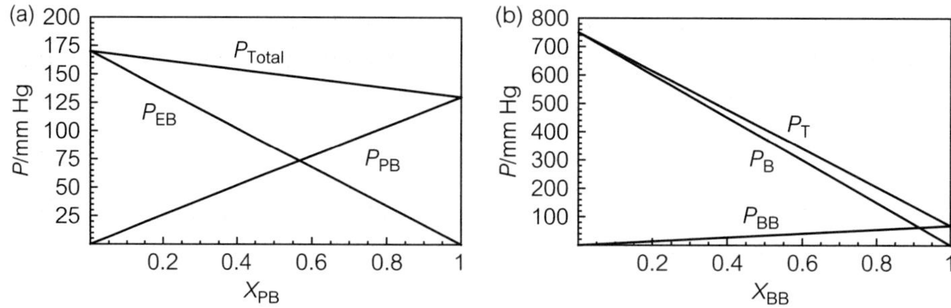

Figure 5.2 The partial vapor pressures and the total pressure of a system of (a) ethylene bromide (EB) and propylene bromide (PB) as a function of composition x (mole fraction PB) at $T = 85\,°C$, based on data by von Zawidski (1900) cited by Guggenheim (1952); (b) benzene (B) and bromobenzene (BB) at $80\,°C$. Note that although the two components have widely different vapor pressures the mixture is nearly SI. (Based on data from McGlashan and Wingrove 1956.)

to the assertion that the two components are *similar* in the sense discussed below. A second example is the mixture of benzene (B) and bromobenzene (BB). These two components have very different vapor pressures in their pure states, yet their mixture is nearly SI in the entire range of composition. Figure 5.2b shows the vapor pressures for this system, based on data by McGlashan and Wingrove (1956).

We shall now show that indeed a much weaker condition is required for SI, which turns out to be both a sufficient and a necessary condition.

We first discuss a two-component system of A and B at T, P and x_A. Since μ_i^P is independent of x_A, differentiation of μ_A in (5.35) gives

$$\left(\frac{\partial \mu_A}{\partial x_A}\right)_{T,P} = \frac{kT}{x_A}, \quad (0 \le x_A \le 1). \tag{5.36}$$

Clearly, (5.35) and (5.36) are equivalent conditions for SI solutions, in the sense that each one follows from the other.

From the KB theory we have equation (4.73)

$$\left(\frac{\partial \mu_A}{\partial x_A}\right)_{T,P} = kT\left(\frac{1}{x_A} - \frac{\rho_T x_B \Delta_{AB}}{1 + \rho_T x_A x_B \Delta_{AB}}\right) \tag{5.37}$$

where $\rho_T = \rho_A + \rho_B$ is the total number density in the mixture. We now show that at any finite density†, ρ_T, a necessary and a sufficient condition for an SI solution (in a binary system at T, P constant) is

$$\Delta_{AB} \equiv G_{AA} + G_{BB} - 2G_{AB} = 0 \quad \text{for } 0 \le x_A \le 1. \tag{5.38}$$

† Note, however, that here ρ_T is not an independent variable; it is determined by T, P, x_A. Also, we shall assume throughout that all the $G_{\alpha\beta}$ are finite quantities.

The condition (5.38) is clearly a sufficient condition for SI solutions. This follows by substituting $\Delta_{AB}=0$ in (5.37). Conversely, if (5.36) and (5.37) are to be equivalent, we must have

$$x_B \rho_T \Delta_{AB} = 0 \quad \text{for } (0 \le x_A \le 1). \tag{5.39}$$

Since ρ_T is presumed to be nonzero, (5.39) implies (5.38) and hence (5.38) is also a necessary condition for SI solutions.

We note that the necessary and sufficient condition, $\Delta_{AB}=0$, for SI was derived here and is valid for mixtures at constant *P*, *T*. The condition (5.38) is very general for SI solutions. It should be recognized that this condition does not depend on any model assumption for the solution. For instance, within the lattice models of solutions we find a *sufficient* condition for SI solutions of the form (Guggenheim 1952)

$$W = W_{AA} + W_{BB} - 2W_{AB} = 0 \tag{5.40}$$

where $W_{\alpha\beta}$ are the interaction energies between the species α and β situated on adjacent lattice points.

We now define the concept of *similarity* between two components *A* and *B* whenever they fulfill condition (5.38). We shall soon see that the concept of "similarity" defined here implies a far less stringent requirement on the two components than does the concept of "very similar" as defined in the previous section.

The concept of "very similar" was defined by the requirement that all intermolecular interactions be the same. For instance, for simple particles, we require

$$U_{AA} = U_{AB} = U_{BA} = U_{BB}. \tag{5.41}$$

We have seen in section (5.17) that (5.41) is a *sufficient* condition for an SI solution.

We now show that condition (5.38) is weaker than (5.41). Let us examine the following series of conditions:

$$\text{(a)}: \quad U_{AA} = U_{AB} = U_{BA} = U_{BB} \tag{5.42a}$$

$$\text{(b)}: \quad g_{AA} = g_{AB} = g_{BA} = g_{BB} \tag{5.42b}$$

$$\text{(c)}: \quad G_{AA} = G_{AB} = G_{BB} \tag{5.42c}$$

$$\text{(d)}: \quad G_{AA} + G_{BB} - 2G_{AB} = 0 \quad (0 \le x_A \le 1). \tag{5.42d}$$

Clearly, each of the conditions in (5.42) follows from its predecessor. Symbolically, we can write

$$\text{(a)} \Rightarrow \text{(b)} \Rightarrow \text{(c)} \Rightarrow \text{(d)}. \tag{5.43}$$

The first relation, (a) ⇒ (b), follows directly from the formal definition of the pair correlation function, assuming pairwise additivity of the total potential energy. The second relation, (b) ⇒ (c), follows from the definition of $G_{\alpha\beta}$, and the third relation (c) ⇒ (d), is obvious.

Since we have shown that condition (d) is a *sufficient* condition for SI solutions, any condition that precedes (d) will also be a sufficient condition. In general, the arrows in (5.43) may not be reversed. For instance, the condition (c) implies an equality of the integrals, which is a far weaker requirement than equality of the integrands $g_{\alpha\beta}$. It is also obvious that (d) is much weaker than (c); i.e., the $G_{\alpha\beta}$ may be quite different and yet fulfill (d). It is not clear whether relation (a) follows from (b). If we require the condition (b) to hold for *all* compositions, and all P, T it is likely that (a) will follow.

We now elaborate on the meaning of condition (5.38) for "similarity" between two components on a molecular level. First, we note that the concept of "very similarity" as defined in section (5.2.1) is independent of temperature or pressure. This is not the case, however, for the concept of "similarity"; Δ_{AB} can be equal to zero at some P, T but different from zero at another P, T.

We recall the definition of $G_{\alpha\beta}$ in equation 4.1 of section 4.1. Suppose we pick up an A molecule and observe the local density in spherical shells around this molecule. The local density of, say, B molecules at a distance R is $\rho_B g_{BA}(R)$; hence, the average number of B particles in a spherical shell of width dR at distance R from an A particle is $\rho_B g_{BA}(R) 4\pi R^2 dR$. On the other hand, $\rho_B 4\pi R^2 dR$ is the average number of B particles in the same spherical shell, the origin of which has been chosen at random. Therefore, the quantity $\rho_B[g_{BA}(R) - 1]4\pi R^2 dR$ measures the excess (or deficiency) in the number of B particles in a spherical shell of volume $4\pi R^2 dR$ centered at the center of an A molecule, relative to the number that would have been measured there using the bulk density ρ_B. Hence, the quantity $\rho_B G_{BA}$ is the average excess of the number of B particles around A. Similarly, $\rho_A G_{AB}$ is the average excess of the number of A particles around B[†]. Thus, G_{AB} is the average excess of A (or B) particles around B (or A) per unit density of A (or B). It is therefore appropriate to refer to G_{AB} as a measure of the *affinity* of A toward B (and vice versa). A similar meaning is ascribed to G_{AA} and G_{BB}.

Note that the aforementioned meaning ascribed to $G_{\alpha\beta}$ is valid only when these quantities have been defined in an *open* system. In a closed system, if we place an A at the origin of our coordinate system, the total deficiency of A's in

[†] Note that $\rho_A G_{AB}$ can be positive or negative. In the latter case we can say that the "excess" is negative, i.e., there is an average deficiency of the number of A particles around B.

the *entire volume* is exactly -1. The total deficiency of B's in the entire volume is exactly zero. In both cases, the correlation due to *intermolecular interactions* extend to a distance of a few molecular diameters. Denote the correlation distance by R_c, we can write the local change in the number of A's around an A at the origin by

$$\Delta N_{AA}(R_c) = \rho_A \int_0^{R_c} [g_{AA}(R) - 1] 4\pi R^2 dR. \tag{5.44}$$

In the open system, $g_{AA}(R)$ is practically *unity* for $R > R_c$, therefore, the local change ΔN_{AA} can be equated to the global change $\rho_A G_{AA}$ in the entire volume. On the other hand, if ΔN_{AA} is defined in the *closed* system, it still has the meaning of the local change in the number of A's in the sphere of radius R_c. It is also true that this is the change due to *molecular interactions* in the system. However, this meaning cannot be retained if we extend the upper limit of the integral from R_c to infinity. The reason is the same as in the one-component system, as discussed in chapter 3. In a closed system, there is a long-range correlation of the form $1 - N^{-1}$ due to the *closure* condition with respect to the number of particles. Therefore, if we extend the limit of integration from R_c to infinity, we add to the integral a finite quantity, the result of which is that ρG_{AA} would not be a measure of the *local* change in the number of A's around an A.

Thus, in a closed system, we have the equality $\rho_A G_{AA}^C = \rho_B G_{BB}^C = -1$. If $\rho_A = \rho_B$, then it follows that $G_{AA}^C = G_{BB}^C$. This is an exact result for a closed system. From this equality, one *cannot* deduce anything regarding the relative affinities between the AA and BB pairs.

In terms of affinities, the condition (5.38) for the SI solutions, $\Delta_{AB} = 0$, is equivalent to the statement that the affinity of A toward B is the arithmetic average of the affinities of A toward A, and B toward B. This is true for all compositions $0 \le x_A \le 1$ at a given T, P. We shall see in chapter 8 another interpretation of Δ_{AB} in terms of preferential solvation.

We end this section by considering the *phenomenological* characterization of SI solutions in terms of their partial molar entropies and enthalpies. If one assumes that equation (5.35) is valid for a finite interval of temperatures and pressures, then, by differentiation, we obtain

$$\overline{S}_i = -\left(\frac{\partial \mu_i}{\partial T}\right)_P = S_i^p - k \ln x_i \tag{5.45}$$

$$\overline{H}_i = \mu_i + T\overline{S}_i = H_i^p \tag{5.46}$$

$$\overline{V}_i = \left(\frac{\partial \mu_i}{\partial P}\right)_T = V_i^p \tag{5.47}$$

where S_i^p, H_i^p and V_i^p are the molar quantities of pure *i*, and $\overline{S}_i, \overline{H}_i, \overline{V}_i$, are the corresponding partial molar quantities. An alternative, although equivalent way of describing SI solutions phenomenologically is by introducing the excess thermodynamic functions defined by

$$G^{\mathrm{EX}} = G - G^{\mathrm{ideal}} = G - \left[\sum_{i=1}^{c} N_i\left(\mu_i^p + kT\ln x_i\right)\right] \tag{5.48}$$

$$S^{\mathrm{EX}} = S - S^{\mathrm{ideal}} = S - \left[\sum_{i=1}^{c} N_i\left(S_i^p - kT\ln x_i\right)\right] \tag{5.49}$$

$$V^{\mathrm{EX}} = V - V^{\mathrm{ideal}} = V - \sum_{i=1}^{c} N_i V_i^p \tag{5.50}$$

$$H^{\mathrm{EX}} = H - H^{\mathrm{ideal}} = H - \sum_{i=1}^{c} N_i H_i^p. \tag{5.51}$$

The SI solutions are characterized by zero excess thermodynamic functions. Clearly, the phenomenological characterization of SI requires stronger assumptions than the condition (5.38).

Finally, we note that the condition (5.38) for SI solutions applies only for systems at constant *P, T*. This condition does not apply to systems at constant volume.

We have discussed in this section the condition for SI solutions for a two-component system. One can show that similar conditions for SI solutions apply to multicomponent systems. We can prove, based on the KB theory, and with a great deal of algebra, that in three- and four-component systems, a necessary and sufficient condition for SI solutions, in the sense of (5.35) for all *i*, is

$$\Delta_{\alpha\beta} = G_{\alpha\alpha} + G_{\beta\beta} - 2G_{\alpha\beta} = 0 \tag{5.52}$$

for all compositions and all pairs of different species $\alpha \neq \beta$. The proof for the general case is relatively easy if we use the expression (4.91). We defer to Appendix K the proof of this contention for the general case.

5.3 **Dilute ideal solutions**

We now turn to a different class of ideal solutions which has been of central importance in the study of solvation thermodynamics. We shall refer to a dilute ideal (DI) solution whenever one of the components is *very dilute* in the solvent. The term "very dilute" depends on the system under consideration, and we shall define it more precisely in what follows. The solvent may be a single component or a mixture of several components. Here, however, we confine ourselves to two-component systems. The solute, say A, is the component diluted in the solvent B. The generalization to multicomponent systems is quite straightforward.

The very fact that we make a distinction between the solute A and the solvent B means that the system is treated *unsymmetrically* with respect to A and B. This is in sharp contrast to the behavior of *symmetrical* ideal solutions.

The characterization of a DI solution can be carried out along different but equivalent routes. Here we have chosen the Kirkwood–Buff theory to provide the basic relations from which we derive the limiting behavior of DI solutions. The appropriate relations needed are (4.23), (4.24), and (4.62) which, when specialized to a two-component system, can be rewritten as

$$\left(\frac{\partial \mu_A}{\partial \rho_A}\right)_{T,\mu_B} = \frac{kT}{\rho_A^2 G_{AA} + \rho_A} = kT\left(\frac{1}{\rho_A} - \frac{G_{AA}}{1+\rho_A G_{AA}}\right) \tag{5.53}$$

$$\left(\frac{\partial \mu_A}{\partial \rho_A}\right)_{T,P} = kT\left(\frac{1}{\rho_A} - \frac{G_{AA}-G_{AB}}{1+\rho_A(G_{AA}-G_{AB})}\right) \tag{5.54}$$

$$\left(\frac{\partial \mu_A}{\partial \rho_A}\right)_{T,\rho_B} = kT\left(\frac{1}{\rho_A} - \frac{G_{AA}+\rho_B(G_{AA}G_{BB}-G_{AB}^2)}{1+\rho_A G_{AA}+\rho_B G_{BB}+\rho_A\rho_B(G_{AA}G_{BB}-G_{AB}^2)}\right). \tag{5.55}$$

Since we are interested in the limiting behavior $\rho_A \to 0$, we have separated the singular part ρ_A^{-1} as a first term on the rhs; this term leads to the divergence of the chemical potential as $\rho_A \to 0$.

Note that the response of the chemical potential to variations in the density ρ_A is different for each set of thermodynamic variables. The three derivatives in (5.53)–(5.55) correspond to three *different* processes. The first corresponds to a process in which the chemical potential of the solvent is kept constant (the temperature being constant in all three cases) and therefore is useful in the study of osmotic experiments. This is the simplest expression of the three and it should be noted that if we simply drop the condition of constant μ_B, we get the appropriate derivative for pure A. This is not an accidental result; in fact, this is the case where a strong resemblance exists between the behavior of the solute A in a solvent B under constant μ_B and a system A in a vacuum which replaces the solvent. We shall return to this analogy in chapter 6.

The second derivative (5.54) is the most important one from the practical point of view since it is concerned with a system under constant temperature and pressure. The third relation (5.55) is concerned with a system under constant volume which is rarely used in practice.

A common feature of all the derivatives (5.53)–(5.55) is the ρ_A^{-1} divergence as $\rho_A \to 0$ (note that we always assume that all the $G_{\alpha\beta}$ are finite quantities). For sufficiently low solute density, $\rho_A \to 0$, the first term on the rhs of each equation (5.53)–(5.55) becomes the dominant one; hence, we get the limiting form of these equations:

$$\left(\frac{\partial \mu_A}{\partial \rho_A}\right)_{T,\mu_B} = \left(\frac{\partial \mu_A}{\partial \rho_A}\right)_{T,P} = \left(\frac{\partial \mu_A}{\partial \rho_A}\right)_{T,\rho_B} = \frac{kT}{\rho_A}, \quad \rho_A \to 0 \tag{5.56}$$

which, upon integration, yields

$$\begin{aligned} \mu_A(T,\mu_B,\rho_A) &= \mu_A^0(T,\mu_B) + kT \ln \rho_A \\ \mu_A(T,P,\rho_A) &= \mu_A^0(T,P) + kT \ln \rho_A, \quad \rho_A \to 0 \\ \mu_A(T,\rho_B,\rho_A) &= \mu_A^0(T,\rho_B) + kT \ln \rho_A. \end{aligned} \tag{5.57}$$

We see that the general dependence on ρ_A for $\rho_A \to 0$ is the same for the three cases. In (5.57) we have used the notation μ_A^0 for the standard chemical potential of A. A few comments regarding equations (5.57) are now in order.

(1) The precise condition that ρ_A must satisfy to attain the limiting behavior depends on the independent variables we have chosen to describe the system. For instance, if $\rho_A G_{AA} \ll 1$, then we may assume the validity of (5.56). The corresponding requirement for (5.54) is that $\rho_A(G_{AA} - G_{AB}) \ll 1$, which is clearly different from the previous condition, if only because the latter depends on G_{AA} as well as on G_{AB}. Similarly, the precise condition under which the limiting behavior of (5.56) is obtained from (5.55) involves all three $G_{\alpha\beta}$. We can define a DI solution for each case, whenever ρ_A is sufficiently small, so that the limiting behavior of either (5.56) or (5.57) is valid.

(2) Once the limiting behavior (5.57) has been attained, we see that all three equations have the same formal form, i.e., a constant of integration, independent of ρ_A, and a term of the form $kT \ln \rho_A$. This is quite a remarkable observation, which holds only in this limiting case. This uniformity of the behavior of the chemical potential already disappears in the first-order deviation from a DI solution, a topic discussed in the next chapter.

(3) The quantities μ_A^0 which appear in (5.57) are constants of integration, and as such depend on the thermodynamic variables we use to describe the system. They are referred to as the "standard chemical potentials" of A in the corresponding set of thermodynamic variables. It is important to realize that these quantities, in contrast to μ_A^p of the previous section, are *not* the chemical potentials of A in any real system. Therefore, it is preferable to refer to μ_A^0 merely as a constant of integration.

(4) Instead of starting with relations (5.53)–(5.55), we could have started from relation (4.73) of the Kirkwood–Buff theory, namely

$$\left(\frac{\partial \mu_A}{\partial x_A}\right)_{T,P} = kT\left(\frac{1}{x_A} - \frac{\rho_B \Delta_{AB}}{1 + \rho_B x_A \Delta_{AB}}\right). \tag{5.58}$$

A limiting behavior of (5.58) is obtained for $x_A \to 0$, in which case we have

$$\left(\frac{\partial \mu_A}{\partial x_A}\right)_{T,P} = \frac{kT}{x_A}, \quad x_A \to 0 \tag{5.59}$$

which, upon integration in the region for which (5.59) is valid, yields

$$\mu_A(T, P, x_A) = \mu_A^{0x}(T, P) + kT \ln x_A, \quad x_A \to 0. \tag{5.60}$$

Again, $\mu_A^{0x}(T, P)$ is merely a constant of integration. It is different from $\mu_A^0(T, P)$ in (5.57), and therefore it is wise to use a different superscript to stress this difference. The exact relation between μ_A^0 and μ_A^{0x} can be obtained by noting that $x_A = \rho_A/\rho_T$, where ρ_T is the total density of the solution. Hence, from (5.60) we get

$$\begin{aligned}\mu_A &= \mu_A^{0x}(T, P) + kT \ln \rho_A - kT \ln \rho_T \\ &= \mu_A^0(T, P) + kT \ln \rho_A.\end{aligned} \tag{5.61}$$

Hence,

$$\mu_A^0(T, P) = \mu_A^{0x}(T, P) - kT \ln \rho_B^P, \quad (\rho_A \to 0) \tag{5.62}$$

where ρ_B^P is the density of pure B. The replacement of ρ_T by ρ_B^P is permissible since $\rho_T \to \rho_B^P$ as $\rho_A \to 0$.

In actual applications, it is sometimes convenient to use the molality scale. Instead of ρ_A or x_A as a concentration variable, one uses the molality of A, which is related to x_A (for dilute solutions) by

$$m_A = \frac{1000 x_A}{M_B} \tag{5.63}$$

M_B being the molecular weight of B. From (5.63) and (5.60), we get

$$\begin{aligned}\mu_A &= \mu_A^{0x}(T, P) + kT \ln\left(\frac{M_B m_A}{1000}\right) \\ &= \left[\mu_A^{0x}(T, P) + kT \ln\left(\frac{M_B}{1000}\right)\right] + kT \ln m_A, \quad m_A \to 0 \\ &= \mu_A^{0m}(T, P) + kT \ln m_A,\end{aligned} \tag{5.64}$$

where we have introduced a new standard chemical potential μ_A^{0m}, which is different from both μ_A^0 and μ_A^{0x}.

We now discuss briefly the behavior of the solvent in a DI solution of a two-component system. The simplest way of obtaining the chemical potential of the solvent B is to apply the Gibbs–Duhem relation, which, in combination with (5.59), yields

$$x_B\left(\frac{\partial \mu_B}{\partial x_A}\right)_{T,P} + x_A\left(\frac{\partial \mu_A}{\partial x_A}\right)_{T,P} = -x_B\left(\frac{\partial \mu_B}{\partial x_B}\right)_{T,P} + kT = 0. \tag{5.65}$$

From (5.65), upon integration in the region for which (5.65) is valid, we get

$$\mu_B(T,P,x_B) = C(T,P) + kT \ln x_B, \quad x_A \to 0. \tag{5.66}$$

Since the condition $x_A \to 0$ is equivalent to the condition $x_B \to 1$, we can substitute $x_B = 1$ in (5.66) to identify the constant of integration as the chemical potential of pure B at the same T and P, i.e.,

$$\mu_B(T,P,x_B) = \mu_B^P(T,P) + kT \ln x_B, \quad x_B \to 1. \tag{5.67}$$

Note that (5.67) has the same form as, say, (5.35), except for the restriction $x_B \to 1$ in the former.

5.4 **Summary**

In this chapter, we have discussed three types of "ideal" solutions. We stress here that the sources of ideality are different for each case. All of the three cases can be derived from the KB theory, specifically, from the relation (5.58), which we rewrite as

$$\left(\frac{\partial \mu_A}{\partial x_A}\right)_{T,P} = kT\left(\frac{1}{x_A} - \frac{\rho_T x_B \Delta_{AB}}{1 + \rho_T x_A x_B \Delta_{AB}}\right). \tag{5.68}$$

An *ideal gas* (IG) mixture is obtained from (5.68) either (theoretically) when no interactions exist, hence all $G_{ij} = 0$, hence $\Delta_{AB} = 0$, hence

$$\left(\frac{\partial \mu_A}{\partial x_A}\right)_{T,P} = \frac{kT}{x_A}; \tag{5.69}$$

or when $\rho_T \to 0$ for which we again obtain (5.69) from (5.68), but now Δ_{AB} is finite.

The *symmetric ideal* (SI) solution is obtained for *similar* components in the sense that $\Delta_{AB} = 0$ for all compositions $0 \le x_A \le 1$, which again leads to (5.69) for all compositions $0 \leq x_A \leq 1$. Clearly, the interactions could be strong and the total density could be large; hence, this case is conceptually very different from the ideal-gas case. The third case is the *diluted ideal* (DI) case obtained whenever x_A (or ρ_A) is very small, so that $(x_A)^{-1}$ becomes very large, hence the dominating term on the rhs of (5.68). Again, we note that there is no limitation on the interactions in the system, nor on the total density of the system.

Clearly, the three types of ideal behavior are quite different. As we shall see in the next chapter, deviations from each of these ideal behaviors occurs for different reasons.

Note that some authors refer to the "symmetrical convention" as the limiting behavior of $\gamma_i \rightarrow 1$ as $x_i \rightarrow 1$, see, for example, Prausnitz et al. (1986). However, this limiting behavior is manifested for *any* mixture, when one of its mole fractions approaches unity. Similarly, for any mixture, when $x_i \rightarrow 0$, we have $\gamma_i \rightarrow 1$. The concept of SI solution is very different from these limiting behaviors.

SIX

Deviations from ideal solutions

In the previous chapter, we described three types of ideal solution, and the conditions under which these behaviors are attained. In this chapter, we discuss deviations from these three types of ideality. The nature of the deviation from ideality is different for each case. The first is due to turning on the interactions in the system. The second arises when the components do not subscribe to the condition of similarity. The third is due to increasing the solute density (the solute being that component which is very diluted in the system).

The deviations from ideality can be expressed either as activity coefficients or as excess functions. Care must be exercised both in notation and in the interpretation of these activity coefficients or the excess functions. Failure to do so is a major source of confusion and erroneous statements regarding these quantities. As before, we treat only two-component systems of A's and B's. The generalization to multicomponent systems is quite straightforward.

6.1 Deviations from ideal-gas mixtures

We start with the KB result, equation (4.73) from chapter 4,

$$\left(\frac{\partial \mu_A}{\partial x_A}\right)_{T,P} = kT\left(\frac{1}{x_A} - \frac{\rho_T x_B \Delta_{AB}}{1 + \rho_T x_A x_B \Delta_{AB}}\right). \tag{6.1}$$

Integrating over x_B, and noting that $dx_A = -dx_B$, we have

$$\mu_A(T, P, x_A) = \mu_A^p(T, P) + kT \ln x_A + kT \int_0^{x_B} \frac{\rho_T x'_B \Delta_{AB}}{1 + \rho_T x'_A x'_B \Delta_{AB}}\, dx'_B \tag{6.2}$$

where $\mu_A^p(T, P)$ is the chemical potential of pure A at the same P, T. This has the general form

$$\mu_A^p(T, P) = W(A|A) + kT \ln \rho_A^p(T, P)\Lambda_A^3 q_A^{-1} \tag{6.3}$$

where $W(A \mid A)$ is the coupling work of A against a surrounding of pure A, and ρ_A^p is the density of pure A, as determined by T and P.

When there are no interactions in the system (theoretical ideal gas), we must have

$$W(A|A) = 0, \quad \Delta_{AB} = 0, \quad \rho_A^p(T, P) = \beta P. \tag{6.4}$$

Hence, the chemical potential in the ideal-gas reference state is

$$\mu_A^{\mathrm{IG}} = kT \ln(\beta P \Lambda_A^3 q_A^{-1}) + kT \ln x_A \tag{6.5}$$

which is the well-known expression for the chemical potential in an ideal-gas mixture.

The excess chemical potential and the corresponding activity coefficient are now defined by

$$\begin{aligned}\mu_A^{\mathrm{EX,IG}} &= \mu_A(T, P, x_A) - \mu_A^{\mathrm{IG}}(T, P, x_A)\\ &= W(A|A) + kT \ln(\rho_A^P(T, P)/\beta P) + kT \int_0^{x_B} \frac{\rho_T x_B' \Delta_{AB}}{1 + \rho_T x_A' x_B' \Delta_{AB}} dx_B'\end{aligned} \tag{6.6}$$

$$\mu_A^{\mathrm{EX,IG}} = kT \ln \gamma_A^{\mathrm{IG}}. \tag{6.7}$$

When we turn off all the interactions, all the three terms on the rhs of (6.6) become zero. Hence, $\mu_A^{\mathrm{EX,IG}}$ measures the deviations of the chemical potential in a real system from the corresponding chemical potential in an ideal-gas system due to turning on all the interactions in the system. The same is true when $\rho_T \to 0$.

We now evaluate the first-order deviations from ideal gas behavior

$$\rho_A^p(T, P) = \beta P - B_{AA}(\beta P)^2 + \cdots \tag{6.8}$$

and

$$kT \ln(\rho_A^P(T, P)/\beta P) = -B_{AA} P + \cdots \tag{6.9}$$

$$W(A|A) = 2kT B_{AA} \rho_A^p(T, P) = 2B_{AA} P + \cdots \tag{6.10}$$

$$\begin{aligned}kT \int_0^{x_B} \beta P x_B' \Delta_{AB}^{00}\, dx_B' &= P \frac{x_B^2}{2} \left[G_{AA}^{00} + G_{BB}^{00} - 2G_{AB}^{00}\right]\\ &= -x_B^2 P[B_{AA} + B_{BB} - 2B_{AB}]\end{aligned} \tag{6.11}$$

where we used the definition of the virial coefficients

$$B_{\alpha\beta} = -\frac{1}{2}\int_0^\infty \{\exp[-\beta U_{\alpha\beta}(R)] - 1\}4\pi R^2\, dR = -\frac{1}{2}G^{00}_{\alpha\beta}. \qquad (6.12)$$

$G^{00}_{\alpha\beta}$ are the limiting values of $G_{\alpha\beta}$ as $P \to 0$ or $\rho_T \to 0$.

The excess chemical potential, and the corresponding activity coefficient, for this case are

$$\begin{aligned}\mu_A^{\mathrm{EX,\,IG}} &= 2B_{AA}P - B_{AA}P - x_B^2 P(B_{AA} + B_{BB} - 2B_{AB})\\ &= B_{AA}P - x_B^2 P(B_{AA} + B_{BB} - 2B_{AB}) = kT\ln\gamma_A^{\mathrm{IG}}.\end{aligned} \qquad (6.13)$$

Note that the Gibbs–Duhem relation determines the behavior of μ_B once the behavior of μ_A is known. The result for B is thus

$$\mu_B^{\mathrm{EX,\,IG}} = B_{BB}P - x_A^2 P(B_{AA} + B_{BB} - 2B_{AB}) = kT\ln\gamma_B^{\mathrm{IG}}. \qquad (6.14)$$

6.2 **Deviations from SI behavior**

Again, we use the *KB* result (6.1) and integrate to obtain (6.2). In the SI solution, by definition $\Delta_{AB}=0$, hence

$$\mu_A^{\mathrm{SI}} = \mu_A^P + kT\ln x_A, \qquad 0 \le x_A \le 1. \qquad (6.15)$$

The excess chemical potential, due to deviations from symmetrical ideal behavior, is thus

$$\begin{aligned}\mu_A^{\mathrm{EX,\,SI}} &= \mu_A(T,P,x_A) - \mu_A^{\mathrm{SI}}(T,P,x_A)\\ &= kT\int_0^{x_B}\frac{\rho_T x_B' \Delta_{AB}}{1+\rho_T x_A' x_B' \Delta_{AB}}\,dx_B'\\ &= kT\ln\gamma_A^{\mathrm{SI}}\end{aligned} \qquad (6.16)$$

Here, both the excess chemical potential, and the activity coefficient measure the deviations from *similarity* of the two quantities. This is fundamentally different from the deviations from the ideal-gas behavior (i.e., total lack of interactions), discussed in section 6.1. Here the limiting behavior of the activity coefficient is

$$\lim_{\Delta_{AB}\to 0}\gamma_A^{\mathrm{SI}} = 1 \qquad (\text{at } T, P, x_A \text{ constants}). \qquad (6.17)$$

The general case (6.16) is not very useful since in general, we do not know the dependence of Δ_{AB} on the composition. However, if the two components deviate only slightly from similarity (i.e., from $\Delta_{AB}=0$), we can assume that

$$\rho_T x_A x_B \Delta_{AB} \ll 1 \qquad (0 \leq x_A \leq 1). \tag{6.18}$$

Note that since here we exclude the case of an ideal-gas mixture, ρ_T is finite, and the condition (6.18) is essentially a condition on Δ_{AB}. If we further assume that $\rho_T \Delta_{AB}$ is independent of the composition, then we can integrate (6.16) to obtain the first-order deviations from SI behavior, namely

$$\mu_A^{\text{EX, SI}} \approx kT\rho_T \Delta_{AB} x_B^2/2. \tag{6.19}$$

In the phenomenological characterization of small deviations from SI solutions, the concepts of regular and athermal solutions were introduced. Normally, the theoretical treatment of these two cases was discussed within the lattice theories of solutions[†]. Here, we discuss only the very general conditions for these two deviations to occur. First, when $\rho_{\text{T}}\Delta_{\text{AB}}$ does not depend on temperature, we can differentiate (6.19) with respect to T to obtain

$$S_A^{\text{EX,SI}} = -\left(\frac{\partial \mu^{\text{EX,SI}}}{\partial T}\right)_{P,x_A} = -k\rho_T \Delta_{AB} x_B^2/2 \tag{6.20}$$

and

$$H_A^{\text{EX,SI}} = \mu_A^{\text{EX,SI}} + TS_A^{\text{EX,SI}} = 0. \tag{6.21}$$

This is the case of *athermal* solution, i.e., no excess enthalpy but finite excess entropy, both with respect to SI solutions.

The second case is when $T\rho_T \Delta_{AB}$ is independent of temperature, in which case

$$S_A^{\text{EX,SI}} = -\left(\frac{\partial \mu^{\text{EX,SI}}}{\partial T}\right)_{P,x_A} = 0 \tag{6.22}$$

$$H_A^{\text{EX,SI}} = \mu_A^{\text{EX,SI}} + TS_A^{\text{EX,SI}} = kT\rho_T \Delta_{AB} x_B^2/2; \tag{6.23}$$

no excess entropy, but finite excess enthalpy. Again, the excesses are with respect to SI solutions-This is the case of regular solutions. We have shown here the theoretical requirements for obtaining the *athermal* and regular solution behavior. The phenomenological characterization is made through the experimental excess entropies and enthalpies of the various components as in equations (6.20)–(6.23). We shall show a possible molecular reason for these two cases in section 6.4.

[†] The term "regular solutions" was first coined by Hildebrand (1929). It was characterized phenomenoligically in terms of the excess entropy of mixing. It was later used in the context of lattice theory of mixtures mainly by Guggenheim (1952). It should be stressed that in both the phenomenological and the lattice theory approaches, the "regular solution" concept applies to deviations from SI solutions. (see also Appendix M).

6.3 **Deviations from dilute ideal solutions**

In the previous two sections we have discussed deviations from ideal-gas and symmetrical ideal solutions. We have discussed deviations occurring at fixed temperature and pressure. There has not been much discussion of these ideal cases in systems at constant volume or of constant chemical potential. The case of dilute solutions is different. Both constant, T, P and constant T, μ_B (osmotic system), and somewhat less constant, T, V have been used. It is also of theoretical interest to see how deviations from dilute ideal (DI) behavior depends on the thermodynamic variable we hold fixed. Therefore in this section, we shall discuss all of these three cases.

We have already seen that the form of the expression for the chemical potential in the limit of dilute ideal solutions is the *same* in the three cases; equations (5.56) and (5.57).

In all cases the limiting behavior has the form

$$\mu_A = \mu_A^0 + kT \ln \rho_A, \qquad (\rho_A \to 0) \tag{6.24}$$

where μ_A^0 is a constant, independent of ρ_A.

Deviations from DI are observed whenever we increase the density of the *solute* ρ_A, beyond the range for which (6.24) is valid. The *extent* of the deviations will, of course, depend on *how*, i.e., under which conditions, we add the solute. Keeping T, P or T, μ_B or T, V constant will result in different deviations. The general cases may be obtained by integrating equations (5.53)–(5.55) under the different conditions T, μ_B, or T, P, or T, V constant. For instance, in the open system (with respect to B) we have

$$\mu_A(T, \mu_B, \rho_A) = \mu_A^{0\rho}(T, \mu_B) + kT \ln \rho_A \gamma_A^{\mathrm{DI}}(T, \mu_B). \tag{6.25}$$

Hence, the excess chemical potential for this case is

$$\begin{aligned} \mu_A^{\mathrm{EX,DI}} &= \mu_A(T, \mu_B, \rho_A) - \mu_A^{\mathrm{DI}}(T, \mu_B, \rho_A) \\ &= kT \ln \gamma_A^{\mathrm{DI}}(T, \mu_B, \rho_A) \\ &= kT \int_0^{\rho_A} \frac{-G_{AA}}{1 + \rho'_A G_{AA}} d\rho'_A. \end{aligned} \tag{6.26}$$

Similar but more complicated expressions may be obtained from (5.54) and (5.55). Clearly, since we do not know the dependence of $G_{\alpha\beta}$ on ρ_A we cannot perform these integrations.

From hereon, we shall discuss only *small* deviations from DI solutions. By expanding the nondivergent parts in equations (5.53)–(5.55), and retaining the first order in ρ_A, we obtain

$$\left(\frac{\partial \mu_A}{\partial \rho_A}\right)_{T,\mu_B} = kT\left(\frac{1}{\rho_A} - G_{AA}^0 + \cdots\right) \tag{6.27}$$

$$\left(\frac{\partial \mu_A}{\partial \rho_A}\right)_{T,P} = kT\left[\frac{1}{\rho_A} - (G_{AA}^0 - G_{AB}^0) + \cdots\right] \tag{6.28}$$

$$\left(\frac{\partial \mu_A}{\partial \rho_A}\right)_{T,\rho_B} = kT\left\{\frac{1}{\rho_A} - \frac{G_{AA}^0 + \rho_B^0[G_{AA}^0 G_{BB}^0 - (G_{AB}^0)^2]}{1 + \rho_B^0 G_{BB}^0} + \cdots\right\}. \tag{6.29}$$

The superscript zero in (6.27)–(6.29) stands for the limiting value of the corresponding quantity as $\rho_A \to 0$. Note that the limit $\rho_A \to 0$ is taken under different conditions in each case, i.e., T and μ_B are constants in the first, T and P are constants in the second, and T and ρ_B in the third.

Since all the $G_{\alpha\beta}^0$ in (6.27)–(6.29) are independent of ρ_A, we can integrate equations (6.27)–(6.29) in the region of ρ_A, for which the first-order expansion is valid, to obtain the first-order correction to DI solutions. These are:

$$\mu_A(T, \mu_B, \rho_A) = \mu_A^0(T, \mu_B) + kT \ln \rho_A - kTG_{AA}^0 \rho_A + \cdots \tag{6.30}$$

$$\mu_A(T, P, \rho_A) = \mu_A^0(T, P) + kT \ln \rho_A - kT(G_{AA}^0 - G_{AB}^0)\rho_A + \cdots \tag{6.31}$$

$$\mu_A(T,\rho_B,\rho_A) = \mu_A^0(T,\rho_B) + kT\ln\rho_A - kT\left[G_{AA}^0 - \frac{\rho_B^0 (G_{AB}^0)^2}{1+\rho_B^0 G_{BB}^0}\right]\rho_A + \cdots \tag{6.32}$$

It is instructive to compare these relations with (5.57). The most important difference between equations (6.30)–(6.32) and (5.57) is that the uniformity shown in the limit of $\rho_A \to 0$ breaks down once we consider deviations from DI behavior. The first-order deviations from DI solutions depend on the thermodynamic variables we choose to describe our system. We can now introduce the excess functions and the activity coefficients corresponding to the first-order deviations from the DI behavior. These are defined by

$$\mu_A^{\text{EX, DI}}(T, \mu_B, \rho_A) = kT \ln \gamma_A^{\text{DI}}(T, \mu_B, \rho_A) = -kTG_{AA}^0 \rho_A \tag{6.33}$$

$$\mu_A^{\text{EX, DI}}(T, P, \rho_A) = kT \ln \gamma_A^{\text{DI}}(T, P, \rho_A) = -kT(G_{AA}^0 - G_{AB}^0)\rho_A \tag{6.34}$$

$$\mu_A^{\text{EX,DI}}(T,\rho_B,\rho_A)=kT\ln\gamma_A^{\text{DI}}(T,\rho_B,\rho_A)=-kT\left(G_{AA}^0-\frac{\rho_B^0(G_{AB}^0)^2}{1+\rho_B^0G_{BB}^0}\right)\rho_A. \quad (6.35)$$

Hence, equations (6.30)–(6.32) can now be written as

$$\mu_A(T,\mu_B,\rho_A) = \mu_A^0(T,\mu_B) + kT\ln[\rho_A\gamma_A^{\text{DI}}(T,\mu_B,\rho_A)] \quad (6.36)$$

$$\mu_A(T,P,\rho_A) = \mu_A^0(T,P) + kT\ln[\rho_A\gamma_A^{\text{DI}}(T,P,\rho_A)] \quad (6.37)$$

$$\mu_A(T,\rho_B,\rho_A) = \mu_A^0(T,\rho_B) + kT\ln[\rho_A\gamma_A^{\text{DI}}(T,\rho_B,\rho_A)]. \quad (6.38)$$

It is clearly observed that the activity coefficients in (6.33)–(6.35) *differ fundamentally* from the activity coefficient introduced in sections 6.1 and 6.2. To stress this difference, we have used the superscript DI to denote deviations from DI behavior. Furthermore, each of the activity coefficients defined in (6.33)–(6.35) depends on the thermodynamic variables, say T and μ_B, or T and P, or T and ρ_B. This has also been indicated in the notation. In practical applications, however, one usually knows which variables have been chosen, in which case one can drop the arguments in the notation for γ_A^{DI}.

The limiting behavior of the activity coefficients defined in (6.33)–(6.35) is, for example,

$$\lim_{\rho_A\to 0}\gamma_A^{\text{DI}}(T,P,\rho_A) = 1, \qquad T,P \text{ constant}. \quad (6.39)$$

Consider next the content of the first-order contribution to the activity coefficients in (6.33)–(6.35). Note that all of these contain the quantity G_{AA}^0. Recall that G_{AA}^0 is a measure of the solute–solute affinity. In the limit of DI, the quantity G_{AA}^0 is still finite, but its effect on the activity coefficient vanishes in the limit $\rho_A \to 0$. It is quite clear on qualitative grounds that the standard chemical potential is determined by the solvent–solvent and solvent–solute affinities (this will be shown more explicitly in the next section). Thus, the effect of solute–solute affinity becomes operative only when we increase the *solute concentration* so that the solute molecules "see" each other, which is the reason for the appearance of G_{AA}^0 in (6.33)–(6.35). In addition to G_{AA}^0, relation (6.34) also includes G_{AB}^0 and relation (6.35) also includes G_{BB}^0.

The quantity G_{AA} (or G_{AA}^0) is often referred to as representing the solute–solute *interaction*. In this book, we reserve the term "interaction" for the *direct intermolecular interaction* operating between two particles. For instance, two hard-sphere solutes of diameter σ do not *interact* with each other at a distance $R>\sigma$, yet the solute–solute *affinity* conveyed by G_{AA} may be different from zero. Therefore, care must be exercised in identifying DI solutions as arising from the absence of solute–solute *interactions*.

Another very common misinterpretation of experimental results is the following. Suppose we measure deviations from a DI solution in a T, P, N_A, N_B system. The corresponding activity coefficient is given by (6.34); the same quantity is often referred to as the excess chemical potential of the solute. One then expands the activity coefficient (or the excess chemical potential) to first order in ρ_A and interprets the first coefficient as a measure of the extent of "solute–solute interaction." Clearly, such an interpretation is valid for an osmotic system provided we understand "interaction" in the sense of affinity, as pointed out above. However, in the T, P, N_A, N_B system, the first-order coefficient depends on the difference $G^0_{AA} - G^0_{AB}$. It is in principle possible that G^0_{AA} be, say, positive, whereas the first-order coefficient in (6.34) can be positive, negative, or zero. This clearly invalidates the interpretation of the first-order coefficient in (6.34) in terms of solute–solute correlation. Similar expansions are common for the excess enthalpies and entropies where the first-order coefficient in the density expansion is not known explicitly.

In practice, the most important set of thermodynamic variables is of course T, P, ρ_A, employed in (6.34). However, relation (6.33) is also useful and has enjoyed considerable attention in osmotic experiments where μ_B is kept constant. This set of variables provides relations which bear a remarkable analogy to the virial expansion of various quantities of real gases. We demonstrate this point by extracting the first-order expansion of the osmotic pressure π in the solute density ρ_A. This can be obtained by the use of the thermodynamic relation

$$\left(\frac{\partial \pi}{\partial \rho_A}\right)_{T,\mu_B} = \rho_A \left(\frac{\partial \mu_A}{\partial \rho_A}\right)_{T,\mu_B}. \tag{6.40}$$

Using (6.27) in (6.40), we get

$$\left(\frac{\partial \pi}{\partial \rho_A}\right)_{T,\mu_B} = \frac{kT}{\rho_A G_{AA} + 1} \xrightarrow{\rho_A \to 0} kT(1 - G^0_{AA}\rho_A + \cdots). \tag{6.41}$$

This may be integrated to obtain

$$\frac{\pi}{kT} = \rho_A - \tfrac{1}{2} G^0_{AA}\rho_A^2 + \cdots \tag{6.42}$$

This expansion is known in the more familiar form

$$\frac{\pi}{kT} = \rho_A + B_2^* \rho_A^2 + \cdots \tag{6.43}$$

where B_2^* is the analog of the second virial coefficient in the density expansion of the pressure (see section 6.5)

$$\frac{P}{kT} = \rho + B_2 \rho^2 + \cdots \tag{6.44}$$

Thus, the virial coefficient B_2 in (6.44) depends on the pair potential. The virial coefficient B_2^* depends on the pair correlation function (or equivalently on the potential of the mean force).

Next, we turn to the chemical potential of the solvent B for a system deviating slightly from DI behavior. The simplest way of doing this is to use relation (4.73) from the Kirkwood–Buff theory, which when written for the B component and expanded to first order in x_A, yields

$$\left(\frac{\partial \mu_B}{\partial x_B}\right)_{T,P} = kT\left(\frac{1}{x_B} - \rho_T^0 \Delta_{AB}^0 x_A + \cdots\right), \qquad x_A \to 0, \tag{6.45}$$

where ρ_T^0 and Δ_{AB}^0 are the limiting values of ρ_T and Δ_{AB} as $x_A \to 0$. Integrating (6.45) yields

$$\mu_B(T,P,x_B) = \mu_B^P(T,P) + kT\ln x_B + \int_0^{x_A} kT\rho_T^0 \Delta_{AB}^0 x'_A \, dx'_A, \quad x_A \to 0 \tag{6.46}$$

Here, $\rho_T^0 \Delta_{AB}^0$ is independent of composition; hence, we can integrate (6.46) to obtain

$$\mu_B(T,P,x_A) = \mu_B^P(T,P) + kT\ln x_B + \tfrac{1}{2} kT\rho_B^0 \Delta_{AB}^0 x_A^2, \quad x_A \to 0. \tag{6.47}$$

Clearly, in the limit $\rho_A \to 0$, we can replace ρ_T^0 by ρ_B^0, the density of the pure solvent B at the given T, P. Note, however, the difference between the excess chemical potential in (6.47), i.e., the last term on the rhs of (6.47), and relation (6.19). These look very similar. Here the expansion is valid for *small* x_A but otherwise the value of $\rho_B^0 \Delta_{AB}^0$ is unrestricted. In (6.19), on the other hand, we have a first-order expansion in Δ_{AB}, which is required to hold for *all compositions* $0 \leq x_A \leq 1$.

6.4 **Explicit expressions for the deviations from IG, SI, and DI behavior**

This section is devoted to illustrating explicitly the three fundamentally different types of ideal mixtures. The first and simplest case is that of the ideal-gas (IG) mixtures, which, as in the case of an ideal gas, are characterized by the complete absence (or neglect) of all intermolecular forces. This case is of least importance in the study of solution chemistry.

The second case, referred to as symmetric ideal (SI) solutions, occurs whenever the various components are "similar" to each other. There are no restrictions on the magnitude of the intermolecular forces or on the densities. The third case, dilute ideal (DI) solutions, consist of those solutions for which at least one component is very diluted in the remaining solvent, which may be a one-component or a multicomponent system. Again, there are no restrictions on the strength of the intermolecular forces, the total density, or the degree of similarity between the various components.

Any mixture of two components can be viewed as deviating from one of the ideal reference cases. This can be written symbolically as

$$\begin{aligned}\mu_A &= c_1 + kT\ln(x_A\gamma_A^{\mathrm{IG}})\\ &= c_2 + kT\ln(x_A\gamma_A^{\mathrm{SI}})\\ &= c_3 + kT\ln(x_A\gamma_A^{\mathrm{DI}})\end{aligned} \tag{6.48}$$

where the constants c_i are independent of x_A. Here, $\gamma_A^{\mathrm{IG}}, \gamma_A^{\mathrm{SI}}, \gamma_A^{\mathrm{DI}}$ are the activity coefficients that incorporate the correction due to non-ideality[†].

There are several ways of reporting experimental data on the deviations from ideal behavior. The most common ones are either the activity coefficients of each component, or the total excess Gibbs energy of the system. If the vapor above the liquid -mixture can be assumed to be an ideal gas, then it is also convenient to plot P_A/P_A^p as a function of x_A where P_A and P_A^p are the partial pressure and the vapor pressure of A, respectively.

Figure 6.1 shows such curves for mixtures of carbon disulphide and acetone, and the second for mixtures of chloroform and acetone. The first shows positive deviations from SI behavior in the entire range of compositions; the second shows negative deviations from SI solution.

We now consider two particular examples of a system which, on the one hand, are not trivial, since interactions between particles are taken into account, yet are sufficiently simple that all three activity coefficients can be written in an explicit form.

6.4.1 **First-order deviations from ideal-gas mixtures**

We choose a two-component system for which the pressure (or the total density) is sufficiently low such that the pair correlation function for each pair of species has the form (see section 2.5)

$$g_{\alpha\beta}(R) = \exp[-\beta U_{\alpha\beta}(R)]. \tag{6.49}$$

[†] For the purpose of demonstration, we have chosen T, P, x_A as the thermodynamic variables. A parallel treatment can be carried out for any other set of thermodynamic variables.

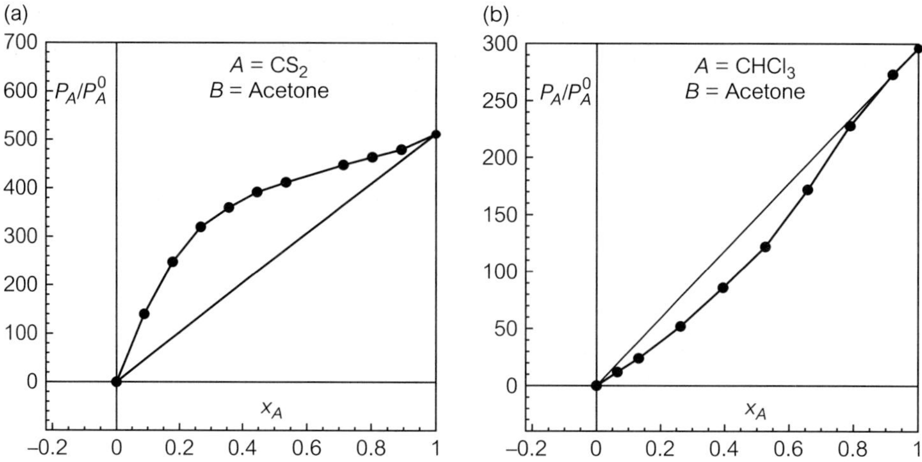

Figure 6.1. (a) Reduced partial pressure (P_A/P_A^0) of carbon disulfide $A = (CS_2)$ as a function of the mole fraction x_A in mixtures of CS_2 and B = acetone, at 35.2 °C. (b) Reduced partial pressure (P_A/P_A^0) of chloroform $A = (CHCl_3)$ as a function of the mole fraction x_A in mixtures of $CHCl_3$ and B = acetone at 35.2 °C.

For simplicity, we have assumed that all the pair potentials are spherically symmetrical, and that all the internal partition functions are unity. The general expression for the chemical potential of, say, A in this system is obtained by a simple extension of the one-component expression given in chapter 3.

$$\begin{aligned}\mu_A &= kT \ln(\rho_A \Lambda_A^3) + \rho_A \int_0^1 d\xi \int_0^\infty U_{AA}(R) g_{AA}(R, \xi) 4\pi R^2\, dR \\ &\quad + \rho_B \int_0^1 d\xi \int_0^\infty U_{AB}(R) g_{AB}(R, \xi) 4\pi R^2\, dR \\ &= kT \ln(\rho_A \Lambda_A^3) + 2kTB_{AA}\rho_A + 2kTB_{AB}\rho_B,\end{aligned} \tag{6.50}$$

where we have used expression (6.49) with $\xi U_{\alpha\beta}(R)$ replacing $U_{\alpha\beta}(R)$ so that integrating over ξ becomes immediate. Also, we have used the more familiar notation

$$B_{a\beta} = -\frac{1}{2}\int_0^\infty \{\exp[-\beta U_{\alpha\beta}(R)] - 1\} 4\pi R^2\, dR. \tag{6.51}$$

We now analyze equation (6.50) with respect to the various kinds of ideality. For the purpose of this section, it is preferable to transform (6.50) so that μ_A is expressed as a function of T, P, and x_A. To do this, we use the analog of the virial expansion for mixtures, which reads

$$\beta P = (\rho_A + \rho_B) + [x_A^2 B_{AA} + 2x_A x_B B_{AB} + x_B^2 B_{BB}](\rho_A + \rho_B)^2 + \cdots \tag{6.52}$$

The factor in the square brackets can be viewed as an "average" virial coefficient for the mixture of two components. We now invert this relation by assuming an expansion of the total density $\rho = \rho_T$ in the form

$$\rho = \rho_A + \rho_B = \beta P + CP^2 + \cdots \tag{6.53}$$

This is substituted into the rhs of (6.52). On equating coefficients of equal powers of P, we get

$$\rho = \rho_A + \rho_B = \beta P - (x_A^2 B_{AA} + 2x_A x_B B_{AB} + x_B^2 B_{BB})(\beta P)^2 + \cdots \tag{6.54}$$

Transforming $\rho_A = x_A \rho$ and $\rho_B = x_B \rho$ in (6.50), and using the expansion (6.54) for ρ, we get the final form of the chemical potential:

$$\begin{aligned}
\mu_A(T,P,x_A) &= kT\ln(x_A\Lambda_A^3) + kT\ln\beta P \\
&\quad + kT\ln[1-(x_A^2 B_{AA}+2x_A x_B B_{AB}+x_B^2 B_{BB})\beta P] \\
&\quad + (2kTB_{AA}x_A + 2kTB_{AB}x_B)\times[\beta P-(x_A^2 B_{AA}+2x_A x_B B_{AB}+x_B^2 B_{BB})(\beta P)^2] \\
&= kT\ln(x_A\Lambda_A^3) + kT\ln\beta P + PB_{AA} - Px_B^2(B_{AA}+B_{BB}-2B_{AB}).
\end{aligned} \tag{6.55}$$

In the last form of (6.55), we have retained only first-order terms in the pressure (except for the logarithmic term). We now view expression (6.55) in various ways, according to the choice of the reference ideal state. Essentially, we shall rewrite the same equation in three different ways, each viewed as deviating from a different reference ideal state.

(1) *Ideal-gas mixture as a reference system.* For $P \to 0$ (or if no interactions exist, so that $B_{\alpha\beta} = 0$), (6.55) reduces to

$$\begin{aligned}
\mu_A^{IG} &= kT\ln(x_A\Lambda_A^3) + kT\ln\beta P \\
&= \mu_A^{0g}(T,P) + kT\ln x_A
\end{aligned} \tag{6.56}$$

where μ_A^{0g} (T, P) is defined in (6.4.9) as the IG standard chemical potential of A (note its dependence on both T and P but not on x_A).

Comparing (6.55) with (6.56), we find the correction due to deviations from the IG mixture. The corresponding activity coefficient is

$$kT\ln\gamma_A^{IG} = PB_{AA} - Px_B^2(B_{AA}+B_{BB}-2B_{AB}), \tag{6.57}$$

and hence

$$\mu_A(T,P,x_A) = \mu_A^{0g}(T,P) + kT\ln(x_A\gamma_A^{IG}). \tag{6.58}$$

It is clear that γ_A^{IG} measures deviations from ideal-gas behavior due to all interactions between the two species. In equation (6.58), we have rewritten the chemical potential by grouping terms which are included in μ_A^{0g} (T, P) and in γ_A^{IG}.

(2) *Symmetric ideal solution as a reference system.* In the next case we assume that the two components A and B are "similar" in the sense of section 5.2, which means that

$$B_{AA} + B_{BB} - 2B_{AB} = 0. \tag{6.59}$$

This is the condition of an SI solution. Substituting in (6.55), we obtain

$$\begin{aligned}\mu_A^{\mathrm{SI}} &= kT\ln(x_A\Lambda_A^3) + kT\ln(\beta P) + PB_{AA}\\ &= \mu_A^P(T, P) + kT\ln x_A.\end{aligned} \tag{6.60}$$

Clearly, μ_A^P is the chemical potential of pure A at this particular T and P, and therefore it depends only on the $A-A$ interactions through B_{AA}. If, on the other hand, the system is not SI, then we define the activity coefficient, for this case, as

$$kT\ln\gamma_A^{\mathrm{SI}} = -Px_B^2(B_{AA} + B_{BB} - 2B_{AB}), \tag{6.61}$$

and (6.55) can be rewritten as

$$\mu_A(T, P, x_A) = \mu_A^P + kT\ln(x_A\gamma_A^{\mathrm{SI}}), \tag{6.62}$$

where γ_A^{SI} as defined in (6.62) is a measure of the *deviations* due to the *dissimilarity* between the two components. In (6.62), we rewrote (6.55) again by regrouping the various terms in (6.55).

(3) *Dilute ideal solution as a reference system.* In this case, we assume that A is very diluted in B, i.e., $x_A \to 0$, or $x_B \to 1$. This is the case of a DI solution. Equation (6.55) reduces to

$$\begin{aligned}\mu_A^{\mathrm{DI}} &= kT\ln(x_A\Lambda_A^3) + kT\ln(\beta P) + P(2B_{AB} - B_{BB})\\ &= \mu_A^{0x} + kT\ln x_A,\end{aligned} \tag{6.63}$$

where μ_A^{0x} is *defined* in (6.63). We can easily transform (6.63) by substituting $x_A = \rho_A/\rho$ in (6.63) and using (6.54) to obtain

$$\mu_A^{\mathrm{DI}} = kT\ln(\rho_A\Lambda_A^3) + 2PB_{AB} = \mu_A^{0\rho} + kT\ln\rho_A \tag{6.64}$$

where $\mu_A^{0\rho}$ is *defined* in (6.64). Note that μ_A^{0x} and $\mu_A^{0\rho}$ do not include the term B_{AA}, which is a measure of the solute–solute interactions. The word "interaction" is appropriate in the present context since in the present limiting case, we know that $g_{\alpha\beta}(R)$ depends only on the *direct* interaction between the pair of species α and β, as we have assumed in (6.49).

The activity coefficients corresponding to the two representations (6.63) and (6.64) are obtained by comparison with (6.55), i.e.,

$$kT \ln \gamma_A^{\mathrm{DI},x} = 2x_A P(B_{AA} + B_{BB} - 2B_{AB}) \tag{6.65}$$

$$kT \ln \gamma_A^{\mathrm{DI},\rho} = 2kT\rho_A(B_{AA} - B_{AB}). \tag{6.66}$$

In (6.65) and (6.66), we have retained only the first-order terms in x_A and in ρ_A. Using these activity coefficients, the chemical potential in (6.55) can be rewritten in two alternative forms:

$$\mu_A(T, P, x_A) = \mu_A^{0x} + kT \ln(x_A \gamma_A^{\mathrm{DI},x}) \tag{6.67}$$

$$\mu_A(T, P, \rho_A) = \mu_A^{0\rho} + kT \ln(\rho_A \gamma_A^{\mathrm{DI},\rho}). \tag{6.68}$$

The different notations $\gamma_A^{\mathrm{DI},x}$ and $\gamma_A^{\mathrm{DI},\rho}$ have been introduced to distinguish between the two cases.

6.4.2 One-dimensional model for mixtures of hard "spheres"

The second example where we can write the exact expression for the chemical potential is a one-dimensional mixture of hard rods (i.e., hard spheres in one-dimensional system). For a system of N_A rods of length (diameter) σ_A and N_B rods of length σ_B, in a "volume" L at temperature T, the canonical partition function is well known† (Ben-Naim 1992).

$$Q(T, L, N_A, N_B) = \frac{(L - N_A\sigma_A - N_B\sigma_B)^{N_A+N_B}}{\Lambda_A^{N_A}\Lambda_B^{N_B} N_A! N_B!} \tag{6.69}$$

from which one can derive the exact expression for the chemical potential

$$\begin{aligned}\mu_A(T, \rho_A, \rho_B) &= kT \ln \rho_A\Lambda_A - kT \ln(1 - \rho_A\sigma_A - \rho_B\sigma_B) \\ &\quad + \frac{kT(\rho_A + \rho_B)\sigma_A}{1 - \rho_A\sigma_A - \rho_B\sigma_B}\end{aligned} \tag{6.70}$$

and in terms of T, P, ρ_A we have‡

$$\mu_A(T, P, \rho_A) = kT \ln \rho_A\Lambda_A + \sigma_A P - kT \ln\left[\frac{1 + \rho_A(\sigma_B - \sigma_A)}{1 + \beta P\sigma_B}\right]. \tag{6.71}$$

† Note that in the one-dimensional case, the canonical partition function has the form of $Q = V_f^N/N!\Lambda$ where V_f^N is the *free volume*. In this case, the quantity V_f is indeed the volume unoccupied by particles. In the "free volume" theories of liquids, this form of the partition function was *assumed* to hold for a three-dimensional liquid.

‡ This is obtained from eliminating ρ_B from the equation of state (6.72) and substituting in (6.70).

Also, the equation of state, and the virial expansion can easily be written for this model:

$$\begin{aligned}\beta P &= \frac{\rho_A + \rho_B}{1 - \rho_A\sigma_A - \rho_B\sigma_B} \\ &= \frac{\rho_T}{1 - \rho_T(x_A\sigma_A + x_B\sigma_B)} \\ &= \rho_T + B_2\rho_T^2 + B_3\rho_T^3 + \cdots \end{aligned} \tag{6.72}$$

where $x_A = \rho_A/\rho_T$ and $\rho_T = \rho_A + \rho_B$ is the total density. The virial coefficients are given by

$$B_K = (x_A\sigma_A + x_B\sigma_B)^{K-1}. \tag{6.73}$$

Note that in this model we always have $\sigma_{AB} = (\sigma_A + \sigma_B)/2$. Transforming into variables T, P, x_A, we obtain the expression for the chemical potential

$$\mu_A(T, P, x_A) = kT \ln x_A\Lambda_A + kT \ln \beta P + \sigma_A P. \tag{6.74}$$

We now use the last expression to derive the three deviations, or the excess chemical potential, with respect to the three ideal behaviors. As in the previous case, we shall rewrite expression (6.74) in three different forms, as follows:

(1) *Deviation from* SI *behavior.* For $x_A = 1$, we obtain the chemical potential of pure A at the same T, P, i.e.,

$$\mu_A^P(T, P) = kT \ln \beta P\Lambda_A + \sigma_A P. \tag{6.75}$$

Substituting in (6.49), we obtain

$$\mu_A(T, P, x_A) = \mu_A^P(T, P) + kT \ln x_A. \tag{6.76}$$

This means that a mixture of A and B differing in size will always behave as an SI solution. Hence, the corresponding excess function is zero:

$$\mu_A^{\text{EX,SI}} = 0 \tag{6.77}$$

This result holds for any multicomponent one-dimensional system. The reason that we observe SI behavior in the one-dimensional system of hard rods, but not for systems of hard disks or hard spheres, is as follows.

The condition for SI behavior is that the coupling work of each molecule is independent of the composition (in the P, T, N system). In the one-dimensional system, each particle "sees" only two "hard points" (the surfaces) in its neighborhood, one in front and one in its back. Hence, the average interaction free energy is *independent* of the *sizes* of its neighbors. This property is particular to the one-dimensional system.

The SI behavior is also consistent with the condition $\Delta_{AB} = G_{AA} + G_{BB} - 2G_{AB} = 0$. In this model, it is easy to compute each of the KB integrals.† The results are

$$G_{ij} = -\sigma_i - \sigma_j + \rho_i \sigma_i^2 + \rho_j \sigma_j^2 \tag{6.78}$$

where σ_i is the length (diameter) of the rod of species i.

Hence, the condition $\Delta_{AB} = 0$ is fulfilled for all compositions.

(2) *Deviations from* DI *behavior.* Taking the limiting behavior, $\rho_A \to 0$, in equations (6.70) and (6.71), we obtain

$$\begin{aligned}\mu_A^{\text{DI}} &= kT \ln \rho_A \Lambda_A - kT \ln(1 - \rho_B \sigma_B) + \frac{kT \rho_B \sigma_A}{1 - \rho_B \sigma_B} \\ &= kT \ln \rho_A \Lambda_A + kT \ln(1 + \beta P \sigma_B) + \sigma_A P.\end{aligned} \tag{6.79}$$

The excess chemical potential with respect to the DI solution is obtained from (6.71) and (6.79).

$$\begin{aligned}\mu_A^{\text{EX,DI}} &= \mu_A - \mu_A^{DI} \\ &= -kT \ln[1 + \rho_A(\sigma_B - \sigma_A)] \\ &= kT \rho_A(\sigma_A - \sigma_B) + \tfrac{1}{2} kT (\sigma_A - \sigma_B)^2 \rho_A^2 + \cdots\end{aligned} \tag{6.80}$$

(3) *Deviations from ideal gas* (IG) *behavior.* Taking the limit $P \to 0$, or $\rho_T \to 0$, we have the ideal-gas chemical potential

$$\mu^{\text{IG}} = kT \ln \rho_A \Lambda_A. \tag{6.81}$$

Hence, from (6.71) and (6.81) we obtain

$$\begin{aligned}\mu_A^{\text{EX,IG}} &= \mu_A - \mu_A^{\text{IG}} \\ &= kT \ln\left[\frac{1 + \beta P \sigma_B}{1 + \rho_A(\sigma_B - \sigma_A)}\right] + \sigma_A P \\ &= -kT \ln(1 - \rho_A \sigma_A - \sigma_B \sigma_B) + \frac{kT(\rho_A + \rho_B)\sigma_A}{1 - \rho_A \sigma_A - \rho_B \sigma_B}.\end{aligned} \tag{6.82}$$

We have thus expressed the three forms of the excess chemical potentials corresponding to the three cases of ideal behaviors.

6.5 The McMillan–Mayer theory of solutions

The McMillan–Mayer (MM) theory is essentially a formal generalization of the theory of real gases. In the theory of real gases we have an expansion of the

† Since we have an explicit expression for the chemical potential, we can also write explicit expressions for the derivatives with respect to the various densities. Using the KB theory one can solve for all G_{ij}. The procedure is lengthy but quite straightforward. Note also that for a one-component system, one can get G directly from the compressibility expression, i.e., $G = -2\sigma + \rho\sigma^2$.

pressure in power series in the density (see chapter 1)

$$\beta P = \rho + B_2(T)\rho^2 + B_3(T)\rho^3 + \cdots \tag{6.83}$$

where B_j is the *j*th virial coefficient. Note that $B_j(T)$ is a function of temperature. The MM theory provides an expansion of the osmotic pressure in power series in the solute density. For a two-component system of A diluted in B, the analogue of the virial expansion is

$$\beta\pi = \rho_A + B_2^*(T, \lambda_B)\rho_A^2 + B_3^*(T, \lambda_B)\rho_A^3 + \cdots \tag{6.84}$$

In this case the coefficients B_j^* are called the virial coefficients of the osmotic pressure. Note that these virial coefficients depend on both the temperature and the solvent activity λ_B, or the solvent chemical potential $\lambda_B = \exp(\beta\mu_B)$.

In the case of real gases, the terms in the expansion (6.83) correspond to successive corrections to the ideal-gas behavior, due to interactions among pairs, triplets, quadruplets, etc., of particles. One of the most remarkable features of the theory is that the coefficients B_j depend on the properties of a system containing *exactly j* particles. For instance, $B_2(T)$ can be computed from a system of two particles in a volume V at temperature T.

Similarly, the coefficients B_j^* are expressed in terms of the properties of exactly j *solute* particles in a pure solvent B at a given solvent activity λ_B and temperature T. When $\lambda_B \to 0$, we recover the expansion (6.83) from (6.84). In this case we may say that the vacuum "fills" the space between the particles. Thus, in (6.83) we have a special (and the simplest) "solvent". In this sense the expansion (6.84) is a generalization of (6.83).

In the MM theory, there is the distinction between the solute and the solvent. The most useful case is the expansion up to ρ_A^2, i.e., the first-order deviation from the dilute ideal behavior. Higher-order corrections to the ideal-gas equations are sometimes useful if we know the interaction energy among j particles. This is not so in the case of the higher-order corrections to the dilute ideal behavior. We shall soon see that B_2^* is an integral over the pair correlation function for two solutes in a pure solvent. Similarly, B_3^* requires knowledge of the triplet correlation function for three solutes in pure solvent. Since we know almost nothing of the triplet (and higher) correlation functions, the expansion (6.84) is useful in actual applications, up to the second-order term in the solute density. For this limiting behavior the result we obtain from the MM theory is identical with the result obtained in section 6.3, from the KB theory. It is in this sense that the KB is a more general theory than the MM theory.

We shall now derive the main result of the MM theory. More detailed derivations may be found elsewhere (Hill 1956: Ben-Naim 1992).

Consider a two-component system of a solute A and a solvent B at a given temperature and activities λ_A and λ_B, respectively. The grand partition function of such a system is defined by

$$\begin{aligned}\Xi(T,V,\lambda_B,\lambda_A) &= \sum_{N_A\geq 0}\sum_{N_B\geq 0} Q(T,V,N_B,N_A)\lambda_A^{N_A}\lambda_B^{N_B} \\ &= \sum_{N_A\geq 0}\sum_{N_B\geq 0}\frac{z_A^{N_A}z_B^{N_B}}{N_A!N_B!}\int \exp[-\beta U(N_B,N_A)]\,d\mathbf{X}^{N_B}\,d\mathbf{X}^{N_A} \\ &= \sum_{N_A\geq 0}\frac{z_A^{N_A}}{N_A!}\int d\mathbf{X}^{N_A}\left\{\sum_{N_B\geq 0}\frac{z_B^{N_B}}{N_B!}\int \exp[-\beta U(N_B,N_A)]d\mathbf{X}^{N_B}\right\} \end{aligned} \tag{6.85}$$

where we have denoted

$$z_\alpha = \frac{\lambda_\alpha q_\alpha}{\Lambda_a^3(8\pi^2)}. \tag{6.86}$$

Also, we have used the shorthand notation $U(N_B,N_A)$ for the total interaction energy among the $N_A + N_B$ particles in the system.

In the limit of the dilute ideal solution, we already know the relation between λ_A and ρ_A. This is

$$\mu_A = W(A|B) + kT\ln \rho_A\Lambda_A^3 q_A^{-1}, \tag{6.87}$$

where $W(A\,|\,B)$ is the coupling work of A against pure B, or equivalently

$$\lambda_A = \exp(\beta\mu_A) = \rho_A\Lambda_A^3 q_A^{-1}\exp[\beta W(A|B)]. \tag{6.88}$$

The limiting behavior of z_A is obtained from (6.86) and (6.87), i.e.,

$$\gamma_A^0 = \lim_{\rho_A\to 0}\frac{z_A 8\pi^2}{\rho_A} = \exp[\beta W(A|B)]. \tag{6.89}$$

Thus, γ_A^0, defined in (6.89), is related to the coupling work of A against *pure B*.

The general definition of the molecular distribution function of n_A solute particles in an open system (chapter 2) is

$$\begin{aligned}\rho^{(n_A)}(\mathbf{X}^{n_A}) &= \Xi^{-1}\sum_{N_A\geq n_A}\sum_{N_B\geq 0}\frac{z_A^{N_A}z_B^{N_B}}{(N_A-n_A)!N_B!}\int \exp[-\beta U(N_B,N_A)]\,d\mathbf{X}^{N_A-n_A}\,d\mathbf{X}^{N_B} \\ &= z_A^{n_A}\Xi^{-1}\sum_{N_A'\geq 0}\sum_{N_B\geq 0}\frac{z_A^{N_A'}z_B^{N_B}}{N_A'!N_B!}\int \exp[-\beta U(N_B,N_A'+n_A)]\,d\mathbf{X}^{N_A'}\,d\mathbf{X}^{N_B}\end{aligned} \tag{6.90}$$

where $\Xi = \Xi(T,\ V,\ \lambda_B,\ \lambda_A)$. In the second form on the rhs of (6.90), we have changed variables from $N_A \geq n_A$ to $N_A' = N_A - n_A \geq 0$. $\rho^{n_A}(\mathbf{X}^{n_A})$ is the probability density of finding n_A particles at a configuration $\boldsymbol{X}^{n_A} = \boldsymbol{X}_1, \ldots, \boldsymbol{X}_{n_A}$ in an open system characterized by the variables T, V, λ_B, λ_A. The correlation function for n_A is defined by

$$g^{(n_A)}(\boldsymbol{X}^{n_A}) = \frac{\rho^{(n_A)}(\boldsymbol{X}^{n_A})}{[\rho_A^{(1)}(\boldsymbol{X})]^{n_A}}$$
$$= \left(\frac{z_A 8\pi^2}{\rho_A}\right)^{n_A} \Xi^{-1} \sum_{N_A \geq 0} \sum_{N_B \geq 0} \frac{z_A^{N_A} z_B^{N_B}}{N_A! N_B!}$$
$$\times \int \exp[-\beta U(N_B, N_A + n_A)] \, d\boldsymbol{X}^{N_A} \, d\boldsymbol{X}^{N_B} \qquad (6.91)$$

where ρ_A is the average density of A (in the open system), and we have replaced N'_A by N_A; otherwise, the sum in (6.90) is the same as the sum in (6.91).

We now take the limit $\rho_A \to 0$ or ($z_A \to 0$; the two are proportional to each other at low density). In this limit, $z_A 8\pi^2/\rho_A \to \gamma_A^0$, and all the terms in the sum (6.91) are zero except those for which $N_A = 0$. Therefore, denoting

$$g^{(n_A)}(\boldsymbol{X}^{n_A}; z_A = 0) = \lim_{\rho_A \to 0} g^{(n_A)}(\boldsymbol{X}^{n_A}, z_A), \qquad (6.92)$$

we obtain from (6.91) the expression

$$g^{(n_A)}(\boldsymbol{X}^{n_A}; z_A = 0) = (\gamma_A^0)^{n_A} \Xi(T, V, \lambda_B)^{-1}$$
$$\times \left\{ \sum_{N_B \geq 0} \frac{z_B^{N_B}}{N_B!} \int \exp[-\beta U(N_B, n_A)] \, d\boldsymbol{X}^{N_B} \right\}. \qquad (6.93)$$

Clearly, for pure solvent B ($\rho_A = 0$, or $z_A = 0$) there are no solute particles and therefore one cannot define the correlation function among the solute particles. The quantity defined in (6.92) is the correlation function among n_A solute particles when the density of all the *remaining* solute particles becomes zero. In other words, $g^{(n_A)}(\boldsymbol{X}^{n_A}; z_A = 0)$ is the correlation function for *exactly* n_A solute particles at configuration $\boldsymbol{X}^{n_A}$ in a *pure* solvent B.

Next we note that the expression in curly brackets in (6.93) is the same as in (6.85) except for the replacement of N_A by n_A. Hence, we rewrite (6.85) using (6.93) as

$$\Xi(T, V, \lambda_B, \lambda_A) = \sum_{N_A \geq 0} \frac{(z_A/\gamma_A^0)^{N_A}}{N_A!} \Xi(T, V, \lambda_B) \int g^{(N_A)}(\boldsymbol{X}^{N_A}; z_A = 0) \, d\boldsymbol{X}^{N_A}. \qquad (6.94)$$

We now define the potential of mean force for N_A, A molecules in a pure solvent B, by

$$W(\boldsymbol{X}^{N_A}; z_A = 0) = -kT \ln g^{(N_A)}(\boldsymbol{X}^{N_A}; z_A = 0). \qquad (6.95)$$

We also recall the fundamental relation between the pressure of a system and the grand partition function

$$P(T, V, \lambda_B, \lambda_A)V = -kT \ln \Xi(T, V, \lambda_B, \lambda_A) \tag{6.96}$$

and for pure B

$$P(T, V, \lambda_B)V = -kT \ln \Xi(T, V, \lambda_B). \tag{6.97}$$

In (6.96), $P(T, V, \lambda_B, \lambda_A)$ is the pressure of a system characterized by the variables T, V, λ_B, λ_A whereas $P(T, V, \lambda_B)$ in (6.97) is the corresponding pressure of the pure solvent B at T, V, λ_B. The difference between these two pressures is, by definition, the osmotic pressure, thus

$$\begin{aligned} \pi V &= [P(T, V, \lambda_B, \lambda_A) - P(T, V, \lambda_B)]V \\ &= -kT \ln \left[\frac{\Xi(T, V, \lambda_B, \lambda_A)}{\Xi(T, V, \lambda_B)} \right]. \end{aligned} \tag{6.98}$$

Denoting

$$\zeta_A = \frac{z_A}{\gamma_A^0}, \tag{6.99}$$

we can rewrite (6.94) as

$$\exp(\beta \pi V) = \sum_{N_A \geq 0} \frac{\zeta_A^{N_A}}{N_A!} \int \exp[-\beta W(\mathbf{X}^{N_A}; z_A = 0)]\, d\mathbf{X}^{N_A}. \tag{6.100}$$

This result should be compared with the corresponding expression for the *one-component* system

$$\exp(\beta P V) = \sum_{N \geq 0} \frac{z^N}{N!} \int \exp[-\beta U(\mathbf{X}^N)]\, d\mathbf{X}^N. \tag{6.101}$$

In (6.101), we have expressed the pressure of a one-component open system as integrals over the *potentials* of N particles $U(\mathbf{X}^N)$. Similarly, in (6.100) the osmotic pressure is related to integrals over the *potentials of mean force* of N_A solute particles, $W(\mathbf{X}^{N_A}; z_A = 0)$ in a pure solvent B.

The virial expansion of the osmotic pressure, although formally exact, is not very useful beyond the first-order correction to the DI limiting case. Higher-order correction terms involve higher-order potentials of mean force about which very little is known.

The derivation of the virial expansion of the osmotic pressure is quite lengthy. We present only the final result for the second virial coefficient in the expansion of the osmotic pressure

$$\beta \pi = \rho_A + B_2^* \rho_A^2 + \cdots \tag{6.102}$$

which is essentially the same as the first-order expansion of the pressure, except for the replacement of the virial coefficients B_i by B_i^*.

The explicit form of the second virial coefficient in the expansion (6.102) is

$$B_2^* = -\frac{1}{2}\int_0^\infty \left\{\exp[-\beta W_{AA}^{(2)}(R)] - 1\right\}4\pi R^2\, dR, \tag{6.103}$$

where $W_{AA}^{(2)}(R)$ is the potential of the mean force between two solutes A at a distance R, in pure solvent B at a given T and μ_B. As expected, the coefficient B_2^* is related to $W_{AA}^{(2)}(R)$ in the same manner as B_2 is related to $U(R)$ for real gases. When we let $\lambda_B \to 0$ in (6.103), $W_{AA}^{(2)}(R) \to U_{AA}(R)$ and $B_2^* \to B_2$.

Using the notation of the KB theory, we arrive at the relation between B_2^* and the KBI

$$B_2^* = -\tfrac{1}{2}G_{AA}^0, \tag{6.104}$$

where G_{AA}^0 is the limit of G_{AA} when $\rho_A \to 0$.

We conclude this section by noting that until the 1980s the MM theory has enjoyed far more attention than the KB theory. This is quite strange in view of the fact that the KB theory is easier to derive and easier to use. Its scope of application is far more wide and general, and its interpretive power is greater.

6.6 Stability condition and miscibility based on first-order deviations from SI solutions

In this section we discuss the stability conditions of a mixture with respect to material flow. There are several ways of expressing the condition of stability. The simplest is in terms of the derivatives of the chemical potential.

Consider a mixture of two components A and B at some given T and P. Let $x_A = N_A/(N_A + N_B)$ be the mole fraction of A. The condition of stability is

$$\left(\frac{\partial \mu_A}{\partial x_A}\right)_{P,T} > 0. \tag{6.105}$$

Basically, this condition means that if a fluctuation occurs at some region in the system such that x_A increases relative to the bulk composition, then the corresponding chemical potential in that region must increase. As a result of this, A will flow out of the region, and the equilibrium composition in this region will be restored.

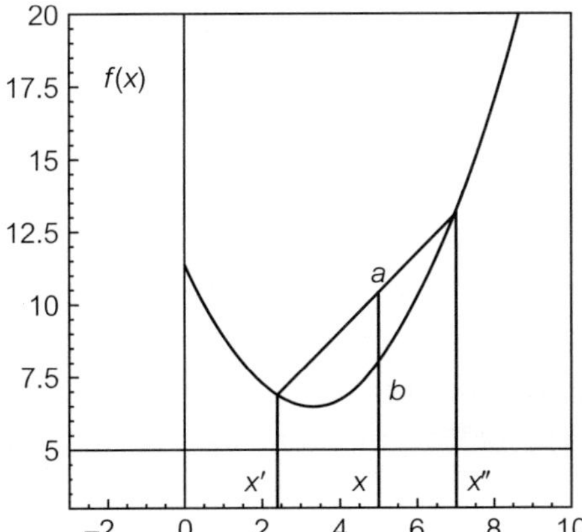

Figure 6.2. An upper concave function $f(x)$. All the points on a straight line connecting any two points on the curve must lie above the curve.

Another way of expressing the stability condition is in terms of the Gibbs energy of the system. If G is the Gibbs energy of the mixture, then

$$G = N_A\mu_A + N_A\mu_B. \tag{6.106}$$

Taking the second derivative with respect to x_A and using the Gibbs–Duhem relation, we obtain

$$\left(\frac{\partial^2 g}{\partial x_A^2}\right)_{P,T} = \frac{1}{x_B}\left(\frac{\partial \mu_A}{\partial x_A}\right)_{P,T} = \frac{1}{(1-x_B)}\left(\frac{\partial \mu_B}{\partial x_B}\right)_{P,T} > 0 \tag{6.107}$$

where $g = G/(N_A + N_B)$. Thus the condition of stability is equivalent to a positive curvature of g (or G) when plotted as a function of x_A.

Geometrically, a positive curvature means that the curve is upward concave. For concave functions the following theorem holds. If a function $f(x)$ is upward concave, i.e., if $\partial^2 f/\partial x^2$ is positive in some region say between $x' \le x \le x''$, then any point on the straight line connecting $f(x')$ and $f(x'')$ (point a in figure 6.2) must lie *above* the point $f(x)$ on the curve (point b in figure 6.2). The physical significance of this theorem is the following.

If $g(x)$ has a positive curvature in some region $x' \le x \le x''$, then a single-phase mixture of composition x is always more stable than any pair of phases with compositions x' and x'' at equilibrium.† When $g(x)$ has a negative curvature, the single phase with composition x is less stable than a pair of mixtures of compositions x' and x''. In figure (6.3), we show $g(x)$ that is a downward concave in the region (x', x''). Any point on the curve $g(x)$ has a higher Gibbs energy than a pair of mixtures with compositions x' and x'', and overall composition x. Therefore, the single phase will split into two phases. The total quantities of the two phases can be calculated as follows.

† In this section, $g(x)$ is the Gibbs energy per mole of the mixture, and x is used for x_A.

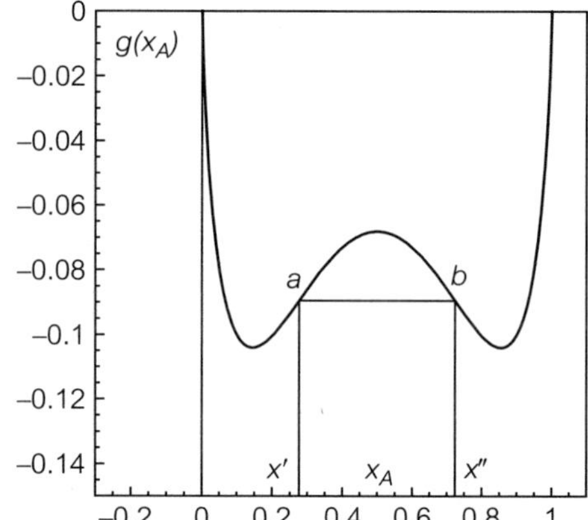

Figure 6.3. The function $g(x)$ with a region (x', x'') where the curve is downward concave (negative curvature) hence the system is unstable in this region.

The conservation of the total number of A's and B's in the system is

$$N_A = N'_A + N''_A, \qquad N_B = N'_B + N''_B. \tag{6.108}$$

The corresponding mole fractions are

$$x' = N'_A/(N'_A + N'_B), \qquad x'' = N''_A/(N''_A + N''_B). \tag{6.109}$$

Hence, the overall composition x may be expressed in terms of x' and x'' as follows

$$\begin{aligned} x &= \frac{N_A}{N_A + N_B} = \frac{N'_A}{N'_A + N'_B}\frac{N'_A + N'_B}{N_A + N_B} + \frac{N''_B}{N''_A + N''_B}\frac{N''_A + N''_B}{N_A + N_B} \\ &= \alpha x' + (1 - \alpha)x'' \end{aligned} \tag{6.110}$$

where $0 \leq \alpha \leq 1$ is

$$\alpha = \frac{N'_A + N'_B}{N_A + N_B}. \tag{6.111}$$

Thus, in this case we have

$$g(x) > \alpha g(x') + (1 - \alpha)g(x''). \tag{6.112}$$

Hence, the single phase at x will split into two phases with compositions x' and x'', with proportional quantities α and $(1 - \alpha)$.

When the system is SI then

$$g = x_A(\mu^p_A + kT \ln x_A) + x_B(\mu^p_B + kT \ln x_B) \tag{6.113}$$

and

$$\left(\frac{\partial^2 g}{\partial x_A^2}\right)_{P,T} = \frac{kT}{x_A x_B} > 0 \quad \text{for all } x_A, \ 0 \leq x_A \leq 1. \tag{6.114}$$

This guarantees that there are no two mixtures of composition x' and x'' that are more stable than the mixture with composition x_A. In other words, the two components A and B are miscible in the entire region $0 \leq x_A \leq 1$.

If the system is not SI, then there could be regions for which the curvature of $g(x)$ is negative. First note that negative curvature cannot be realized in the *entire* range of composition $0 \le x_A \le 1$. If this happens then for each mixture with composition x, there are two *pure* phases of A and B with overall composition x such that the two pure phases are more stable than the one-phase solution. Clearly this cannot happen for all x_A. We know that at very dilute solution, the system obeys Henry's law which is equivalent to

$$\left(\frac{\partial^2 g}{\partial x_A^2}\right)_{P,T} = \frac{kT}{x_A x_B} > 0 \qquad (\text{at } x_A \approx 0). \tag{6.115}$$

Thus, whenever either $x_A \to 0$ or $x_B \to 0$, we must have *positive* curvature of $g(x)$. Physically this means that the two strictly pure phases cannot exist at equilibrium when at contact. Since the chemical potential of say A in pure B will be $-\infty$, this will produce infinite driving force for A to flow into pure B, and the same applies for B to flow into pure A. Of course, this theoretical condition might not be realized in practice. In extreme cases, the solubility could be less than, say 10^{-30} mol/cm^3 which means that on average one cannot find even one A molecule in one mole of pure B. In general, the curve of $g(x)$ is similar to the one drawn in figure 6.3, i.e., the one-phase mixture is stable near $x=0$ and $x=1$ but in the "inner" region $0 < x' \le x \le x'' < 1$, there are two phases with compositions x' and x'', the combined Gibbs energy of which is lower than the Gibbs energy of the one-phase mixture at x. In this region, no single phase exists with composition x, $x' \le x \le x''$.

We now turn to examine the molecular origin of this kind of instability. We first use the first-order deviation from SI solution (see section 6.2)

$$\mu_A(T, P, x_A) = \mu_A^P + kT \ln x_A + \tfrac{1}{2} kT \rho_T \Delta_{AB} x_B^2. \tag{6.116}$$

This is equivalent to

$$g = x_A \mu_A^P + x_B \mu_B^P + kT x_A \ln x_A + kT x_B \ln x_B + \tfrac{1}{2} kT \rho_T x_A x_B \Delta_{AB}. \tag{6.117}$$

If $\rho_T \Delta_{AB}$ is independent of x_A, then

$$\left(\frac{\partial^2 g}{\partial x_A^2}\right)_{P,T} = \frac{kT}{x_A x_B} - kT \rho_T \Delta_{AB}. \tag{6.118}$$

The following discussion is the "traditional" way of examining the condition of stability. It was originally discussed in the context of the lattice model of mixtures (see below).

First, when $\Delta_{AB} < 0$, the curvature is always positive hence the one-phase system is always stable. A negative Δ_{AB} means that $G_{AA} + G_{BB} < 2G_{AB}$, i.e., the affinity between A–B pairs is larger than the average of affinities between A–A

and B–B pairs. This makes sense since if A "prefers" to be surrounded by B more than by A, and B "prefers" to be surrounded by A rather than by B, the mixture will always be more stable than two separate mixtures. It should be stressed however, that this is true only within the *first*-order deviations, as expressed in 6.116–6.118. See also section 6.7.

Second, when $\Delta_{AB} > 0$, but small, i.e.[†],

$$0 < \rho\Delta_{AB} < \frac{1}{x_A x_B} \leq 4 \tag{6.119}$$

we shall still have a stable one-phase system. However, when$\Delta_{AB} > 0$ becomes very large, such that $\rho\Delta_{AB} > 4$ then the curvature changes sign and becomes *negative*. Hence, the one-phase system is not stable. A large positive value of Δ_{AB} means that $G_{AA} + G_{BB} > 2G_{AB}$, meaning that the average affinities between A–A and B–B is larger than the affinity G_{AB}. Again, it makes sense to expect instability in this case but one must be careful in reaching any conclusion regarding the stability for $|\rho\Delta| \geq 4$ since in this case the first-order expansion we have used (equation 6.116) might not be valid. See also section 6.7, and Appendix M.

In figure 6.4, we present a few examples of negative deviations from Raoult's law, i.e., $P_A^* = P_A/P_A^0 < x_A$, for different values of $\rho\Delta_{AB} = -5, -4, -3, -2, -1, 0$. It is seen that the system is stable in the entire range of compositions. Note that since ρ is always positive, the sign of the deviation is determined by Δ_{AB}.

Figure 6.5 shows some examples of *small* positive deviations from Raoult's law, $\rho\Delta_{AB} = 0$, 1, 2, 3, 4. Again in this case, the system is stable in the entire range of compositions. Note that when $\rho\Delta_{AB} = 4$, the curve for $P_A^* = P_A/P_A^0$ has an inflection point at $x_A = \frac{1}{2}$. Figure 6.6 shows the behavior for *large* values of $\rho\Delta_{AB} = 4$, 6, 8. Here for each value of $\rho\Delta_{AB} > 4$, we have a region of instability where $\partial\mu_A/\partial x_A$ becomes negative, or equivalently the excess of Gibbs energy becomes concave downward. In all of these cases, the instability region is around the center $x_A = \frac{1}{2}$, and there are always two regions of stability near the edges $x_A \approx 0$ and $x_A \approx 1$. The latter becomes narrower as we increase $\rho\Delta_{AB}$. A summary of the regions of stability and instability is shown in figure 6.7.

We now turn to some simple examples where Δ_{AB} can be calculated. We still stay within the first-order deviation from ideal symmetrical solution.

(1) *Mixture of gases at very low Pressure.* In the limit of $P \to 0$ or $\rho_T \to 0$, we have

$$G_{ij} = \int_0^\infty \{\exp[-\beta U_{ij}] - 1\} 4\pi R^2 \, dR. \tag{6.120}$$

[†] In the following we use ρ instead of ρ_T.

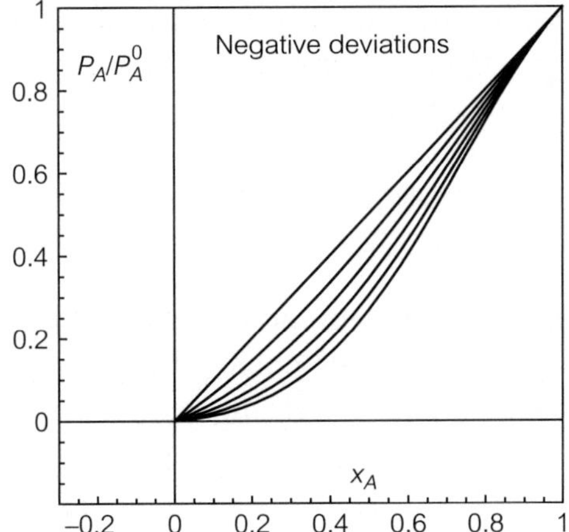

Figure 6.4. Negative deviations from Raoult's law. Plots of P_A/P_A^0 (or the activity) as a function of x_A for different values of $\rho\Delta_{AB} = -5, -4, -3, -2, -1, 0$. The larger $|\rho\Delta_{AB}|$ the farther the curve from the diagonal line.

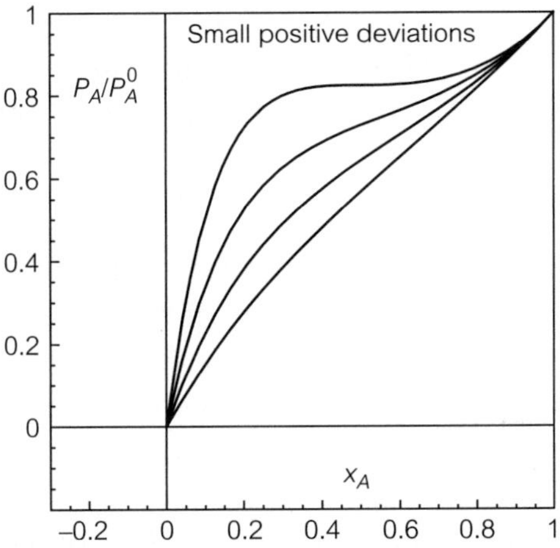

Figure 6.5. Small positive deviations. Same as in figure 6.4 but with $\rho\Delta_{AB} = 0, 1, 2, 3, 4$.

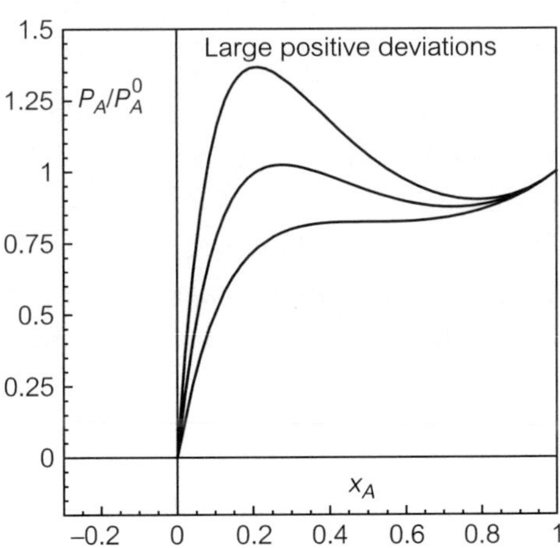

Figure 6.6 Large positive deviations. Same as in figure 6.4 but with $\rho\Delta_{AB} = 4, 6, 8$.

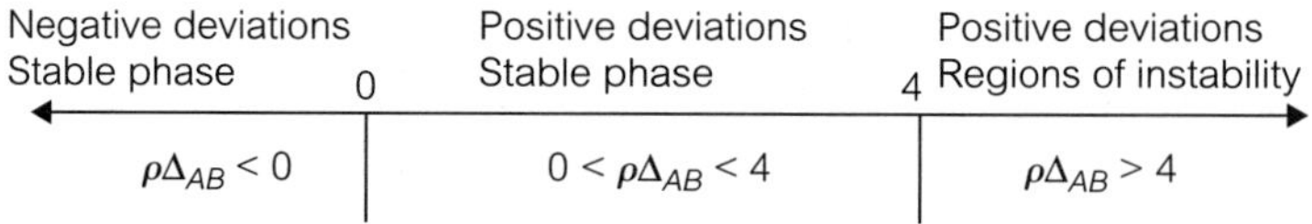

Figure 6.7. Region of stability and instability as determined by $\rho\Delta_{AB}$, based on first-order deviation from SI solutions.

Note that this is *not* a theoretical ideal gas. We take the limit $P \to 0$ for a real gas mixture. $U_{ij}(R)$ is the pair potential for the ij pair. We assume that the pair potential has a square-well form, i.e.,

$$U_{ij}(R) = \begin{Bmatrix} \infty & \text{for} & R < \sigma_{ij} \\ -\varepsilon_{ij} & \text{for} & \sigma_{ij} \leq R \leq \sigma_{ij} + \delta_{ij} \\ 0 & \text{for} & R > \sigma_{ij} + \delta_{ij} \end{Bmatrix} \tag{6.121}$$

where δ_{ij} is the range of the potential. If all $\beta\varepsilon_{ij}$ are small, we can write (6.120) as

$$G_{ij} = \int_0^{\sigma_{ij}} (-1)4\pi R^2\, dR + \int_{\sigma_{ij}}^{\sigma_{ij}+\delta_{ij}} \beta\varepsilon_{ij} 4\pi R^2\, dR$$

$$= -\frac{4\pi\sigma_{ij}^3}{3} + \beta\varepsilon_{ij} V_{ij}. \tag{6.122}$$

The first term on the rhs of (6.122) is due to the "hard" part of the interaction; it equals the volume of a sphere of radius σ_{ij}. The second is due to the "soft" part of the interaction; here, V_{ij} is the region where the soft part is operative. One can easily see that for hard-sphere mixtures† (i.e., all $\varepsilon_{ij} = 0$) $\Delta_{AB} < 0$, hence the curvature of g is always *positive.*

Next suppose that A and B have the same size and the same interaction range, i.e., $V_{ij} = V_{int}$, but differ in the soft interaction parameter ε_{ij}, in which case

$$\Delta_{AB} = \beta V_{\text{int}}[\varepsilon_{AA} + \varepsilon_{BB} - 2\varepsilon_{AB}]. \tag{6.123}$$

Here we obtain $\Delta_{AB} > 0$ whenever $\varepsilon_{AA} + \varepsilon_{BB} > 2\varepsilon_{AB}$ and $\Delta_{AB} < 0$ whenever $\varepsilon_{AA} + \varepsilon_{BB} < 2\varepsilon_{AB}$. It is tempting to conclude that in these two cases we shall obtain stable and unstable mixtures, respectively. However, we must remember that in the limit of $P \to 0$, $\rho_T = \beta P$ and $\rho_T\Delta_{AB} = \beta^2 P V_{\text{int}}[\varepsilon_{AA} + \varepsilon_{BB} - 2\varepsilon_{AB}]$ must be small. Therefore, one *cannot* predict the behavior at large values of $|\rho_T\Delta_{AB}|$.

Thus, from the above discussion one can predict the occurrence of positive or negative deviations from SI solutions. But since in this limit $P \to 0$, and $\rho_T \to 0$ we also expect $|\rho_T\Delta_{AB}|$ to be small, therefore we must have miscibility in the entire region of compositions.

† Note that this is true for hard spheres which obeys $\sigma_{AB} = (\sigma_{AA} + \sigma_{AB})/2$. In a one-dimensional system, $\Delta_{AB} = 0$, see section 6.4.

(2) *Lattice model of solutions.* In a lattice model of mixtures (Guggenheim 1952) the size of particles are assumed to be nearly the same. Hence A and B can occupy the same lattice sites. It is well known that in this case the deviations from SI solutions are expressed in the form (see Appendix M)

$$\mu_A = \mu_A^P + kT \ln x_A - \frac{zWx_B^2}{2} \tag{6.124}$$

where $W = E_{AA} + E_{BB} - 2E_{AB}$ is the so-called exchange energy. $E_{ij} < 0$ are the interaction energies between i and j on adjacent sites, and z is the coordination number or the number of nearest neighbors to any site. The condition for instability is when W is large and negative, i.e.,

$$W = E_{AA} + E_{BB} - 2E_{AB} < 0 \tag{6.125}$$

(note that $E_{ij} < 0$, whereas ε_{ij} in (6.123) are positive).

Most of the analysis of the conditions for stability were carried out using equation (6.124) (see Guggenheim 1952; Denbigh 1966; Prausnitz et al. 1986). Unfortunately, it was not explicitly recognized that the expression (6.124) is the *first-order* deviation from SI. However, going through the arguments reveals that at some point one uses the approximation

$$\langle e^{-X} \rangle \approx e^{-\langle X \rangle}. \tag{6.126}$$

This "elevation" of the average sign to the exponent is valid only when X is small compared to unity. For more details, see Appendix M. Failing to recognize the "first-orderness" of the expression (6.124) has misled almost two generations of scientists to conclude that large *positive* deviations in the sense of $\rho_T \Delta_{AB} > 0$ lead to instability, but, large negative deviations do not produce any instability. It should be stressed however, that either (6.116) or (6.124) are valid only for small values of $|\rho \Delta_{AB}|$ or $|zW|$, respectively. Therefore, no conclusions should be reached for large deviations from SI behavior. This point will be further examined in the next section.

6.7 **Analysis of the stability condition based on the Kirkwood–Buff theory**

In section 6.6, we have examined the conditions for stability using the first-order deviations from SI solutions. The regions of stability and instability were summarized in figure 6.7.

We noted in section 6.6 that care must be exercised when examining the conditions for stability, based on first-order deviations. The reason is simple – the first-order expansion is valid only for *small* values of $\rho_T\Delta_{AB}$. Clearly, one cannot use the first-order expansion to examine the behavior of these solutions at large values of $|\rho_T\Delta_{AB}|$. To the best of the author's knowledge the molecular conditions for stability were never studied beyond the first-order correction to SI solutions[†]. This is true also within the lattice theories of solutions where the conditions for stability were examined in terms of the exchange energy W, in equation (6.124).

In this section, we examine the conditions for stability using the exact result of the KB theory

$$\left(\frac{\partial^2 g}{\partial x_A^2}\right) = \frac{1}{x_B}\left(\frac{\partial \mu_A}{\partial x_A}\right)_{P,T} = \frac{kT}{x_B}\left(\frac{1}{x_A} - \frac{x_B\rho\Delta_{AB}}{1 + x_A x_B \rho\Delta_{AB}}\right) = \frac{kT}{x_B x_A(1 + x_A x_B \rho\Delta_{AB})}. \tag{6.127}$$

As in section 6.6, we shall examine the conditions under which the derivative in equation (6.127) is positive, i.e., when the free energy is concave upwards. The examination of the excess chemical potential, or the activity coefficient, is less convenient in this case since it requires integration of equation (6.127). In most of what follows we assume that Δ_{AB} is independent of the composition. We shall examine the conditions under which the rhs of (6.127) changes sign.

First, it is clear that for any positive values of $\rho\Delta_{AB}$, the system is stable everywhere, i.e., the Gibbs energy of the system is concave upwards, in the *entire* range of compositions. This is in sharp contrast to the conclusion based on the first-order expansion, where we found that instability ensues when $\rho\Delta_{AB}$ became large and positive beyond $\rho\Delta_{AB} \geq 4$. See figure 6.7.

Second, for $\rho\Delta_{AB} < 0$ we have two regions.

(1) For $-4 < \rho\Delta_B < 0$ (in the entire range of compositions) the rhs of (6.127) is positive, and the system is stable.

(2) At $\rho\Delta_B = -4$ the derivative diverges at $x_A = x_B = \frac{1}{2}$ and for $\rho\Delta_B < -4$, we have regions of stability at the edges, i.e., near $x_A \approx 0$ and $x_A \approx 1$ but instabilities in the center of the composition range.

This is summarized in figure 6.8 which should be compared with figure 6.7 of the previous section. Note that in the region where $|\rho\Delta_{AB}| < 4$, the two

[†] Much work has been done on the regions of stability and instability using empirical expression for the excess free energy; see, for example, Prausnitz et al. (1968) and Novák et al. (1987).

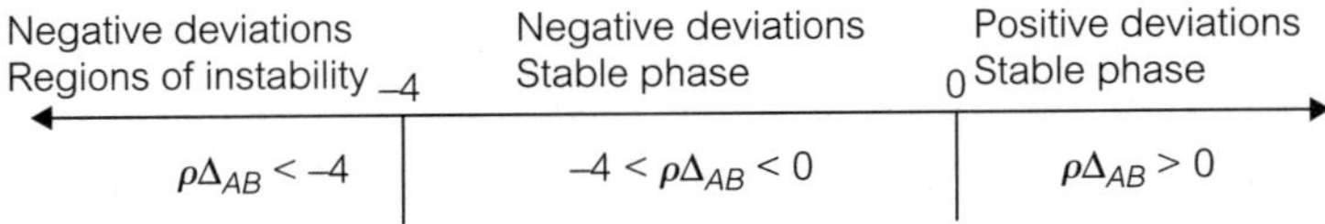

Figure 6.8. Region of stability and instability based on the exact *KB* expression (6.127).

figures agree. They do not agree for larger values of $|\rho\Delta_{AB}| \geq 4$. For larger values of $|\rho\Delta_{AB}|$ the use of the *first-order* expansion should not be trusted.

The reason for this discrepancy is quite simple. If we expand the rhs of equation (6.127) in a power series about the point $\rho\Delta_{AB}=0$, we have

$$\left(\frac{\partial^2 g}{\partial x_A^2}\right)_{P,T} = \frac{kT}{x_A x_B}\left[1 - x_A x_B \rho\Delta_{AB} + (x_A x_B \rho\Delta_{AB})^2 - (x_A x_B \rho\Delta_{AB})^3 + \cdots\right]. \quad (6.128)$$

Clearly, we can trust the first-order expansion only for small values of $|\rho\Delta_{AB}|$. As we have seen in section 6.6, positive values of $\rho\Delta_{AB}$ lead to *positive* deviations from Raoult's law, and negative values of $\rho\Delta_{AB}$ lead to negative deviations from Raoult's law. However, once we get beyond the first-order deviations from SI solutions, one must use higher-order terms in the expansion, or better, the exact analytical expression (6.127) which is valid for all values of $\rho\Delta_{AB}$ except for values of $\rho\Delta_{AB}$ where the denominator is zero.

We next turn to some numerical illustrations of the behavior of the derivative (6.127) for different values of $\rho\Delta_{AB}$.

In figure 6.9, we plot the second derivative of g with respect to x_A, for a few values of $\rho\Delta_{AB}$. First, for values of $\rho\Delta_{AB}=0$, 1, 2, 3, 4 we have a stable phase in agreement with the case of the first-order deviations. For $\rho\Delta_{AB}=4$, 6, 8 we find a stable phase, in disagreement with the case of the first-order deviations. For $\rho\Delta_{AB}= -0.35$, -0.36, -0.37, -0.38, -0.39 we still have a stable phase, but as $\rho\Delta_{AB}$ approaches $\rho\Delta_{AB}= -4$, the second derivative of g diverges at $x_A=x_B=\frac{1}{2}$.

Figure 6.10 shows what happens when $\rho\Delta_{AB}$ decreases beyond -4. Recall that within the first-order deviations, we found that the system is stable. Here, however, once we pass below $\rho\Delta_{AB}< -4$ we have instability in the "inner" region of compositions (around $x_A=\frac{1}{2}$), but stability in the "outer" region, i.e., near the edges of $x_A\approx 0$ and $x_A\approx 1$. Clearly as $\rho\Delta_{AB}$ becomes more negative, the region of instability widens and the regions of stability are pushed towards the edges. This again is in contrast to the conclusions reached in section 6.6. For each value of $\rho\Delta_{AB}< -4$ there are two points at which the rhs of equation (6.127) diverges. The two points are the solution of the equation $1+x(1-x)\rho\Delta_{\mathrm{AB}}=0$.

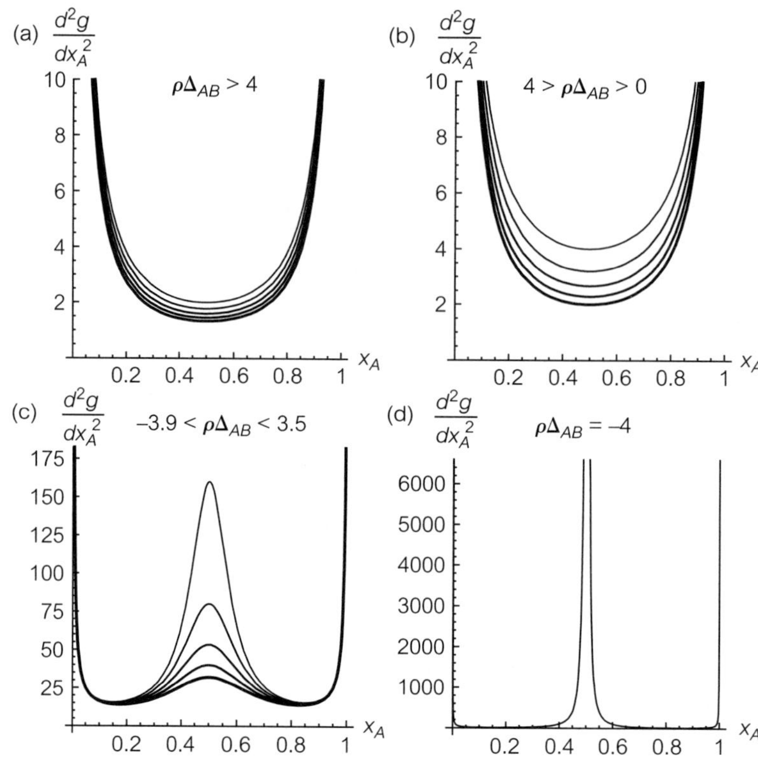

Figure 6.9. The second derivative of g (the Gibbs energy of the system per mole of the mixture) as a function of x_A for various values of $\rho\Delta_{AB}$: (a) small positive values of $\rho\Delta_{AB}=0, 1, 2, 3, 4$. The upper curve is for $\rho\Delta_{AB}=0$, the lower for $\rho\Delta_{AB}=4$; (b) large positive values of $\rho\Delta_{AB}=4, 6, 8$ (the upper curve for $\rho\Delta_{AB}=4$, the lower for $\rho\Delta_{AB}=8$); (c) small negative values of $\rho\Delta_{AB}= -0.35, -0.36, -0.37, -0.38$ and -0.39; (d) divergence of the derivative at $x_A= 1/2$ for the case $\rho\Delta_{AB}= -4$.

In figure 6.11, we plot the pair of compositions (x', x'') between which the system is unstable. In this particular example all the pairs (x', x'') are symmetrical about $x_A=\frac{1}{2}$. This is a result of the choice of a constant value (i.e., independent of composition) of $\rho\Delta_{AB}$. Clearly for $\rho\Delta_{AB}> 4$, there are no real solutions to the equation $1+x(1-x)\rho\Delta_{AB}=0$. For $\rho\Delta_{AB}<-4$ there are two solutions x', x''. These are plotted for different values of $\rho\Delta_{AB}$.

Finally, we note first that in all our calculations, we have assumed that $\rho\Delta_{AB}$ is constant in the entire range of composition. In real systems we should expect that $\rho\Delta_{AB}$ will change with composition. This will affect the details of the regions of stability and instability but grossly the qualitative behavior should not be much different. In figure 6.12, we present two examples of the behavior of the second derivative of g for two cases when $\rho\Delta_{AB}$ depends linearly on x_A. Second, we recall that the KB theory and the

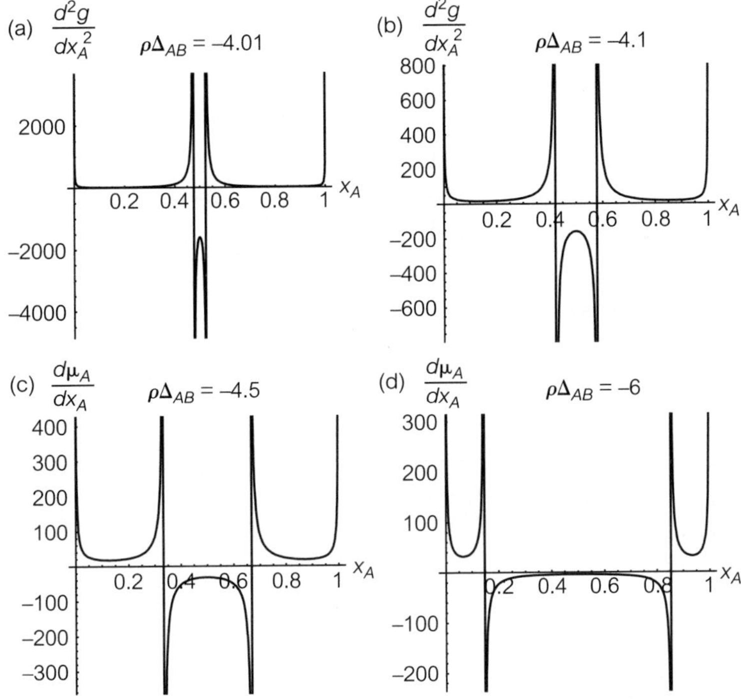

Figure 6.10. The second derivative of g as a function of x_A for large negative values of $\rho\Delta_{AB} = -4.1$, -4.5, -6, -10. In each case there are two points x_A (symmetrical about $x_A = 1/2$) at which the curve diverges.

result we have used in equation (6.127) is valid for stable mixture. This was explicitly assumed in chapter 4 when we inverted the matrix $\boldsymbol{B}$. The existence of the inverse matrix is equivalent to the assumption of stability. Therefore, one cannot use the KB theory to study systems that are unstable with respect to composition. In Appendix P we further examine the relation between deviations from SI as measured by $\rho\Delta_{AB}$, and experimental deviations from Raoult's law as measured by P_A/P_A^0 or by the activity coefficient γ_A^{SI}.

6.8 The temperature dependence of the region of instability: Upper and lower critical solution temperatures

In sections 6.6 and 6.7 we have analyzed the region of stability for a constant value of $\rho\Delta_{AB}$. We have seen that when the "magic" value of -4 is crossed, we

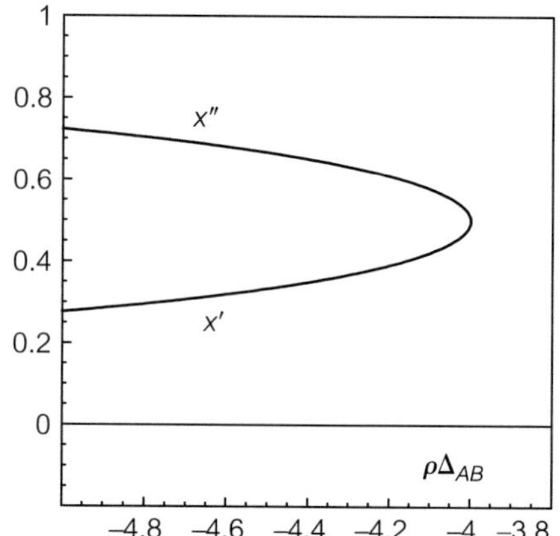

Figure 6.11. The pair of compositions x' and x'' at which the second derivative of the Gibbs energy diverges, as a function of $\rho\Delta_{AB}$.

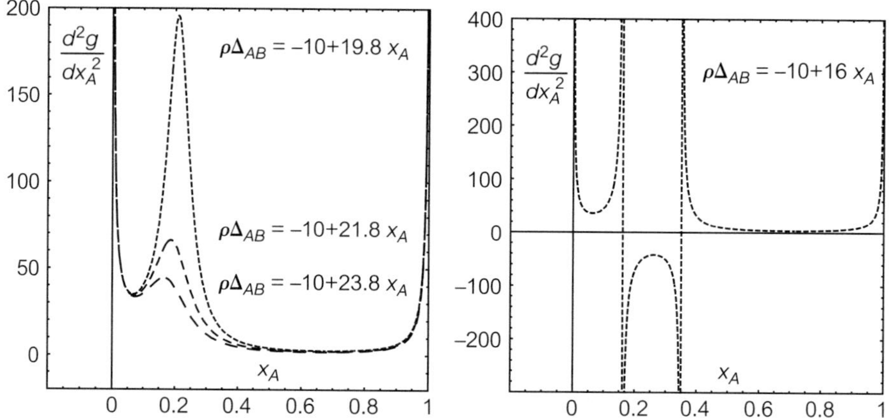

Figure 6.12. As in figure 6.10 but now we assume that $\rho\Delta_{AB}$ is a function of composition. This particular illustration is for $\rho\Delta_{AB} = -10 + ax$, with different values of a as indicated.

pass from stable to unstable mixtures.[†] In this section we examine the conditions under which we get upper and lower critical solution temperatures (UCST and LCST, respectively). These are obtained when, as a result of change in the temperature (or the pressure) we cross the "magic" border of $\rho\Delta_{AB} = -4$. We shall still assume that $\rho\Delta_{AB}$ is independent of composition, but assume that it is a function of temperature. To obtain a UCST or LCST we must cross the borderline of $\rho\Delta_{AB} = -4$. We now examine a few possible cases of the temperature dependence of $\rho\Delta_{AB} = f(T)$. In all cases, in order to obtain either a UCST or a LCST, the function $f(T)$ must cross the borderline of $f(T) = -4$ at least once.

[†] The unique value of $\rho\Delta_{AB} = -4$ is simply a result of the fact that the product $x_A(1 - x_A)$ has a maximum value of 1/4 at $x_A = \frac{1}{2}$. Hence, whenever $\rho\Delta_{AB} = -4$, the denominator of equation (6.127) becomes zero. The divergence of the derivative at exactly $x_A = \frac{1}{2}$ is a result of the assumption of $\rho\Delta_{AB}$ being independent of composition.

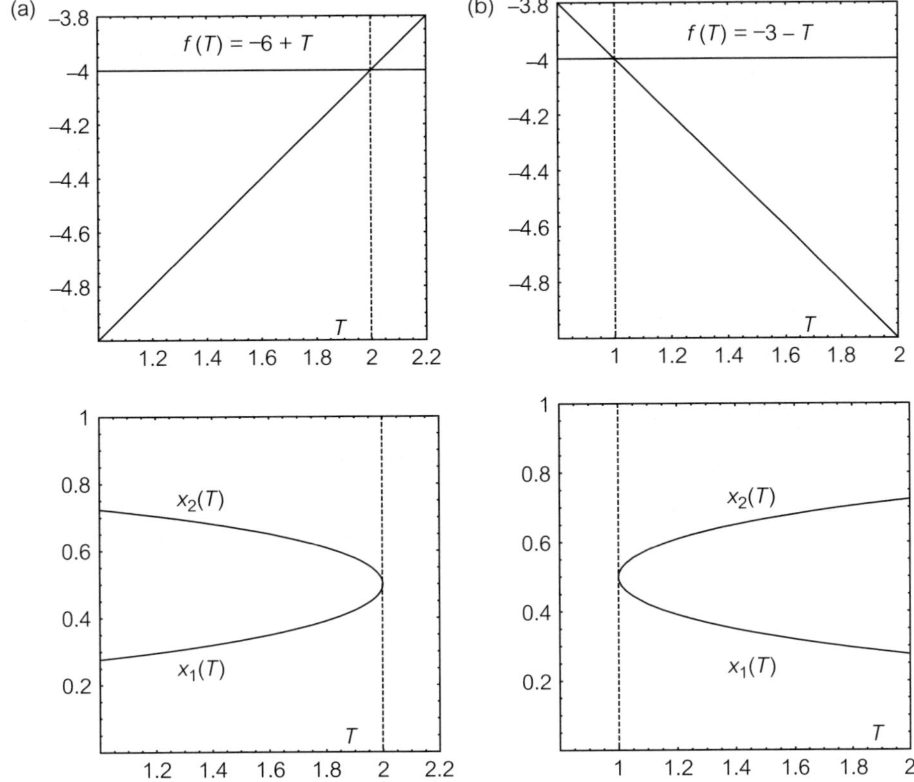

Figure 6.13. (a) A $T(x)$ diagram for a linearly increasing function of $f(T)$. In this particular illustration $f(T) = -6 + 2T$. (b) As in (a) but for a linear decreasing function $f(T)$. Specifically $f(T) = -3 - T$.

(1) *A linearly increasing function of T.* Figure 6.13a shows a linearly increasing function $f(T)$ that crosses the line at -4. Clearly, as long as $f(T)$ is below -4, we have a region of instability with a width $(x_1(T), x_2(T))$. As the temperature increases, the width of the instability region decreases and at the temperature where $f(T) = -4$, the region of instability shrinks to zero. Beyond $f(T) = -4$ we have a single stable phase. This is the case of a UCST. In figure 6.13 we show both the function $f(T)$ and the corresponding $T(x)$ diagram.

An example of a system with a UCST is a mixture of n-hexane and nitrobenzene. At one atmospheric pressure, this system has a critical temperature at about 19 °C.

(2) *A linearly decreasing function of T.* Figure 6.13b shows a simple linearly decreasing function $f(T)$ that starts above the line -4 and cross downward below -4. As long as $f(T)$ is above -4 we have a single and stable phase.

Once we cross the borderline, we have an incipient instability – the system splits into two phases with compositions $(x_1(T), x_2(T))$. At the temperature

T_C when we cross the line -4, we obtain a LCST. An example of the system with a LCST is a mixture of water and diethylamine, which has a critical temperature at about 143.5 °C.

In both of the examples above, we have assumed that $\rho\Delta_{AB}$ is independent of composition and that the dependence of $\rho\Delta_{AB}$ on the temperature is linear. Clearly if the temperature dependence is nonlinear, but still monotonically increasing or decreasing functions of T, we shall obtain UCST and LCST, respectively. If, on the other hand, $\rho\Delta_{AB}$ is also a function of x_A, then we lose the symmetrical behavior of the $T(x)$ diagram, but still the general phenomenon is the same.

(3) *A function f(T) that crosses the borderline twice; upwards then downwards.* Figure 6.14a shows an example of a function $f(T)$, here chosen as a parabolic function that starts at low temperature, below -4, at higher temperature cross

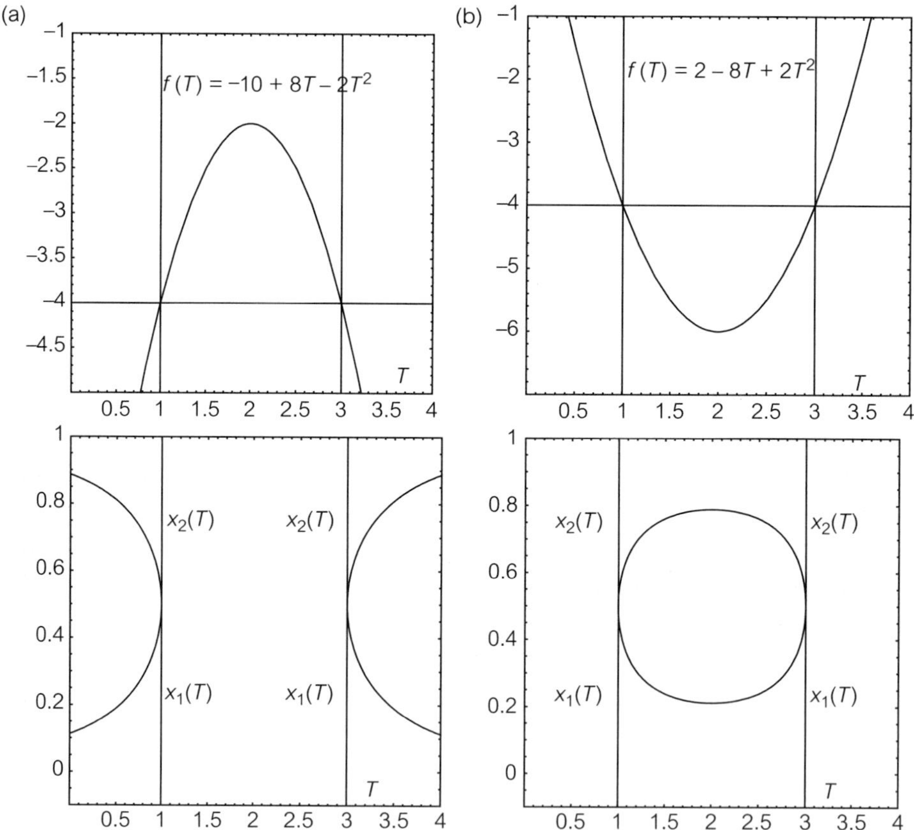

Figure 6.14. (a) As in figure 6.13 but with a parabolic function $f(T)$. In this illustration $f(T) = -10+8T-2T^2$. (b) Here again we have a parabolic function but concave downwards, specifically $f(T) = 2-8T+2T^2$.

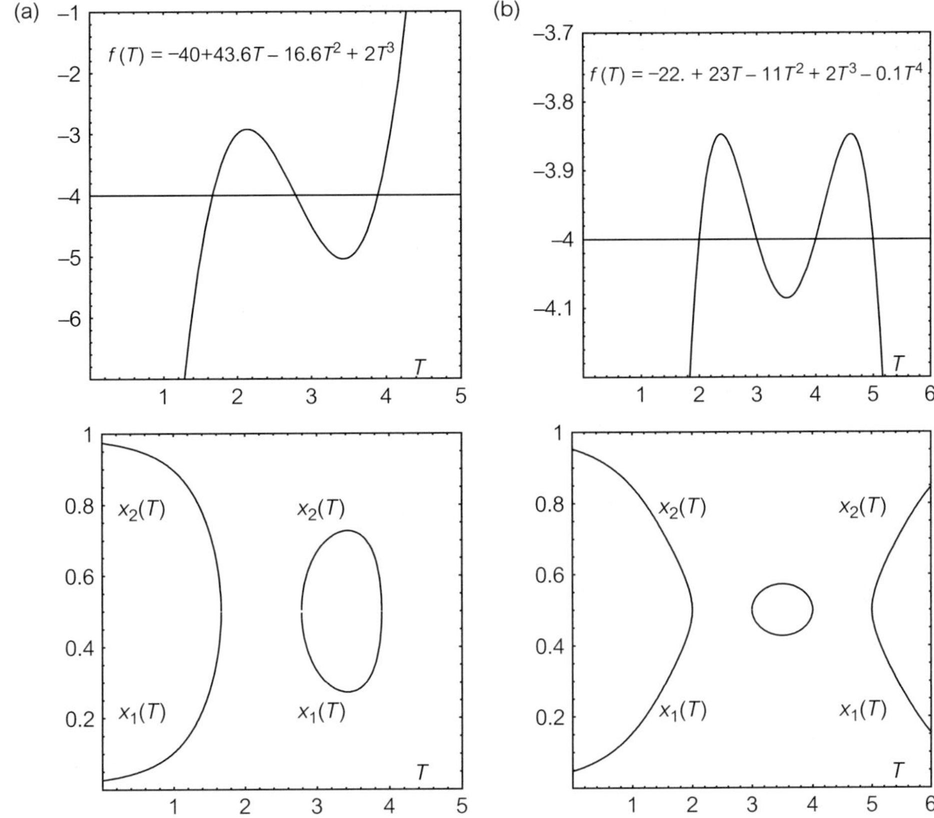

Figure 6.15. (a) A possible system with three critical temperatures, first UCST then LCST, then UCST again with $f(T) = -40 + 43.6T - 16.6T^2 + 2T^3$. (b) A possible system with four critical temperatures, first UCST followed by LCST, UCST, and LCST, with $f(T) = -22 + 23\text{T} - 11T^2 + 2T^3 - 0.1T^4$.

the line -4 upwards, then at even higher temperature crosses again the borderline -4 downwards. The corresponding behavior of the $T(x_A)$ diagram is simply a combination of the two cases (a) and (b) of figure 6.13. Initially, when $\rho\Delta_{AB} < -4$ we have instability in the region $(x_1(T), x_2\ (T))$, at crossing the border line at -4, we enter into a single-phase region, and stay there as long as $f(T)$ is above -4. Once the function $f(T)$ crosses the borderline again we enter into an instability region. The $T(x_A)$ will show first a UCST and then a LCST. An example of such a system where two-phase regions are separated by a one-phase region is carbon dioxide and ethane. This kind of behavior is observed in solutions of polymers (Sandler 1994).

(4) *A function $f(T)$ that crossed the borderline twice; downwards then upwards.* Figure 6.14b shows what happens when the function $f(T)$ crosses the borderline twice, first downward then upwards. In this case we get first a LCST at low

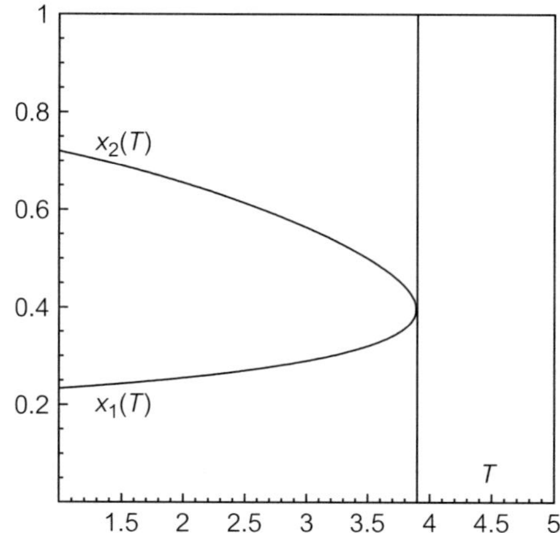

Figure 6.16. An asymmetric $T(x)$ curve obtained for a system for which $\rho\Delta_{AB}$ is a function of T and x. $\rho\Delta_{AB} = (-6 + 0.5x) + (0.1 + 0.8x)T$. This system shows a UCST at a mole fraction of about $x = 0.4$.

temperature, then a closed loop of instability region at higher temperature and eventually reaching a UCST at higher temperature. Examples of such a closed loop with UCST and LCST are mixtures of nicotine and water, and of glycerol and water.

(5) *A function that crosses the borderline thrice.* Figures 6.15 shows two possible cases. In (a) the function $f(T)$ crosses the line -4 three times; first starts lower than -4, crosses the borderline upwards, then crosses downwards and then crosses upward again. A second possibility is shown in (b) where the function $f(T)$ crosses the line -4 four times.

One can go on and examine more complex functions $f(T)$ that cross the borderline many times; the corresponding $T(x)$ diagram would have as many LCST and UCST as the number of times the function crosses the borderline. Theoretically, there is no limit to the number of LCST and UCST, the only question is whether there are real mixtures for which $\rho\Delta_{AB}$ will have this kind of complex behavior.

Figure 6.16 shows an example for which $\rho\Delta_{AB}$ is a function of both T and x: $\rho\Delta_{AB} = (-6 + 0.5x) + (0.1 + 0.8x)T$. This system shows a UCST at a mole fraction of about $x = 0.4$.

SEVEN
Solvation thermodynamics

The study of solvation thermodynamics has a very long history. Almost any experiment carried out in a solution necessarily involves solvation. Prior to the 1970s there were several quantities referred to as standard Gibbs energies of solution (or hydration, when the solvent is water). All of these were defined in terms of a process of transferring of a solute from a specified standard state in the gaseous phase into some other standard state in the liquid phase. It was not at all clear which of these quantities is truly a measure of the Gibbs energy of interaction between the solute and the solvent. It is no wonder that Lewis and Randall (1961) have written on this matter: "Of all the applications of thermodynamics to chemistry, none has in the past presented greater difficulty, or been subject to more misunderstanding."

Traditionally, solvation was studied within the context of thermodynamics. In this context, it could be studied only in the limit of very dilute solution, i.e., in the concentration range when Henry's law is obeyed.

In 1978, a new process of solvation was introduced along with the corresponding thermodynamic quantities (Ben-Naim 1978[†], 1987). Initially the new measure was referred to as "nonconventional," "generalized", and "local" quantities. It was only much later, after I had been convinced that these quantities are the only *bonafide* measure of solvation, that I claimed the already used term "solvation thermodynamic quantities" to the newly introduced quantities. With this new concept, the study of solvation became a powerful tool to probe the extent of interaction between the solute and the solvent, and the effect of the solute on various molecular distribution functions in the solvent. We shall study these aspects of solvation in the following sections. Before doing that, we present here several situations in which the concept of solvation Gibbs energy arise "naturally," hence justifies its study. The rest of the chapter is devoted to the study of solvation of different systems with increasing degree of complexity.

† This paper was entitled: "Standard thermodynamics of transfer. Uses and misuses." Later in 1987, a monograph was published where all the advantageous aspects of the new quantities were spelled out in great detail.

7.1 **Why do we need solvation thermodynamics?**

Consider a general chemical reaction in the gaseous phase, written symbolically as

$$R \rightarrow P \tag{7.1}$$

where R stands for all the reactants, and P for all the products. If the reaction (7.1) is carried out in an ideal-gas phase, then one can compute the equilibrium constant of this reaction, as well as the corresponding standard thermodynamic quantities of this reaction, from the knowledge of the properties of all the molecules involved in the reaction. For many relatively simple reactions, one can actually compute the partition function of each of the species involved in the reaction and from that, the equilibrium constant by the well-known procedure of statistical mechanics.

First, the chemical potential of each species is written as

$$\mu_i = kT \ln \rho_i \Lambda_i^3 q_i^{-1} = \mu_i^{0g} + kT \ln \rho_i \tag{7.2}$$

where ρ_i is the number density, Λ_i^3 the momentum partition function (or the de Broglie thermal wavelength), and q_i include all the internal partition functions of the species i.

At any given P, T and composition $\mathbf{N} = (N_1, \ldots, N_c)$ of the system, the total Gibbs energy is given by

$$G = \sum_{i=1}^{c} N_i \mu_i \tag{7.3}$$

where the sum is over *all* c-components in the system (including all reactants and products). At equilibrium, the Gibbs energy must attain its minimum with respect to the "reaction coordinate;" any infinitesimal change in the composition of the system away from the equilibrium composition must increase the Gibbs function. This leads to the well-known equilibrium condition†

$$\sum_{i=1}^{c} \upsilon_i \mu_i = 0 \tag{7.4}$$

where υ_i is the stoichiometric coefficient for the species i in the reaction (7.1). By convention, υ_i is positive for the products and negative for the reactants. Using the expression (7.2) for each of the species involved in reaction (7.1), we

† See, for example, Prigogine and Defay (1954), or Denbigh (1966).

arrive at the expression for the ideal-gas equilibrium constant

$$\sum_{i=1}^{c} \upsilon_i \mu_i^{0g} + kT \ln \prod_{i=1}^{c} \rho_i^{\upsilon_i} = 0, \tag{7.5}$$

or equivalently,

$$K^{ig} = \prod_{i=1}^{c} \rho_i^{\upsilon_i} = \exp[-\beta \Delta G^{0g}] \tag{7.6}$$

where ΔG^{0g} is the standard Gibbs energy of the reaction in the ideal-gas phase

$$\Delta G^{0g} = \sum_{i=1}^{c} \upsilon_i \mu_i^{0g}. \tag{7.7}$$

As a simple example, suppose two monomers (M) form a dimer (D). The reaction is

$$2M \leftrightarrow D. \tag{7.8}$$

The equilibrium constant, and the corresponding standard Gibbs energy of this reaction, are

$$K^{ig} = \left(\frac{\rho_D}{\rho_M^2}\right)_{eq}^{ig} = \exp[-\beta \Delta G^{0g}] \tag{7.9}$$

and

$$\Delta G^{0g} = \mu_D^{0g} - 2\mu_M^{0g} \tag{7.10}$$

where "eq" indicates that the ratio is evaluated at equilibrium. Now, suppose that we carry out exactly the same process, at the same P, T, but in some solvent l. How is the equilibrium constant (and the corresponding standard Gibbs energy) modified by this transfer? The qualitative answer, as suggested by inspection of figure 7.1, is that each chemical potential in (7.3) is modified by the appropriate coupling work $W(i)$, hence

$$\mu_i = W(i) + kT \ln \rho_i \Lambda_i^3 q_i^{-1}. \tag{7.11}$$

The modified equilibrium constant in the liquid phase is

$$K^l = \left(\frac{\rho_D}{\rho_M^2}\right)_{eq}^{l} = \exp[-\beta \Delta G^{0l}] = K^{ig} \exp[-\beta(\Delta\mu_D^* - 2\Delta\mu_M^*)] \tag{7.12}$$

where $\Delta\mu_i^*$ is the *solvation* Gibbs energy of the component i. This will be defined more precisely in section 7.2. It is clear that in order to obtain the standard

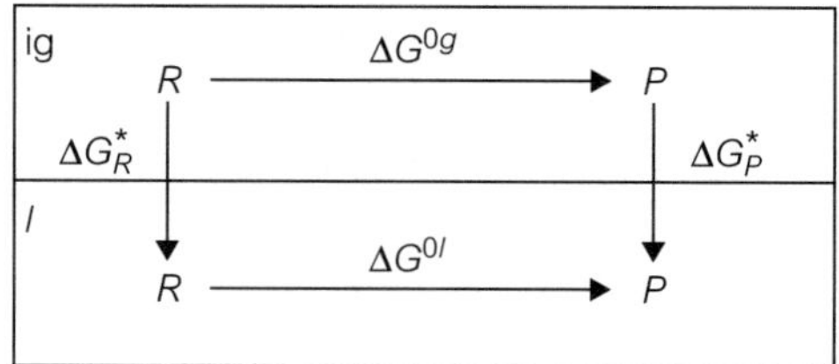

Figure 7.1 Schematic diagram showing the relation between standard free energies of reaction in two phases (ideal gas and a liquid), and the corresponding solvation Gibbs energies of all the species involved in the reaction.

Gibbs energy in the liquid phase, we have to transfer each of the species from the ideal-gas phase into the liquid. This transfer involves the solvation Gibbs energies of all the species participating in the reaction.

The second situation, where the solvation Gibbs energy features naturally, is in the distribution of a solute s between two phases α and β. Let a solute s be distributed between two phases α and β. At equilibrium, we have

$$\mu_s^\alpha = \mu_s^\beta. \tag{7.13}$$

Hence,

$$\left(\frac{\rho_s^\beta}{\rho_s^\alpha}\right)_{\text{eq}} = \exp[-\beta(\Delta\mu_s^{*\beta} - \Delta\mu_s^{*\alpha})]. \tag{7.14}$$

Thus the distribution of s between the two phases α and β is determined by the difference in the solvation Gibbs energies of s in the two phases. As we shall see in the next sections, relation (7.14) is actually used to *measure* the solvation Gibbs energy when one of the phases is chosen to be an ideal-gas phase. Thus, when α is an ideal-gas phase then

$$\left(\frac{\rho_s^l}{\rho_s^{ig}}\right)_{\text{eq}} = \exp[-\beta\Delta\mu_s^{*l})] \tag{7.15}$$

which *defines* $\Delta\mu_s^{*l}$ of s in the liquid l. We shall discuss this definition of the solvation quantities in detail in the next section.

Beyond the importance of the solvation Gibbs energy in calculating equilibrium constants in solutions, or solubilities (which is also an equilibrium constant), the solvation Gibbs energy and its derivatives are themselves valuable tools for studying the interaction between the solute (or any molecule) and the solvent (or any liquid in which the molecule is immersed), and the effect of the inserted molecule on the solvent, such as structural changes in the solvent or the effect of the solute on various molecular distribution functions of the solvent. As we shall see throughout this chapter, solvation quantities are powerful tools for probing the effect of a solute on the solvent at a *local* level, i.e., in the immediate surroundings of the solutes.

In the aforementioned paragraphs, we have referred to the solute and solvent in their traditional meanings, i.e., when one component, the solute, is relatively diluted in the solvent. As we shall see in the next section, the process of solvation and the corresponding thermodynamic quantities can be defined for *any* molecule in any medium. This has infinitely increased the range of systems for which this concept could be defined and applied.

7.2 **Definition of the solvation process and the corresponding solvation thermodynamics**

For the sake of convenience, we *define* the process of solvation using the following simple thought experiment. We transfer a molecule s from a *fixed* position in an ideal gas phase into a *fixed* position in the liquid phase[†], Figure 7.2.

If s is a spherical particle, such as argon, then we place the center of s at the fixed position. When s is not spherical, then we may choose the center of mass of the molecule, or any other convenient point, say the center of the oxygen atom in water, to be the center of the molecule[‡].

Clearly, the process as defined above cannot be carried out experimentally, but since we consider a classical system, it is meaningful to think of a particle lacking any translational degrees of freedom, i.e., at fixed position and zero velocity. As we shall see later in this section, this thought experiment is convenient but not essential for constructing the thermodynamic quantities associated with the process of solvation. Once we have established the meaning of the thermodynamic quantities defined below, we can do away with this thought experiment. Instead, we can just imagine fixing the origin of our coordinate system at the center of the inserted particle.

We start with the general expression for the chemical potential of s in any liquid system which reads

$$\mu_s^l = \mu_s(T, P, \mathbf{N}) = \mu_s^{*l}(T, P, \mathbf{N}) + kT \ln \rho_s \Lambda_s^3. \tag{7.16}$$

We use here the independent variables T, P, $\mathbf{N}$. These are the most common variables that are controlled in solution chemistry. In writing (7.16), we

[†] Since all points in the system are considered equivalent except for some negligible region near the boundaries of our system, we do not need to specify the location of the fixed point.

[‡] We also assume in most of this chapter that the molecule is rigid or nearly rigid. In sections 7.8 and 7.11, we shall also treat molecules with internal rotations, such as butane or a protein. In this case, one must define an average solvation Gibbs energy, averaged over all internal configurations.

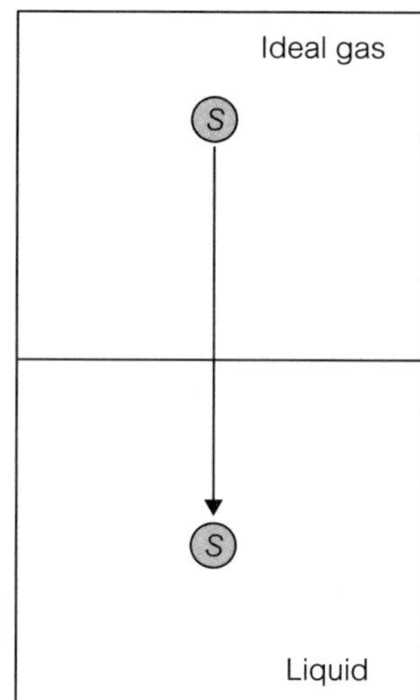

Figure 7.2 Definition of the solvation process. *A* solute *s* is transferred from a fixed position in an ideal gas phase to a fixed position in a liquid.

have already committed ourselves to systems where the translational degrees of freedom are classical (Λ_s^3 arises from the integration over the classical, Maxwell–Boltzmann, distribution of momenta). Any other internal degrees of freedom that s might have can be treated either classically or quantum mechanically. The corresponding *internal* partition function of s is included in q_s, and q_s is absorbed by μ_s^{*l}.

In section 3.4 we also found a convenient interpretation of the two terms in equation (7.16). The process of *adding* one s to the system is performed in two steps, figure 3.1. First, we place s at a fixed position in the system. The corresponding change in Gibbs energy is μ_s^* referred to as the pseudo-chemical potential of s. Second, we release the particle from the constraint of being at a fixed position. We call this the liberation process. The corresponding change in Gibbs energy is $kT \ln \rho_s \Lambda_s^3$. We have also shown that there are three, conceptually different, contributions to the liberation free energy. One is due to the acquiring of momenta; the second is due to gaining access to the entire volume V of the system; the third is due to the process of assimilation of the newly added s particles into the community of the already existing N_s particles of the same species. Altogether, the liberation Gibbs energy must be negative[†].

It is also convenient to introduce the concept of a *solvaton*. The solvaton s is simply that particular particle which we have added to our system. As long as it is outside the system, i.e., before its insertion, or when it is at some fixed position, it is distinguishable from all other particles of the same species in the system; otherwise it is *identical* to all N_s particles. Once we have released the

[†] This is the condition for which the classical partition function is valid.

solvaton from its fixed position, it loses its distinguishability. Alternatively we may think of the solvaton as one selected particle at the center of which we place our origin of the coordinate systems. All the locations of the other particles are defined with respect to this origin.

If s is in an ideal gas phase, then its chemical potential has the form

$$\begin{aligned}\mu_s^{\mathrm{ig}} &= \mu_s^{*\mathrm{ig}} + kT\ln\rho_s\Lambda_s^3 \\ &= -kT\ln q_s + kT\ln\rho_s\Lambda_s^3. \end{aligned} \tag{7.17}$$

Again, we may interpret the first term on the rhs of (7.17) as the change in Gibbs energy for the process of placing s at a fixed position in the ideal-gas phase. Clearly, in this process the particle carries with it all its internal degrees of freedom. By definition of the ideal-gas system, there are no interactions between s and any of the particles in the system.We now define the solvation Gibbs energy associated with the solvation process by the difference

$$\Delta\mu_s^* = \Delta G_s^*(\mathrm{ig}\to\mathrm{l}) = \mu_s^{*l} - \mu_s^{*\mathrm{ig}}. \tag{7.18}$$

Note that in writing (7.16) and (7.17), we have assumed that Λ_s^3 is the same in the two phases. Again, classically speaking, the momentum partition function depends only on the temperature and is not affected by the interaction of the solvaton with the rest of the system. Any other degrees of freedom might or might not be affected by the interactions. In most sections of this chapter, we assume that q_s is unaffected by the interaction of the solvaton with the rest of the system. Hence, in this case, (7.18) reduces to

$$\Delta\mu_s^* = \Delta G_s^*(\mathrm{ig}\to\mathrm{l}) = W(s|l) = -kT\ln\langle\exp[-\beta B_s]\rangle_0 \tag{7.19}$$

i.e., the solvation Gibbs energy is the coupling work of s to the rest of the system. This is the average of the quantity $\exp[-\beta B_s]$, where B_s is the binding energy of s to the rest of the system at some fixed configuration. The average is taken over all the configurations of the molecules in the system, in the T, P, N ensemble, using the distribution function of configurations of all the particles in the system *before* the addition of s; this average is indicated by the symbol $\langle\ \rangle_0$.

In the more general case, when s is inserted into the system, some internal degrees of freedom might change. Hence, $\Delta\mu_s^*$, in general, includes both the coupling work and the effect of the solvation process on the internal degrees of freedom. We shall discuss an example of this in section 7.8, but from hereon we shall assume that q_s (as well as Λ_s^3) is not affected by the process of solvation. To avoid confusion, two comments are in order:

First, $\Delta\mu_s^*$ is in general *not* the excess chemical potential of s, neither with respect to ideal gas, nor with respect to dilute ideal solutions (see sections 6.1 and 6.3, respectively). Unfortunately, $\Delta\mu_s^*$ is sometimes referred to as the excess chemical potential of the solute. The two excess chemical potentials are defined as

$$\mu_s^{\text{EX,IG}} = \mu_s^l - kT \ln \rho_s \Lambda_s^3 q_s^{-1} \tag{7.20}$$

$$\mu_s^{\text{EX,DI}} = \mu_s^l - [\mu_s^{0\rho} + kT \ln \rho_s]. \tag{7.21}$$

When q_s is *unaffected* by the interactions, we have

$$\mu_s^l = W(s|l) + kT \ln \rho_s \Lambda_s^3 q_s^{-1} = \mu_s^* + kT \ln \rho_s \Lambda_s^3. \tag{7.22}$$

Hence, (7.20) and (7.21) reduce to

$$\mu_s^{\text{EX,IG}} = W(s|l) = \Delta\mu_s^{*l} \tag{7.23}$$

$$\mu_s^{\text{EX,DI}} = W(s|l) - \lim_{\rho_s \to 0} W(s|l) = \Delta\mu_s^{*l} - \lim_{\rho_s \to 0} \Delta\mu_s^{*l}. \tag{7.24}$$

Thus, only when q_s is unaffected by the solvation process, $\mu_s^{\text{EX,IG}}$ becomes identical with the solvation Gibbs energy, whereas $\mu_s^{\text{EX,DI}}$ is reduced to the difference in the solvation Gibbs energy in the actual system and in the limit of the dilute ideal system with respect to *s*.

The second comment concerns the choice of *standard states*. Clearly, in defining the process of solvation, one must specify the thermodynamic variables under which the process is carried out. Here we used the temperature *T*, the pressure *P*, and the composition $N_1, \ldots, N_c$ of the system into which we added the solvaton. In the traditional definitions of solvation, one needs to specify, in addition to these variables, a *standard state* for the *solute* in both the ideal gas phase and in the liquid phase. In our definition, there is no need to specify any standard state for the solvaton. This is quite clear from the definition of the solvation process yet there exists some confusion in the literature regarding the "standard state" involved in the definition of the solvation process. The confusion arises from the fact that $\Delta\mu_s^*$ is *determined experimentally* in a similar way as one of the conventional standard Gibbs energy of solvation. The latter does involve a choice of standard state, but the solvation process as defined in this section does not. For more details, see the next two sections.

We have used the variables *T, P,* **N** to define, the chemical potential and the solvation process. These are the most common variables used in practice. However, one can define solvation quantities in any other ensemble. Sometimes it is more convenient in theoretical work to use the *T, V,* **N** ensemble.

Note that $\Delta\mu_s^*$ as defined in (7.19) may be referred to as the *free energy of interaction* of *s* with the system. This should be clearly distinguished from the *average interaction energy between s* and *l.*

Having defined the Gibbs energy of solvation, we can derive all the other thermodynamic quantities of solvation by using standard thermodynamic relationships. The most important quantities are the first derivatives of the Gibbs energy, i.e.,

$$\Delta S_s^* = -\left(\frac{\partial \Delta G_s^*}{\partial T}\right)_P \tag{7.25}$$

$$\Delta H_s^* = \Delta G_s^* + T\Delta S_s^* \tag{7.26}$$

$$\Delta V_s^* = \left(\frac{\partial \Delta G_s^*}{\partial P}\right)_T. \tag{7.27}$$

The Helmholtz energy of solvation is given by $\Delta A_s^* = \Delta G_s^* - P\Delta V_s^*$ and the internal energy of solvation is $\Delta E_s^* = \Delta H_s^* - P\Delta V_s^*$. However, for most of the specific systems discussed in this book, the term $P\Delta V_s^*$ is usually negligible with respect to ΔH_s^* or ΔG_s^*. Therefore, the distinction between ΔG_s^* and ΔA_s^* or between ΔH_s^* and ΔE_s^* is usually insignificant. For more details see Ben-Naim (1987).

Clearly, all the thermodynamic quantities associated with the solvation process, as defined above, pertain to exactly the *same* process. We stress again that the process of solvation is not experimentally feasible, i.e., we cannot carry out this process in the laboratory. For this reason, it cannot be handled within the realm of classical thermodynamics. Fortunately, as we shall see in the following section, statistical mechanics does provide a simple connection between solvation quantities and experimentally measurable quantities.

7.3 Extracting the thermodynamic quantities of solvation from experimental data

In this section we turn to the question of evaluating the pertinent thermodynamic quantities from experimental data. We discuss in this section the case of a solvaton s which does not undergo dissociation in any of the phases. The more complicated case of dissociated solvatons (such as ionic solutes) will be discussed separately in section 7.9.

Consider two phases α and β in which s molecules are distributed. We do not impose any restrictions on the concentration of s in the two phases. At equilibrium, assuming that the two phases are at the same temperature and pressure, we have the following equation for the chemical potential of s in two phases:

$$\mu_s^\alpha = \mu_s^\beta. \tag{7.28}$$

In the traditional thermodynamic treatment, one usually imposes the restriction of a very dilute solution of s in two phases. However, here we shall use the general expression (7.16) for the chemical potential of s in the two phases, to rewrite (7.28) as

$$\mu_s^{*\alpha} + kT \ln \rho_s^\alpha = \mu_s^{*\beta} + kT \ln \rho_s^\beta \tag{7.29}$$

From (7.29) we obtain

$$\begin{aligned}\Delta G_s^{*\beta} - \Delta G_s^{*\alpha} &= (\mu_s^{*\beta} - \mu_s^{*ig}) - (\mu_s^{*\alpha} - \mu_s^{*ig}) \\ &= \mu_s^{*\beta} - \mu_s^{*\alpha} \\ &= kT \ln \left(\frac{\rho_s^\alpha}{\rho_s^\beta} \right)_{eq}\end{aligned} \tag{7.30}$$

where ρ_s^α and ρ_s^β are the number densities of s in the two phases, *at equilibrium* (eq). Relation (7.30) provides a very simple way of computing the difference in the solvation Gibbs energies of s in the two phases α and β, from the measurement of the two densities ρ_s^α and ρ_s^β at equilibrium. A specific case of (7.30) occurs when one of the phases, say a, is an ideal gas. In such a case $\Delta G_s^{*\alpha} = 0$ and relation (7.30) reduces to

$$\Delta \mu_s^{*\beta} = \Delta G_s^{*\beta} = kT \ln \left(\frac{\rho_s^{ig}}{\rho_s^\beta} \right)_{eq}. \tag{7.31}$$

It is important to note that there are no restrictions on the density of s in phase β, but ρ_s^{ig} must be low enough to ensure that this phase is an ideal gas†. A specific example is a liquid–vapor equilibrium in a one-component system. If the vapor pressure is low enough, we may safely assume that the vapor behaves as an ideal gas. In such a case we rewrite equation (7.31) as

$$\Delta \mu_s^{*p} = \Delta G_s^{*p} = kT \ln \left(\frac{\rho_s^{ig}}{\rho_s^p} \right)_{eq} \tag{7.32}$$

where $\Delta \mu_s^{*p}$ or (ΔG_s^{*p}) is the solvation Gibbs energy of s in its own *pure* liquid s. As we shall see in the next section, this quantity cannot be studied within the traditional approach to solvation.

Another limiting case is the very dilute solution of s in phase β, say, argon in water, for which we have the limiting form of equation (7.31) which reads

$$\Delta \mu_s^{*0} = \Delta G_s^{*0} = kT \ln \left(\frac{\rho_s^{ig}}{\rho_s^\beta} \right)_{eq}. \tag{7.33}$$

† We cannot use here the theoretical definition of an ideal gas since we are concerned with experimentally determinable quantities.

This relation is identical in *form* to an equation derived from thermodynamics.

However, we stress that *conceptually*, the two relations differ from each other. Further elaboration on this point is given in the next section.

Relation (7.31) [as well as (7.32) and (7.33)], provide a simple and convenient way of determining $\Delta\mu_s^*$ from experimental data. It is fortunate that statistical mechanical considerations have provided us with means of *measuring* a quantity which pertains to an unfeasible process.

Having obtained the Gibbs energy of solvation through one of the relations cited above, it is a straightforward matter to calculate other thermodynamic quantities of solvation using the standard relations. For instance, the solvation entropy can be calculated from the temperature dependence of the $\Delta G_s^{*\beta}$, i.e.,

$$\Delta S_s^{*\beta} = -\left\{\frac{\partial}{\partial T}\left[kT\ln\left(\frac{\rho_s^{ig}}{\rho_s^{\beta}}\right)_{eq}\right]\right\}_P. \tag{7.34}$$

The differentiation in this equation is carried out at constant pressure P. One must distinguish between this derivative and the derivative along the liquid–vapor equilibrium line. The relation between the two quantities is discussed in section 7.6.

Using standard thermodynamic relationships, we can derive all the thermodynamic quantities of solvation from experimental data using equation (7.31).

7.4 Conventional standard Gibbs energy of solution and the solvation Gibbs energy

In this section, we present a detailed comparison between the *solvation* quantities as defined in section 7.2 and the conventional standard thermodynamic quantities of solutions. The latter are also referred to as solvation quantities. As we shall demonstrate in this section, the conventional quantities are always restrictive measures of solvation quantities, sometimes even inadequate measures of solvation.

There are quite a few conventional quantities that have been employed in the literature. We shall discuss in this section only three of these, which we believe to be the most frequently used. In order to avoid any (understandable) confusion, a special notation will be used for the various *conventional* processes (i.e., x-process, ρ-process, etc.) The superscript asterisk is reserved for the solvation process as defined in section 7.2.

Let C_s be any concentration unit utilized to measure the concentration of s in the system. The most common units are the molarity ρ_s, the molality m_s, and the mole fraction x_s. We shall often refer to ρ_s as either the molar or the number density. The two differ by the Avogadro number. It should be clear from the context as to which of these we are referring to in a particular case.

The conventional thermodynamic approach always applies to the limit of a very dilute solution of s, where the chemical potential of the solute s can be written as

$$\mu_s = \mu_s^{0c} + kT \ln C_s \tag{7.35}$$

where μ_s^{0c} is referred to as the *standard* chemical potential of s, based on the concentration scale of C. This is formally defined as the limit

$$\mu_s^{0c} = \lim_{C_s \to 0} (\mu_s - kT \ln C_s) \tag{7.36}$$

but it is interpreted using (7.35) as the chemical potential of s in a "standard" state where $C_s = 1$. In general, one cannot guarantee that in this standard state, the system is DI[†]. Hence, the meaning assigned to the μ_s^{0c} is dubious.

In all the conventional processes to be discussed below, it is important to bear in mind that the so-called (conventional) standard quantities only apply to *very dilute solutions* of s in the system. This is a very severe restriction on the applicability of the standard quantities defined below.

Let α and β be two phases in which the concentrations of s are C_s^α and C_s^β, respectively. If the limiting expression (7.35) applies to both phases, we may define the Gibbs energy change for the process of transferring *one s* (or one mole of s) from α to β as

$$\Delta G \begin{bmatrix} \alpha \to \beta \\ C_s^\alpha, C_s^\beta \end{bmatrix} = \mu_s^\beta - \mu_s^\alpha = \mu_s^{0c\beta} - \mu_s^{0c\alpha} + kT \ln(C_s^\beta / C_s^\alpha). \tag{7.37}$$

We shall always assume that the temperature and the pressure are the same in the two phases. Hence, these will be omitted from our notation. On the left-hand side of relation (7.37), we do specify the concentrations of s in the two phases. These can be chosen freely as long as they are within the range of validity of equation (7.35).

Next, one makes a choice of a *standard process*. In principle, one can choose any specific values of C_s^α and C_s^β to characterize this standard process. The three most commonly employed choices in the literature are the following.

(1) *The ρ-process.* This is the process of transferring one s molecule from an ideal-gas phase into a dilute ideal solution (Henry's law) at fixed temperature

[†] In fact, one cannot guarantee that such a *physical* state exists at all. I owe this comment to Dr. R.M. Mazo.

and pressure and such that $\rho_s^l = \rho_s^g$. This is essentially a special case of the general process described in (7.37) with the choice of number (or molar) concentration units for C_s i.e., $C_s^\beta = \rho_s^l, C_s^\alpha = \rho_s^g$. The corresponding Gibbs energy change is

$$\Delta G_s^{0\rho} = \Delta G_s\,(\rho\text{-process}) = \mu_s^{0\rho l} - \mu_s^{0\rho g} \tag{7.38}$$

where ΔG_s (ρ-process) is a shorthand notation for

$$\Delta G\begin{pmatrix} g \to l \\ \rho_s^g = \rho_s^l \end{pmatrix}. \tag{7.39}$$

The superscript "$0\rho l$" stands for *standard*, ρ-units, and *liquid phase l.*

The relation between ΔG_s (ρ-process) and the solvation Gibbs energy ΔG_s^* may be obtained by applying relation (7.16) to the two phases:

$$\Delta G_s^{0\rho} = \Delta G_s(\rho\text{-process}) = \mu_s^{*l} - \mu_s^{*ig} = \Delta\mu_s^*. \tag{7.40}$$

Thus, the standard Gibbs energy of solvation $\Delta G_s^{0\rho}$ is equal to the solvation Gibbs energy. It is also determined experimentally in the same way as in (7.31). To see that, we write the chemical potential of s in the two phases in the traditional convention, valid only in the limit of ideal gas and ideal dilute solutions:

$$\mu_s^g = \mu_s^{0\rho g} + kT\ln\rho_s^g \tag{7.41}$$

$$\mu_s^l = \mu_s^{0\rho l} + kT\ln\rho_s^l. \tag{7.42}$$

At equilibrium (7.28) holds, hence from (7.41) and (7.42) it follows that

$$\Delta G_s^{0\rho} = \mu_s^{0\rho l} - \mu_s^{0\rho g} = -kT\ln\left(\rho_s^l/\rho_s^g\right)_{\text{eq}} \tag{7.43}$$

which is exactly the same as relationship as (7.31).

Relation (7.40) is quite remarkable. The apparent identity of the two free energy changes is deceiving, however. One should be careful in interpreting this relation as implying the identity of the two processes.

The reason for misinterpretation, which is commonly committed in the literature, is the following. The ρ-process and the solvation process are two *distinctly different* processes. It so happens that under very special conditions (s is very diluted in the two phases), their Gibbs energy changes are equal in magnitude as stated in (7.40). This has led some authors to identify $\Delta\mu_s^*$ with the Gibbs energy change for the ρ-process and actually refer to $\Delta\mu_s^*$ as the *standard* Gibbs energy of solvation based on the ρ standard states. This is not the case, however. First because $\Delta\mu_s^*$ is applicable to an infinitely larger range of concentrations than the very restricted range of applicability of ΔG_s (ρ-process). The *magnitude* of the two quantities happen to coincide only in the limit of $\rho_s \to 0$ in both phases. Second, some of the other thermodynamic quantities, as we shall see below, have different magnitudes for the two processes even in the limit of DI solutions in both phases. For instance, an equality of the form (7.40) does not exist for the entropy change even in the limit of DI solutions (see below).

(2) *The x-process.* This is defined as the process of transferring an s molecule from an ideal gas at 1 atm pressure to a hypothetical dilute-ideal solution in which the mole fraction of s is unity. (The temperature T and pressure of 1 atm are the same in the two phases).†

The relation between the standard Gibbs energy of the x-process and the solvation Gibbs energy is obtained from the general expression (7.16) for the chemical potential. Thus,

$$\Delta G_s^{0x} = \Delta G_s(x\text{-process}) = [\mu_s^{*l} - \mu_s^{*g} + kT\ln(\rho_s^l/\rho_s^g)]_{\substack{P_s=1\\x_s=1}} \tag{7.44}$$

where we must substitute $P_s = 1$ atm and $x_s = 1$ in equation (7.44). To achieve that, we transform variables as follows. Since the x-process applies for ideal gases, we have

$$\rho_s^g = P_s/kT \tag{7.45}$$

where P_s is the partial pressure of s in the gaseous phase. Furthermore, the x-process applies to dilute-ideal solutions; hence, we may write

$$x_s^l = \frac{\rho_s^l}{\sum \rho_i^l} \approx \frac{\rho_s^l}{\rho_B^l}, \qquad \rho_s \to 0 \tag{7.46}$$

where ρ_B^l is the number density of the solvent B, which in principle can be a mixture of many components.

After this transformation of the variables, we rewrite equation (7.44) in the form

$$\begin{aligned}\Delta G_s^{0x} = \Delta G_s(x\text{-process}) &= [\mu_s^{*l} - \mu_s^{*g} + kT\ln(x_s^l\rho_B^l kT/P_s)]_{\substack{P_s=1\\x_s=1}}\\ &= \mu_s^{*l} - \mu_s^{*g} + kT\ln(kT\rho_B^l)\\ &= \Delta G_s^* + kT\ln(kT\rho_B^l).\end{aligned} \tag{7.47}$$

Note that since we put $P_s = 1$ atm, $kT\rho_B^l$ must also be expressed in units of atmospheres. This renders the argument of the logarithm a dimensionless quantity.

The connection between ΔG_s^{0x} and experimental data is similar to the connection (7.43). One writes the chemical potentials of s in the two phases, in

† The hypothetical dilute ideal state with $x_s = 1$ is awkward, to say the least. More on this can be found in Ben-Naim (1978).

the traditional form

$$\mu_s^g = \mu_s^{0g} + kT \ln P_s \tag{7.48}$$

$$\mu_s^l = \mu_s^{0xl} + kT \ln x_s^l \tag{7.49}$$

and imposes the equilibrium condition (7.28) to obtain

$$\Delta G_s^{0x} = \mu_s^{0xl} - \mu_s^{0g} = -kT \ln\left(\frac{P_s}{x_s}\right)_{\text{eq}}. \tag{7.50}$$

Clearly, in this case, we do not have an equality of the type (7.40). In (7.47), we see that ΔG_s^{0x} and ΔG_s^* are, in general, different quantities, pertaining to two distinctly different processes.

Relationship (7.47) clearly shows that ΔG_s^{0x} could be either larger or smaller than ΔG_s^*, depending on whether $kT\rho_B^l$ is larger or smaller than 1 atm. One could adopt ΔG_s^{0x} as a measure of solvation Gibbs energy shifted by the quantity $kT \ln(kT\rho_B^l)$. This is unacceptable, however, for reasons that could not have been noticed within the traditional approach to solvation. The reason is that in the limit of very small ρ_B^l, $\Delta\mu_s^*$ must tend to zero since in this limit the average of the quantity $\exp[-\beta B_s]$ will tend to unity; hence, $\Delta\mu_s^* \to 0$. On the other hand, ΔG_s^{0x} will diverge to minus infinity. This renders ΔG_s^{0x} invalid as a measure of the solvation Gibbs energy. Unfortunately, this fact was both elusive to the traditional approach to solvation as well as to the most trained eyes of practitioners in this field.

(3) The *m-process.* This is defined as the process of transferring one s molecule from an ideal-gas phase at 1 atm pressure to a hypothetical ideal solution in which the *molality* of s is unity[†]. (The temperature T and the pressure of 1 atm are the same in the two phases.)

The basic connection between the Gibbs energies of the *m-process* and the solvation process is again obtained from equation (7.16). The result is

$$\Delta G_s^{0m} = \Delta G_s(m\text{-process}) = [\mu_s^{*l} - \mu_s^{*g} + kT \ln(\rho_s^l/\rho_s^g)]_{\substack{P_s=1\\ m_s=1}}. \tag{7.51}$$

Assuming that the gas is ideal, we can use the transformation (7.45). Furthermore, for very dilute solutions of s and assuming for simplicity that the solvent B is a one-component liquid with density ρ_B^l and molecular mass M_B, we can write the transformation from ρ_s^l into molality units m_s as follows:

$$\rho_s^l = M_B \rho_B^l m_s/1000. \tag{7.52}$$

[†] As in the previous case, the hypothetical state of dilute ideal solution at 1 molality is awkward since at this concentration, in reality, the system would not be dilute ideal.

Substitution of relations (7.45) and (7.52) into equation (7.51) yields

$$\begin{aligned}\Delta G_s^{0m} &= \Delta G_s(m\text{-process}) \\ &= [\mu_s^{*l} - \mu_s^{*g} + kT\ln(M_B\rho_B^l m_s kT/1000P_s)]_{\substack{P_s=1\\ m_s=1}} \\ &= \Delta G_s^* + kT\ln(M_B\rho_B^l kT/1000)]\end{aligned} \tag{7.53}$$

which is the required relation between the Gibbs energy changes for the m-process and for the solvation process.

We now summarize the three relationships between the various Gibbs energies of solution and the Gibbs energy of solvation:

$$\Delta G_s^{0\rho} = \Delta G_s(\rho\text{-process}) = \Delta G_s^* \tag{7.54}$$

$$\Delta G_s^{0x} = \Delta G_s(x\text{-process}) = \Delta G_s^* + kT\ln(kT\rho_B^l) \tag{7.55}$$

$$\Delta G_s^{0m} = \Delta G_s(m\text{-process}) = \Delta G_s^* + kT\ln(M_B\rho_B^l kT/1000). \tag{7.56}$$

Note that in both (7.55) and (7.56), the argument in brackets under the logarithm sign must be rendered dimensionless, consistent with the substitution of 1 atm in (7.55), and 1 atm and 1 molal in (7.56).

Some general comments are now in order. First, we stress that all the three relations (7.54)–(7.56) are valid for the limiting case when one phase is an ideal gas and one phase is dilute ideal solution with respect to the solute s. It is only in this case that the three traditional standard quantities are defined and applicable. ΔG_s^*, on the other hand, is defined and applicable for all the concentration range of s, from $\rho_s = 0$ up to the concentration of pure s, ρ_s^P.

Second, the equality in (7.54) is somewhat deceptive; it is an equality between two quantities which pertain to two *different* processes and which happen to have the same magnitude at very specific conditions of ideality. For most concentrations of ρ_s in either phase, no such equality exists.

Third, the equality of the Gibbs energies for the ρ-process and the solvation process do not imply equality between any other thermodynamic quantities pertaining to these two processes. One must exercise extreme care in deriving the relations between, say, the standard entropy of the ρ-process and the entropy of solvation; these cannot be obtained by taking the temperature derivative of equation (7.54). As we shall see in the next section, this is a tricky point which has been overlooked even by experts working in this field.

Finally, we note that in both (7.55) and (7.56) the term $kT \ln kT\rho_B$ originates from the liberation Gibbs energy (or the translation Gibbs energy) of the solvent B. The solvation Gibbs energy ΔG_s^* does not include any contribution due to the translational free energy of either the solute or the solvent. From (7.54) it is clear that also $\Delta G_s(\rho\text{-process})$ is devoid of any contribution due to the translational free energy[†]. As we shall soon see, the entropy change corresponding to the ρ-process does include the temperature derivative of the solvent density ρ_B.

In relations (7.55) and (7.56), we see that the difference between the Gibbs energy of the thermodynamic process (either x-process or m-process), and the solvation process is a constant quantity depending on the properties of the solvent. This means that if we wish to compare *various solutes* in the *same* solvent B we may disregard the constant quantities on the rhs of equations (7.55) and (7.57). But what if we wish to study a *single solute* in *various* solvents? Here we encounter a serious problem because of the special way the solvent density features in the quantities on the rhs of equations (7.55) and (7.56)[‡]. To demonstrate the difficulty, suppose we wish to study the solvation Gibbs energy of a given solute s in a series of solvents B having decreasing densities ρ_B. It can easily be shown that ΔG_s^* will tend to zero as $\rho_B^l \rightarrow 0$. This is clearly the behavior we should expect from a quantity that measures the extent of the Gibbs energy of *interaction* between s and its environment. In the extreme case when $\rho_B^l = 0$, we have $\Delta G_s^* = 0$ as it should be! On the other hand, a glance at equations (7.55) and (7.56) show that when $\rho_B^l \rightarrow 0$ both ΔG_s (x-process) and ΔG_s (m-process) will tend to minus infinity[¶]. This behavior is clearly unacceptable for a quantity that is presumed to measure the Gibbs energy of *interaction* between s and its environment. We shall see below that such divergent behavior is exhibited by other thermodynamic quantities corresponding to the x-process and the m-process. It is for this reason that both of these quantities cannot, in principle, serve as *bona fide* measures of the solvation Gibbs energy.

[†] In discussing the various standard states, Friedman and Krishnan (1973) commented that "At an elementary level, the choice of the standard state in equation (16) (referring to the ρ-process) eliminates the translational entropy contribution to $\Delta G(\rho\text{-process})$, but a *deeper* analysis shows that this is not really so." In spite of my correspondence with these authors, I still do not know what that "deeper analysis" is and why that comment has been made.

[‡] Arnett and McKelvey (1969) found that the standard free energy of transferring propane from H_2O to D_2O have different *signs* if calculated using the mole fraction or molality scale. They referred to this finding as a "shocking example," and indeed it is. Thus, within the conventional standard quantities of solvation, one could not tell even the sign of the change when passing from one solvent to another.

[¶] It is ironic to note that Tanford, who has enthusiastically advocated the use of the mole fraction scale (Tanford 1973; Reynolds et al. 1974), reacted to my publication (Ben-Naim 1978), by saying that "... those who dismiss work of this kind on the basis of *second-order terms* in theoretical equations ..." (See Tanford 1979, and Ben-Naim 1979). Divergence to infinity is deemed to be "second order terms!"

7.5 **Other thermodynamic quantities of solvation**

In this section, we derive some more relations between the solvation quantities and standard thermodynamic quantities of solution.

7.5.1 **Entropy**

As before, our starting point is the general equation (7.16) for the chemical potential. The partial molar (or molecular) entropy of s is

$$\overline{S}_s = \frac{-\partial \mu_s}{\partial T} = \frac{-\partial \mu_s^*}{\partial T} - k \ln \rho_s \Lambda_s^3 - \frac{kT \partial \ln(\rho_s \Lambda_s^3)}{\partial T}. \tag{7.57}$$

All the differentiations are taken at constant pressure and composition of the system.

We denote by S_s^* the entropy change corresponding to the process of adding one s molecule to a fixed position in the system. On performing the differentiation with respect to temperature in equation (7.57), we obtain

$$\overline{S}_s = S_s^* - k \ln \rho_s \Lambda_s^3 + kT\alpha_p + \tfrac{3}{2}k \tag{7.58}$$

where $\alpha_p = V^{-1}\partial V/\partial T$ is the thermal expansion coefficient of the system at constant pressure.

Applying equation (7.58) for an ideal-gas phase and for an ideal dilute solution, we may derive the entropy changes associated with the *standard* processes as defined above. For the solvation process we simply have the relation

$$\Delta S_s^* = S_s^{*l} - S_s^{*g} = \frac{-\partial \Delta G_s^*}{\partial T}. \tag{7.59}$$

(In most cases, the superscript g for "gas" is understood to stand for ideal gas. If the gaseous phase is not ideal, then ΔS_s^* is the difference in the solvation entropy between the two phases.)

For the ρ-process, we have

$$\begin{aligned} \Delta S_s(\rho\text{-process}) &= [\overline{S}_s^l - \overline{S}_s^g]_{\rho_s^l=\rho_s^g} \\ &= S_s^{*l} - S_s^{*g} + kT\alpha_p^l - kT\alpha_p^g \\ &= \Delta S_s^* + kT\alpha_p^l - k \end{aligned} \tag{7.60}$$

where $\alpha_p^g = T^{-1}$ for the ideal-gas phase.

Similarly, for the x-process and the m-process we have, respectively.

$$\begin{aligned} \Delta S_s(x\text{-process}) &= [\overline{S}_s^l - \overline{S}_s^g]_{\substack{P_s=1 \\ x_s=1}} = S_s^{*l} - S_s^{*g} - k\ln(kT\rho_B^l) + kT\alpha_p^l - k \\ &= \Delta S_s^* - k\ln(kT\rho_B^l) + kT\alpha_p^l - k \end{aligned} \tag{7.61}$$

and

$$\Delta S_s(m\text{-process}) = [\overline{S}_s^l - \overline{S}_s^g]_{\substack{P_s=1\\ m_s=1}}$$
$$= \Delta S_s^* - k \ln(M_B \rho_B^l kT/1000) + kT\alpha_p^l - k. \tag{7.62}$$

We see that in contrast to the case of the solvation Gibbs energies where we encountered only *three* different quantities which correspond to *four* distinctly different processes, here we have *four* different quantities. The entropy change for the ρ-process is *unequal* to the entropy of solvation. Hence, one *cannot* identify the solvation process with the ρ-process. Unfortunately, such an identification is frequently made in the literature. We also note that ΔS_s (ρ-process) cannot be obtained by direct differentiation of $\Delta G_s(\rho\text{-process})$ with respect to the temperature. When we take the temperature derivative of $\Delta G_s^{0\rho}$ we *do not* get the entropy change for the ρ-process. The reason is quite subtle and has to do with the choice of standard states. We shall elaborate further on this aspect.

We recall that the pseudo-chemical potential was defined as the Gibbs energy change for the process of inserting s at a fixed position. Hence, the temperature derivative gives the entropy change for the same process, i.e.,

$$S_s^* = -\left(\frac{\partial \mu_s^*}{\partial T}\right)_{P,N}. \tag{7.63}$$

When we take the temperature derivative of μ_s^l with respect to temperature, we obtain

$$\overline{S}_s = -\left(\frac{\partial \mu_s}{\partial T}\right)_{P,N} = S_s^* - k \ln \rho_s \Lambda_s^3 - kT\frac{\partial \ln(\rho_s \Lambda_s^3)}{\partial T}. \tag{7.64}$$

We can read this equation as follows. The entropy change for adding one s particle to the system is composed of two parts: the entropy change associated with the process of placing the added solute s at a fixed position; this is S_s^*, a second term due to the liberation of the particle. The latter consists of the two terms on the rhs of (7.64). Equation (7.64) and the interpretation of the two contributions to the entropy just mentioned *holds true for any* ρ_s.

The situation is different when we write the expression for the chemical potential in the traditional convention, either (7.41) or (7.42). For example, when we take the temperature derivative of μ_s^l we obtain

$$\overline{S}_s = -\left(\frac{\partial \mu_s^l}{\partial T}\right)_{P,N} = -\frac{\partial \mu_s^{0\rho l}}{\partial T} - k \ln \rho_s - kT\frac{\partial \ln \rho_s}{\partial T}. \tag{7.65}$$

Normally, $\mu_s^{0\rho l}$ is defined as an integration constant and has no meaning as a chemical potential. However, $\mu_s^{0\rho l}$ has been traditionally interpreted as the chemical potential of s in the hypothetical dilute ideal solutions, for which $\rho_s = 1$. Clearly, in (7.65), we cannot interpret $-(\partial \mu_s^{0\rho l}/\partial T)$ as the entropy change corresponding to the process of adding an s particle to the system at $\rho_s = 1$. To do this, we must substitute $\rho_s = 1$ in (7.65) to obtain the required entropy change which we denote by

$$\overline{S}_s\,(\text{at}\,\rho_s = 1) = -\frac{\partial \mu_s^{0\rho l}}{\partial T} + \frac{kT}{V}\left(\frac{\partial V}{\partial T}\right)_{P,N}. \tag{7.66}$$

In other words, one cannot substitute $\rho_s = 1$ in (7.42) and then take the temperature derivative of $\mu_s^l(\rho_s = 1)$ to obtain the required entropy. Instead, one must *first* take the temperature derivative of (7.42) to obtain (7.65), *thereafter* substitute $\rho_s = 1$ to obtain the required entropy, (7.66). Repeating the same process for the two phases, we obtained relation (7.60). Actually, to obtain (7.60), we do not need to choose $\rho_s = 1$ for *each* phase. It is sufficient to require $\rho_s^l = \rho_s^g$ when we form the quantity ΔS_s (ρ-process). Similar procedures should be taken to obtain (7.61) and (7.62).

A glance at the expressions (7.61) and (7.62) for ΔS_s(x-process) and ΔS_s (m-process) shows that both contain the solvent density ρ_B^l under the logarithm sign. Thus, for a series of solvents with decreasing densities, both ΔS_s (x-process) and ΔS_s(m-process) will diverge to infinity, clearly an undesirable feature for a quantity that is presumed to measure the solvation entropy of a molecule s. On the other hand, ΔS_s^* tends to zero as the solvent density decreases to zero, as it should! In addition to this unacceptable behavior of ΔS_s(x-process) and ΔS_s(m-process), all of these standard entropies of transfer contain the term $kT\alpha_p^l - k$, which is irrelevant to the solvation process of the molecule s.

7.5.2 **Enthalpy**

The enthalpies of the various processes are obtained from the combination $\overline{H}_s = \mu_s + T\overline{S}_s$. The results are

$$\Delta H_s^* = H_s^{*l} - H_s^{*g} = \Delta G_s^* + T\Delta S_s^* \tag{7.67}$$

$$\begin{aligned}\Delta H_s(\rho\text{-process}) &= \Delta H_s(x\text{-process}) \\ &= \Delta H_s(m\text{-process}) \\ &= \Delta H_s^* + kT^2\alpha_p^l - kT.\end{aligned} \tag{7.68}$$

We see that the enthalpy changes for the three standard processes are identical. This follows from the assumption of ideality introduced in the definition of these processes. On the other hand, these three processes produce an enthalpy change which *differs* from the solvation enthalpy by the quantity $kT^2\alpha_p^l - kT$.

Although this quantity is not divergent as $\rho_B^l \to 0$, it is certainly irrelevant to the process of solvation.

7.5.3 **Volume**

Taking the derivative of the general expression for the chemical potential with respect to pressure, we obtain the partial molar (or molecular) volume of *s*, i.e.,

$$\begin{aligned} \overline{V}_s = \left(\frac{\partial \mu_s}{\partial P}\right)_T &= \left(\frac{\partial \mu_s^*}{\partial P}\right)_T + kT\left(\frac{\partial \ln \rho_s}{\partial P}\right)_T \\ &= V_s^* + kT\kappa_T^l \end{aligned} \tag{7.69}$$

where V_s^* is the volume change due to the addition of one *s* molecule at a fixed position in the system and κ_T^l is the isothermal compressibility of the phase *l*, defined as

$$\kappa_T^l = \frac{-1}{V}\left(\frac{\partial V}{\partial P}\right)_T \tag{7.70}$$

which, for an ideal-gas phase, reduces to

$$\kappa_T^{\text{ig}} = 1/P. \tag{7.71}$$

The volume changes for the four processes of interest are

$$\Delta V_s^* = \frac{\partial \Delta G_s^*}{\partial P} = V_s^{*l} - V_s^{*g} \tag{7.72}$$

$$\Delta V_s(\rho\text{-process}) = \Delta V_s^* + [kT(\kappa_T^l - P^{-1})]_{\rho_s^l = \rho_s^g} \tag{7.73}$$

$$\Delta V_s(x\text{-process}) = \Delta V_s(m\text{-process}) = \Delta V_s^* + kT(\kappa_T^l - 1/\text{atm}). \tag{7.74}$$

Note that the volume changes for the last two processes are identical. We note also that for the liquid phases at room temperature κ_T^l is much smaller than 1 atm^{-1} (e.g., for water at 0 °C, $kT\kappa_T \approx 1\,\text{cm}^3\,\text{mol}^{-1}$, $\Delta V_s^* \approx 20\,\text{cm}^3\,\text{mol}^{-1}$, and kT / atm $\approx 2 \times 10^4\,\text{cm}^3\,\text{mol}^{-1}$). Similarly, in equation (7.73) $\kappa_T^l \ll P^{-1}$ (the limit of an ideal-gas phase). Thus, the volume change for the three standard processes is dominated by the terms which originate from the ideal-gas compressibility. Because of this undesirable feature, it is common to abandon these processes when studying the volume of solvation. Almost all researchers who study the solvation phenomena apply one of these standard processes for quantities like the Gibbs energy, entropy and enthalpy of solvation, but for the volume of solvation they switch to partial molar volumes at infinite dilution. The latter clearly corresponds to a *different* process. Similar difficulties are encountered in the study of higher-order derivatives of the free energy.

No such difficulty arises in the study of the *solvation* process as defined in section 7.2. This can be applied *uniformly* to all the thermodynamic quantities of solvation. Obviously, this is a convenient feature of the solvation process which is not shared by the conventional standard processes.

From the quantities derived above, one may construct the internal energy of solvation ($\Delta E_s^* = \Delta H_s^* - P\Delta V_s^*$) and the Helmholtz energy of solvation ($\Delta A_s^* = \Delta G_s^* - P\Delta V_s^*$). As noted in section 7.2, the difference between ΔE_s^* and ΔH_s^* and between ΔA_s^* and ΔG_s^* is usually very small and may be neglected for most systems of interest discussed in this book. For more details see Ben-Naim (1987).

It is straightforward to go beyond first-order derivatives of the Gibbs energy. One can define the compressibility, heat capacity, thermal expansion, and so on, for the process of solvation. These quantities are of potential interest in the study of solvation phenomena.

We now recap the main differences between the two approaches to the study of solvation phenomena. First and foremost is the fact that the solvation process as defined in section 7.2 is the most direct tool of probing the free energy of interaction of a solvaton with its environment. All the thermodynamic quantities of solvation tend to zero when the solvent density goes to zero (i.e., when there are no interactions between the solvaton and its environment). This is not the case for the conventional thermodynamic quantities, some of which even diverge to plus or minus infinity in this limit. Furthermore, by adopting the solvation process we achieve both a generalization and a uniformity in application of this concept. The generalization involves the extension of the range of concentration of s for which the solvation thermodynamics may be studied, from the very dilute s up to pure liquid s. This immensely increases the range of systems which may be studied by means of solvation. The uniformity involves the application of the *same* process for all thermodynamic quantities. This is in sharp contrast to the conventional approach where different processes are used for studying different thermodynamic quantities.

Finally, we may add that once we adopt the definition of the solvation process as given in section 7.2, we can forget about all standard states. This is a drastic simplification compared to the traditional approach where in addition to specifying the thermodynamic variables of the system, one must also choose a standard state.

7.6 Further relationships between solvation thermodynamics and thermodynamic data

In section 7.3, we outlined the fundamental connections between solvation thermodynamics and experimental data. However, in many cases thermodynamic data are available that may be directly converted into solvation thermodynamic quantities without going through the fundamental relationships. Some of these transformations are presented here.

7.6.1 Very dilute solutions of *s* in *l*

These are the only systems for which studies of solvation, in the conventional meaning, have been carried out. For these systems there are numerous publications of tables of thermodynamics of solution (or solvation) which pertain to one of the processes discussed in section 7.4. All the conversion formulas for these cases have already been derived in the previous section. Here, we add one more connection with a very commonly used quantity, the Henry law constant. In its most common form it is defined by

$$K_H = \lim_{x_s^l \to 0} (P_s / x_s^l) \tag{7.75}$$

where P_s is the partial pressure and x_s^l is the mole fraction of s in the system and the limit takes x_s^l into the range where Henry's law becomes valid. The general expression for the solvation Gibbs energy in this case is

$$\Delta G_s^{*0} = kT \ln(\rho_s^{\text{ig}} / \rho_s^l)_{\text{eq}}. \tag{7.76}$$

Assuming that we have a sufficiently dilute solution of s in l such that Henry's law in the form $P_s = K_H x_s^l$ is obeyed, we can transform equation (7.76) into

$$\begin{aligned} \Delta G_s^{*0} &= kT \ln(P_s / kT x_s^l \rho_B^l) \\ &= kT \ln(K_H / kT \rho_B^l). \end{aligned} \tag{7.77}$$

This is a connection between the tabulated values of K_H as defined in (7.75), and the solvation Gibbs energy. (Here we have assumed, for simplicity, that the solvent consists of one component with a number density ρ_B^l.) We also note that from relations (7.52) and (7.77), we also have

$$\Delta G_s(x\text{-process}) = kT \ln K_H \tag{7.78}$$

i.e., information on K_H is essentially equivalent to information on the Gibbs energy change for the x-process.

7.6.2 **Concentrated solutions**

For solutions (or rather mixtures) of s at higher concentrations beyond the realms of Henry's law, we depart from the traditional notion of solvation and must use the definition as presented in section 7.2. There exists a variety of data which measures the extent of deviation from dilute ideal solutions. These include tables of activity coefficients, osmotic coefficients, and excess functions. All of these may be used to compute solvation thermodynamic quantities.

As always, our starting point is the general expression for the chemical potential of s in the liquid phase l, equation (7.16),

$$\mu_s = \mu_s^{*l} + kT \ln \rho_s^l \Lambda_s^3. \tag{7.79}$$

The chemical potential in the same system may be expressed in conventional thermodynamics as

$$\mu_s = \mu_s^{0\rho} + kT \ln \rho_s^l + kT \ln \gamma_s^{D,\rho} \tag{7.80}$$

where $\gamma_s^{D,\rho}$ is the activity coefficient which measures deviations with respect to the *ideal-dilute behavior*, based on the *number density* ρ_s as concentration units; $\mu_s^{0\rho}$ is the conventional standard chemical potential of s in the ρ-concentration scale, and is formally defined by

$$\mu_s^{0\rho} = \lim_{\rho_s^l \to 0} (\mu_s - kT \ln \rho_s^l). \tag{7.81}$$

Substitution of equation (7.79) into (7.81) yields

$$\mu_s^{0\rho} = \lim_{\rho_s^l \to 0} (\mu_s^{*l} + kT \ln \rho_s^l \Lambda_s^3 - kT \ln \rho_s^l) = \mu_s^{*0l} + kT \ln \Lambda_s^3 \tag{7.82}$$

which is the required connection between the *conventional* standard chemical potential $\mu_s^{0\rho}$, and the *pseudo*-chemical potential of s at infinite dilution μ_s^{*0l}. By using relations (7.79), and (7.80) and (7.81), we arrive at the final expression:

$$\mu_s^{*l} - \mu_s^{*0l} = kT \ln \gamma_s^{D,\rho}. \tag{7.83}$$

This quantity is equivalent to the difference between the solvation Gibbs energy of s in the phase l, and the solvation Gibbs energy of s in the same phase except for taking the limit $\rho_s \to 0$. Using the notation of section 7.2, we may rewrite this quantity as

$$\Delta G_s^* - \Delta G_s^{*0} = \mu_s^{*l} - \mu_s^{*0l} = kT \ln \gamma_s^{D,\rho}. \tag{7.84}$$

Thus, from the activity coefficient (based on the ρ-concentration scale), one can compute the solvation Gibbs energy of s in a liquid phase l (containing any quantity of s), relative to the solvation Gibbs energy of s in the same phase but as $\rho_s \to 0$. This quantity is quite useful whenever the infinite dilute limit is already known.

Relation (7.83) holds only for the activity coefficient as defined in equation (7.82), i.e., based on the number density scale. However, it is quite a simple matter to use any other activity coefficient to extract the same information. Let C_s be any other concentration units (e.g., molality, mole fraction, etc). We write the general conversion relation between C_s and ρ_s symbolically as

$$\rho_s = T_s^c C_s. \tag{7.85}$$

Where T_s^c is defined in (7.85)
The general expression of the chemical potential may be expressed in the C_s scale as

$$\mu_s = \mu_s^{*l} + kT \ln T_s^c C_s \Lambda_s^3. \tag{7.86}$$

In the limit of a very dilute solution $\rho_s \to 0$, we have

$$\mu_s = \mu_s^{*0l} + kT \ln T_s^{c,0} C_s \Lambda_s^3 \tag{7.87}$$

where

$$T_s^{c,0} = \lim_{\rho_s \to 0} T_s^c. \tag{7.88}$$

The conventional standard chemical potential in the C_s scale is given by

$$\begin{aligned}
\mu_s^{0c} &= \lim_{\rho_s \to 0} (\mu_s - kT \ln C_s) \\
&= \lim_{\rho_s \to 0} (\mu_s^{0\rho} + kT \ln \rho_s \gamma_s^{D,\rho} - kT \ln C_s) \\
&= \mu_s^{0\rho} + kT \ln T_s^{c,0}.
\end{aligned} \tag{7.89}$$

Hence, the relation between the two activity coefficients is

$$\begin{aligned}
kT \ln \gamma_s^{D,c} &= \mu_s - \mu_s^{0c} - kT \ln C_s = \mu_s - \mu_s^{0\rho} - kT \ln(T_s^{c,0} \rho_s / T_s^c) \\
&= kT \ln \gamma_s^{D,\rho} + kT \ln(T_s^c / T_s^{c,0}).
\end{aligned} \tag{7.90}$$

The connection with the solvation Gibbs energy is

$$\Delta G_s^* - \Delta G_s^{*0} = kT \ln \gamma_s^{D,c} + kT \ln(T_s^c / T_s^{c,0}) \tag{7.91}$$

which may be used when activity coefficients based on any concentration scale are available.

The second source of data available for multicomponent mixtures are the excess thermodynamic quantities. These are equivalent to activity coefficients that measure deviations from *symmetrical ideal solutions* and should be distinguished carefully from activity coefficients which measure deviations from ideal dilute solutions (see chapter 6). In a symmetrical ideal (SI) solution, the chemical potential is written in the form

$$\mu_i = \mu_l^p + kT \ln x_i \tag{7.92}$$

where μ_l^p is the chemical potential of the pure component i at the same temperature and pressure.

The SI behavior is realized by a variety of two-component systems of two "similar" species. Deviations from this behavior may be expressed by introducing either an activity coefficient γ_s^{SI} or an excess function. These are defined as

$$\mu_i = \mu_i^p + kT \ln x_i + kT \ln \gamma_i^{\mathrm{SI}} = \mu_i^p + kT \ln x_i + \mu_i^{\mathrm{EX}}. \tag{7.93}$$

The total excess Gibbs energy of the entire system is defined as

$$G^{\mathrm{EX}} = \sum_i N_i \mu_i^{\mathrm{EX}} \tag{7.94}$$

and the excess Gibbs energy per molecule of the mixture is defined by

$$g^{\mathrm{EX}} = G^{\mathrm{EX}} \Big/ \sum_i N_i = \sum_i x_i \mu_i^{\mathrm{EX}}. \tag{7.95}$$

For some two-component systems, the quantity g^{EX} is available as an analytical function of the composition of the system. Let A and B be the two components, and x_A the mole fraction of A in the system:

$$x_A = \frac{N_A}{N_A + N_B}. \tag{7.96}$$

When g^{EX} is given as a function of x^A in the entire range of compositions, one can recover both μ_A^{EX} and μ_B^{EX} using the following well-known procedure:

$$\begin{aligned} \mu_A^{\mathrm{EX}} &= \left(\frac{\partial G^{\mathrm{EX}}}{\partial N_A}\right)_{P,T,N_B} \\ &= \frac{\partial}{\partial N_A}[(N_A + N_B) g^{\mathrm{EX}}] \\ &= (N_A + N_B)\frac{\partial g^{\mathrm{EX}}}{\partial N_A} + g^{\mathrm{EX}}. \end{aligned} \tag{7.97}$$

Transforming variables from N_A into x_A

$$\frac{\partial}{\partial N_A} = \frac{N_B}{(N_A + N_B)^2} \frac{\partial}{\partial x_A} \tag{7.98}$$

yields

$$\mu_A^{\mathrm{EX}} = g^{\mathrm{EX}} + x_B \frac{\partial g^{\mathrm{EX}}}{\partial x_A} \tag{7.99}$$

and similarly

$$\mu_B^{\rm EX} = g^{\rm EX} + x_A \frac{\partial g^{\rm EX}}{\partial x_B}. \tag{7.100}$$

The connection between excess chemical potentials and solvation Gibbs energies may be obtained as follows. We write both μ_A and μ_A^p using the general expression (7.16), and the expression (7.92) to obtain

$$\mu_A = \mu_A^p + kT \ln x_A + \mu_A^{\rm EX} = \mu_A^{*p} + kT \ln \rho_A^p \Lambda_A^3 x_A + \mu_A^{\rm EX} \tag{7.101}$$

$$\mu_A = \mu_A^* + kT \ln \rho_A \Lambda_A^3. \tag{7.102}$$

By comparing equation (7.100) with (7.101), we arrive at

$$\Delta G_A^* - \Delta G_A^{*p} = \mu_A^* - \mu_A^{*p} = kT \ln(\rho_A^p x_A/\rho_A) + \mu_A^{\rm EX}. \tag{7.103}$$

In (7.103), we obtained the solvation Gibbs energy of A in the mixture ΔG_A^* relative to the solvation of A, in pure A ΔG_A^{*p} (at the same temperature and pressure). This may be computed for any composition from knowledge of the excess chemical potential $\mu_A^{\rm EX}$, and the densities of A in the mixture and in the pure component, ρ_A and ρ_A^p, respectively. A similar expression may be written for the second component B.

Sometimes, the densities ρ_A are not available in the entire range of compositions. Instead, data on excess volume are available. This may be used as follows. The excess volume per molecule of the mixture is given by

$$v^{\rm EX} = \frac{V^{\rm EX}}{N_A + N_B} = \frac{V - N_A V_A^p - N_B V_B^p}{N_A + N_B} = v_m - x_A V_A^p - x_B V_B^p \tag{7.104}$$

where V_A^p and V_B^p are the molar (or molecular) volumes of pure A and B, respectively, and v_m is the volume per molecule of the mixture given by $v_m = V/(N_A + N_B)$. Thus,

$$\frac{\rho_A^p x_A}{\rho_A} = \frac{\rho_A^p}{\rho_A + \rho_B} = \frac{v_m}{V_A^p} = \frac{v^{\rm EX} + x_A V_A^p + x_B V_B^p}{V_A^p}. \tag{7.105}$$

This may be used in (7.103) to calculate the relative solvation Gibbs energy of A.

7.6.3 **Pure liquids**

The extreme limit of high density of s is the pure liquid. Normally, the liquids of interest are either at room temperature and 1 atm pressure or along the liquid–vapor coexistence equilibrium line. Let l and g be the liquid and the gaseous phases of a pure component s at equilibrium. The Gibbs energy of transferring s from a fixed position in g into a fixed position in l is equal to the difference in the solvation Gibbs energies of s in the two phases, i.e.,

$$\mu_s^{*l} - \mu_s^{*g} = \Delta G_s^{*l} - \Delta G_s^{*g} = kT \ln(\rho_s^g/\rho_s^l)_{\text{eq}}. \tag{7.106}$$

Knowing the densities s of in the two phases at equilibrium, gives us only the *difference* in the solvation Gibbs energies of s in two phases. However, in many cases, especially near the triple point, the density of the gaseous phase is quite low, in which case we may assume that $\Delta G_s^{*g} \approx 0$ and therefore relation (7.106) reduces to

$$\begin{aligned} \Delta G_s^{*p} &= kT \ln(\rho_s^{\text{ig}}/\rho_s^l)_{\text{eq}} \\ &= kT \ln(P_s/kT\rho_s^l)_{\text{eq}} \end{aligned} \tag{7.107}$$

where P_s is the vapor pressure of s at temperature T.

When evaluating other thermodynamic quantities of solvation from data along the equilibrium line, care must be exercised to distinguish between derivatives at *constant pressure* and derivatives *along the equilibrium line.* The connection between the two is

$$\left(\frac{d\Delta G_s^*}{dT}\right)_{\text{eq}} = \left(\frac{\partial \Delta G_s^*}{\partial T}\right)_P + \left(\frac{\partial \Delta G_s^*}{\partial P}\right)_T \left(\frac{dP}{dT}\right)_{\text{eq}}. \tag{7.108}$$

Here, we use straight derivatives to indicate differentiation along the equilibrium line. The two derivatives of ΔG_s^* on the rhs of equation (7.108) are identified as the solvation entropy and the solvation volume, respectively; thus,

$$\left(\frac{d\Delta G_s^{*p}}{dT}\right)_{\text{eq}} = -\Delta S_s^{*p} + \Delta V_s^{*p} \left(\frac{dP}{dT}\right)_{\text{eq}}. \tag{7.109}$$

Usually, data are available to evaluate both of the straight derivatives in equation (7.108). This is not sufficient, however, to compute both ΔS_s^{*p} and ΔV_s^{*p}. Fortunately, ΔV_s^{*p} may be obtained directly from data on molar volume and compressibility of the pure liquid. For the pure system s we have

$$\mu_s^p = \mu_s^{*p} + kT \ln \rho_s^p \Lambda_A^3 \tag{7.110}$$

Differentiation with respect to P at constant T yields

$$\overline{V}_s^p = \left(\frac{\partial \mu_s^p}{\partial P}\right)_T = V_s^{*p} + kT\kappa_T^p \tag{7.111}$$

where κ_T^p is the isothermal compressibility of the pure s.

We now write the analog of (7.111) for an ideal gas system as

$$\overline{V}_s^{\text{ig}} = V_s^{*\text{ig}} + kT\kappa_T^{\text{ig}} \tag{7.112}$$

(all quantities are for pure s). Hence, the solvation volume is

$$\Delta V_s^{*p} = \overline{V}_s^p - \overline{V}_s^{\text{ig}} - kT[\kappa_T^p - \kappa_T^{\text{ig}}]. \tag{7.113}$$

Care must be exercised in taking the limit of ideal gas ($\rho \to 0$ or $P \to 0$). In general, one *cannot* take the limiting behavior of $\overline{V}_s^{\text{ig}}$ and κ_T^{ig} as:

$$\overline{V}_s^{\text{ig}} \to \frac{kT}{P}, \qquad \kappa_T^{\text{ig}} \to \frac{1}{P}. \tag{7.114}$$

This would lead to $V_s^{*\text{ig}} = 0$, which in general, is not correct†. Further discussion of this point is presented in Appendix O.

7.7 Stepwise solvation processes

We have defined the solvation process as the process of transfer from a fixed position in an ideal gas phase to a fixed position in a liquid phase. We have seen that if we can neglect the effect of the solvent on the internal partition function of the solvaton s, the Gibbs or the Helmholtz energy of solvation is equal to the coupling work of the solvaton to the solvent (the latter may be a mixture of any number of component, including any concentration of the "solute" s). In actual calculations, or in some theoretical considerations, it is often convenient to carry out the coupling work in steps. The specific steps chosen to carry out the coupling work depend on the way we choose to write the solute–solvent interaction.

For simplicity, we discuss here a solute s in a one-component solvent b, in a system at T, V, N_s, N_b.‡ We assume that the solute–solvent interaction can be written as a sum of two parts, say

$$U_{sb}(R) = U_{sb}^X(R) + U_{sb}^Y(R). \tag{7.115}$$

The coupling work is the same as the work of "turning on" the interaction $U_{sb}(R)$. If $U_{sb}(R)$ has the form (7.115), we can carry out the coupling work in two steps; first we couple $U_{sb}^X(R)$, and then we couple the second part $U_{sb}^Y(R)$. This procedure was found useful in interpreting the solvation quantities of simple solutes in water, the study of hydrophobic hydrophilic interactions and

† For a theoretical ideal gas $V_s^{*\text{ig}} = 0$. But this is not true for a real gas at $\rho \to 0$ or $P \to 0$. This error has been made in Ben-Naim (1987).

‡ Note that we use the terms "solute" and "solvent" in the nontraditional sense. The solute can be any molecule and its concentration in the solvent can either be very low $\rho_s \to 0$ or very high $\rho_s \to \rho_s^P$. In the latter, we have pure liquid s, where the traditional distinction between solute and solvent does not apply.

the solvation of macromolecules, such as proteins (see section 7.11). In all cases we assume some kind of additivity of the intermolecular potential and then derive the corresponding split of solvation Gibbs or Helmholtz energy.

7.7.1 Stepwise coupling of the hard and the soft parts of the potential

One of the earliest attempts to interpret the anomalous properties of aqueous solutions was based on splitting the solvation process into two parts: first, we form a cavity in the solvent, then introduce the solute into the cavity. In the present section, we shall discuss the statistical-mechanical basis for such a split of the solvation process. For more details see Eley (1939, 1944) Ben-Naim (1992).

In essence, the molecular basis for splitting the solvation into two (or more) steps stems from the recognition of the two (or more) parts of the solute–solvent intermolecular potential function, in the present case, the hard and soft parts of the interactions.

For simplicity, we assume that the solute–solvent pair potential is a function of the distance R only, and that this function may be written as

$$U_{sb}(R) = U_{sb}^{H}(R) + U_{sb}^{S}(R) \tag{7.116}$$

where the "hard" part of the potential is defined, somewhat arbitrarily, by choosing an *effective* hard-core diameter for the solute σ_s and the solvent σ_b, respectively, i.e.,

$$U_{sb}^{H}(R) = \begin{cases} \infty & \text{for } R \leq \sigma_{sb} \\ 0 & \text{for } R > \sigma_{sb} \end{cases} \tag{7.117}$$

where $\sigma_{sb} = (\sigma_s + \sigma_b)/2$. In practice, one can always find a small enough value of σ_{sb} such that for $R \leq \sigma_{sb}$, the potential function rises so steeply so as to justify an approximation of the form (7.117). The "soft" part, U_{sb}^{S}, is next defined through equation (7.116). This is illustrated in figure 7.3.

We now write the solvation Helmholtz energy of s as follows

$$\begin{aligned} \Delta A_s^* &= -kT \ln\langle \exp(-\beta B_s)\rangle_0 \\ &= -kT \ln\langle \exp(-\beta B_s^H - \beta B_s^S)\rangle_0 \end{aligned} \tag{7.118}$$

where B_s^H and B_s^S are the "hard" and "soft" parts of the binding energies of s to all the solvent molecules.

The average quantity in (7.118) can be viewed as an average of a product of two functions. If these were independent (in the sense of the independence of two random variables), one could rewrite this average as a product of two

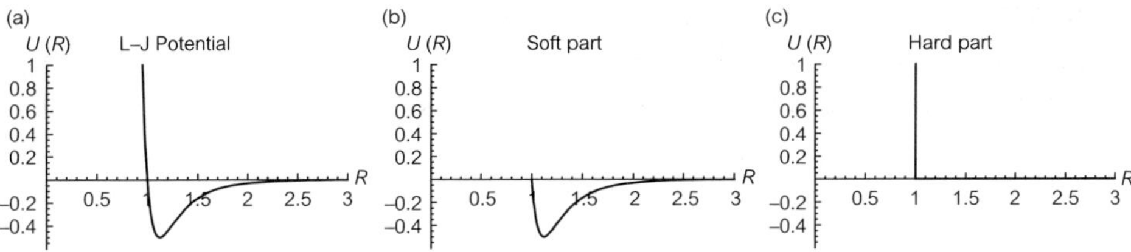

Figure 7.3 A typical Lennard-Jones pair potential (a), is split into two parts: a soft part (b) and a hard part (c).

averages, i.e.,

$$\langle \exp(-\beta B_s^H) \exp(\beta B_s^S) \rangle_0 = \langle \exp(-\beta B_s^H) \rangle_0 \langle \exp(-\beta B_s^S) \rangle_0 \tag{7.119}$$

and ΔA_s^* could be split into a sum of two terms[†]

$$\Delta A_s^* = \Delta A_s^{*H} + \Delta A_s^{*S}. \tag{7.120}$$

However, the factorization in equation (7.119) is in general invalid, and therefore a split of ΔA_s^* in terms of a hard and soft part in the form as in (7.120) is not justified. To see why, we rewrite the average in (7.119) as

$$\begin{aligned}
&\left\langle \exp(-\beta B_s^H) \exp(-\beta B_s^S) \right\rangle_0 \\
&\quad = \frac{\int d\mathbf{X}^N \exp(-\beta U_N - \beta B_s^H) \exp(-\beta B_s^S)}{\int d\mathbf{X}^N \exp(-\beta U_N)} \\
&\quad = \frac{\int d\mathbf{X}^N \exp(-\beta U_N - \beta B_s^H)}{\int d\mathbf{X}^N \exp(-\beta U_N)} \frac{\int d\mathbf{X}^N \exp(-\beta U_N - \beta B_s^H) \exp(-\beta B_s^S)}{\int d\mathbf{X}^N \exp(-\beta U_N - \beta B_s^H)} \\
&\quad = \left\langle \exp(-\beta B_s^H) \right\rangle_0 \left\langle \exp(-\beta B_s^S) \right\rangle_H.
\end{aligned} \tag{7.121}$$

The first average on the rhs of (7.121) is the same as in (7.119); the second average on the rhs of equation (7.121) is a *conditional average*, using the probability distribution function

$$P(\mathbf{X}^N/\mathbf{X}_s^H) = \frac{P(\mathbf{X}^N, \mathbf{X}_s^H)}{P(\mathbf{X}_s^H)} = \frac{\exp(-\beta U_N - \beta B_s^H)}{\int d\mathbf{X}^N \exp(-\beta U_N - \beta B_s^H)} \tag{7.122}$$

where $P(\mathbf{X}^N/\mathbf{X}_s^H)$ is the probability density of finding a configuration $\mathbf{X}^N$ of the N particles, *given* a hard particle at $\mathbf{X}_s^H$. By using the form (7.121), the solvation Helmholtz energy can be written as

$$\Delta A_s^* = \Delta A_s^{*H} + \Delta A_s^{*S/H} \tag{7.123}$$

where ΔA_s^{*H} is the Helmholtz energy of solvation of the *hard* part of the potential. This part is the same as in (7.120). The second term on the rhs of (7.122), $\Delta A_s^{*S/H}$, is the *conditional Helmholtz energy* of solvation of the *soft* part

[†] Note that the subscript s is for *solute*, and the superscript S for the *soft* part of the potential.

of the potential. This is different from the second term in (7.120). It is the work required to couple, or to switch-on, the soft part of the solute–solvent interaction, *given* that the hard part of the potential has already been coupled. Clearly, the procedure outlined above may be generalized to include other contributions to the pair potential. An important generalization of equation (7.116) might be the inclusion of electrostatic interactions, hydrogen bonding, etc. The generalization of equation (7.116) might look like this:

$$U_{sb}(R) = U_{sb}^{H}(R) + U_{sb}^{S}(R) + U_{sb}^{el}(R) + U_{sb}^{HB}(R). \tag{7.124}$$

Correspondingly, the solvation Helmholtz energy will be written in a generalization of expression (7.123) as follows:

$$\Delta A_s^* = \Delta A_s^{*H} + \Delta A_s^{*S/H} + \Delta A_s^{*el/S,H} + \Delta A_s^{HB/el,S,H} \tag{7.125}$$

where $\Delta A_s^{*el/S,H}$ is the conditional Helmholtz energy of solvation of the electrostatic part of the interaction, given that a solvaton with the *soft* and *hard* parts (excluding the electrostatic part) has already been placed at a fixed position in the solvent. Similar interpretation applies to the last term on the rhs of (7.125).

It is important to take note of the order in which we couple each part of the potential. When the *n*th part has already been coupled, it has an effect on the distribution of the configurations of the solvent molecules. Therefore, in the next step we have to replace the distribution function used in the coupling of the *n*th part by a *conditional* distribution function for calculating the average in the $(n+1)$th part.

The solvation Helmholtz energy of a hard particle in a solvent is related to the probability of finding a cavity of suitable size in the liquid (see section 7.11 and appendix N). Hence, (7.123) may be rewritten as

$$\exp(-\beta\Delta A_s^*) = \Pr(\text{cavity})\langle\exp(-\beta B_s^S)\rangle_H \tag{7.126}$$

Equation (7.126) is useful in actual estimation of the solvation Helmholtz energy. The cavity work is usually estimated by the scaled particle theory (Appendix N). If the soft part of the interaction is small, i.e., if $|\beta B_s^S| \ll 1$, then we may estimate

$$\langle\exp(-\beta B_s^S)\rangle_H \approx 1 - \beta\langle\beta B_s^S\rangle_H \tag{7.127}$$

where $\langle\beta B_s^S\rangle_H$ is the *conditional average binding energy* of the soft part of the interaction between the solute *s* and the solvent, given that the hard part has already been coupled.

7.7.2 **Stepwise coupling of groups in a molecule**

In this section, we examine the molecular basis of a group-additivity approach. As we shall see, the problem is essentially the same as treated in the previous section; i.e., it originates from a split of the solute–solvent intermolecular potential function into two or more parts.

Consider a solute of the form $X-Y$, where X and Y are two groups, say CH_3 and CH_3 in ethane, or CH_3 and OH in methanol. We assume that the solute–solvent interaction may be split into two parts as follows:

$$U(X-Y,i) = U(X,i) + U(Y,i) \tag{7.128}$$

where $U(X, i)$ and $U(Y, i)$ are interaction energies between the groups X and Y, and the ith solvent molecule, respectively (figure 7.4).

As in the preceding section, where we had split the interaction energy into a hard and soft part, we also have here an element of ambiguity as to the exact manner in which this split may be achieved. For instance for ethane, we assume that the ethane–water interaction may be written as the sum of the two methyl–water interactions, as schematically depicted in figure 7.4. Next, we proceed to split the total binding energy of the solute $X-Y$ into two parts,

$$B_{X-Y} = \sum_{i=1}^{N} U(X,i) + \sum_{i=1}^{N} U(Y,i) = B_X + B_Y. \tag{7.129}$$

The solvation Helmholtz energy of the solute $X-Y$ is now written as

$$\begin{aligned}\Delta A^*_{X-Y} &= -kT\ln\langle\exp(-\beta B_{X-Y})\rangle_0 \\ &= -kT\ln\langle\exp(-\beta B_X)\exp(-\beta B_Y)\rangle_0.\end{aligned} \tag{7.130}$$

As in the previous section, we have again an average of a product of two functions. This, in general, *may not* be factorized into a product of two average quantities. If this could have been done, then relation (7.130) could have been written as a sum of two terms, i.e.,

$$\begin{aligned}\Delta A^*_{X-Y} &= -kT\ln\langle\exp(-\beta B_X)\rangle_0 - kT\ln\langle\exp(-\beta B_Y)\rangle_0 \\ &= \Delta A^*_X + \Delta A^*_Y\end{aligned} \tag{7.131}$$

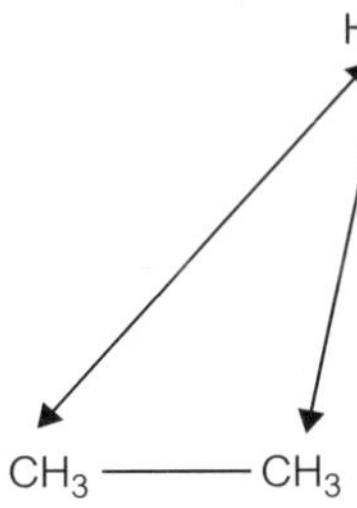

Figure 7.4 Schematic split of the ethane–water interaction into two methyl–water interactions.

Such a group additivity, though very frequently assumed to hold for ΔA^*_{X-Y} (as well as for other thermodynamic quantities), has clearly no justification on a molecular basis, even when the split of the potential function in expression (7.128) is exact.

The reason is quite simple: since the two groups X and Y are very close to each other, their solvation must be correlated, and therefore the average in (7.130) cannot be factored into a product of two averages. In order to achieve some form of "additivity" similar to relation (7.131), we rewrite equation (7.130) as follows:

$$\begin{aligned}
\Delta A^*_{X-Y} &= -kT\ln\langle\exp(-\beta B_X - \beta B_Y)\rangle_0 \\
&= -kT\ln\left[\frac{\int d\mathbf{X}^N \exp(-\beta U_N)\exp(-\beta B_X)\exp(-\beta B_Y)}{\int d\mathbf{X}^N \exp(-\beta U_N)\exp(-\beta B_X)}\right. \\
&\qquad \left.\times\frac{\int d\mathbf{X}^N \exp(-\beta U_N)\exp(-\beta B_X)}{\int d\mathbf{X}^N \exp(-\beta U_N)}\right] \\
&= -kT\ln\left[\int d\mathbf{X}^N P(\mathbf{X}^N/\mathbf{X}_X)\exp(-\beta B_Y)\int d\mathbf{X}^N P(\mathbf{X}^N)\exp(-\beta B_X)\right] \\
&= \Delta A^*_X + \Delta A^*_{Y/X}.
\end{aligned} \tag{7.132}$$

The interpretation of equation (7.132) is the following. The solvation Helmholtz energy of the solute $X-Y$ is written [presuming the validity of relation (7.129)] as a sum of two contributions. First, the solvation Helmholtz energy of the group X, and second, the conditional solvation Helmholtz energy of the group Y, *given* that the group X is already at a fixed point in the liquid. This is schematically shown in figure 7.5.

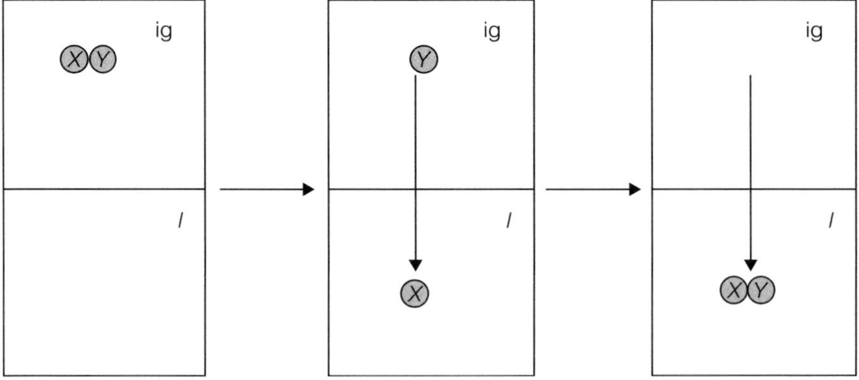

Figure 7.5 Insertion of a solute X–Y in two steps. First, we insert the group X. The resulting solvation Helmholtz energy is ΔA^*_X. Next we insert Y to a fixed position next to X. The conditional solvation Helmholtz energy is $\Delta A^*_{Y/X}$.

Clearly, since the second group Y is brought to a location very near X, one cannot ignore the effect of X on the (conditional) solvation of Y. In other words, in the first step, we insert X in pure solvent. In the second, we insert Y in a solvent which has been perturbed by the presence of X. It is only in very extreme cases when X and Y are very far apart that we could assume that their solvation Helmholtz energies will be strictly additive, as in equation (7.131).

Because of symmetry, we could, of course, replace equation (7.132) by the equivalent relation

$$\Delta A^*_{X-Y} = \Delta A^*_Y + \Delta A^*_{X/Y}. \tag{7.133}$$

We see that relations (7.132) and (7.133) have the same form as relation (7.120). Placing X at a fixed position has an effect on the distribution of the configurations of solvent molecules around X. Therefore, when we couple the group Y at a location near X, the average work is now calculated with the conditional distribution function as shown in (7.132). The procedure outlined above may be generalized to include large molecules with many groups. We shall make use of such a procedure for proteins in section 7.11.

Note that since ΔA^*_{X-Y} applies to the entire molecule $X-Y$, the conditional solvation Helmholtz energy $\Delta A^*_{X/Y}$ does not include the direct bond energy between X and Y. We shall see in the next section the relation between $\Delta A^*_{X/Y}$ and ΔA^*_X, and the correlation function between the two groups.

7.7.3 Conditional solvation and the pair correlation function

In this section, we examine in some more detail the relation between the conditional and the unconditional solvation Helmholtz energies. Consider a solvent l at any given temperature, volume, and composition. Let s be a simple spherical molecule. The solvation Helmholtz energy of s is given by

$$\exp[-\beta\Delta A^*_s(\mathbf{R}_1)] = \langle \exp[-\beta B_s(\mathbf{R}_1)]\rangle_0. \tag{7.134}$$

Note that we have explicitly introduced the location of $\mathbf{R}_1$ at which we have placed the solvation s. This is in general not necessary since all points in the solvent are equivalent. In this section, however, we shall produce inhomogeneity in the solvent by introducing two particles at $\mathbf{R}_1$ and $\mathbf{R}_2$; therefore, the recording of their locations is important. In this section, we assume that we have a dilute solution of s in a solvent. Theoretically, we can think of having just one s in a pure solvent. The symbol $\langle\ \rangle_0$ stands for an average over all the configurations of the solvent molecules in the T, V, $\mathbf{N}$ ensemble, i.e.,

$$\langle \exp[-\beta B_s(R_1)]\rangle_0 = \int\cdots\int d\mathbf{X}^N P_0(\mathbf{X}^N)\exp[-\beta B_s(\mathbf{R}_1, \mathbf{X}^N)] \tag{7.135}$$

where $P_0(\boldsymbol{X}^N)$ is the probability density of finding a configuration $\boldsymbol{X}^N$, i.e.,

$$P_0(\boldsymbol{X}^N) = \frac{\exp[-\beta U(\boldsymbol{X}^N)]}{\int \cdots \int d\boldsymbol{X}^N \exp[-\beta U(\boldsymbol{X}^N)]}. \tag{7.136}$$

Suppose we have one solute at $\boldsymbol{R}_1$ and we introduce a second solute at a different location $\boldsymbol{R}_2$. The corresponding work is obtained by taking the ratio of the two partition functions

$$\begin{aligned}\exp[-\beta \Delta A_s^*(\boldsymbol{R}_2/\boldsymbol{R}_1)] &= \frac{Q(T, V, \boldsymbol{N}; \boldsymbol{R}_1, \boldsymbol{R}_2)}{Q(T, V, \boldsymbol{N}; \boldsymbol{R}_1)} \\ &= \frac{\int \cdots \int d\boldsymbol{X}^N \exp[-\beta U(\boldsymbol{X}^N) - \beta B_s(\boldsymbol{R}_1) - \beta B_s(\boldsymbol{R}_2) - \beta U(\boldsymbol{R}_1, \boldsymbol{R}_2)]}{\int \cdots \int d\boldsymbol{X}^N \exp[-\beta U(\boldsymbol{X}^N) - \beta B_s(\boldsymbol{R}_1)]} \\ &= \exp[-\beta U(\boldsymbol{R}_1, \boldsymbol{R}_2)] \langle \exp[-\beta B_s(\boldsymbol{R}_2)] \rangle_{\boldsymbol{R}_1}. \end{aligned} \tag{7.137}$$

Here, $U(\boldsymbol{R}_1, \boldsymbol{R}_2)$ is the *direct* interaction between the two solutes at $\boldsymbol{R}_1$, $\boldsymbol{R}_2$, and the symbol $\langle\ \rangle_{R_1}$ stands for a conditional average, i.e., an average over all configurations of the solvent molecules, given one solute at $\boldsymbol{R}_1$. The corresponding conditional density is

$$P(\boldsymbol{X}^N/\boldsymbol{R}_1) = \frac{\exp[-\beta U(\boldsymbol{X}^N) - \beta B_s(\boldsymbol{R}_1)]}{\int \cdots \int d\boldsymbol{X}^N \exp[-\beta U(\boldsymbol{X}^N) - \beta B_s(\boldsymbol{R}_1)]} \tag{7.138}$$

Equation (7.137) can be rewritten in another equivalent form as

$$\exp[-\beta \Delta A_s^*(\boldsymbol{R}_2/\boldsymbol{R}_1)] = \langle \exp[-\beta U(\boldsymbol{R}_1, \boldsymbol{R}_2)] \rangle \frac{\langle \exp[-\beta B_{ss}(\boldsymbol{R}_1, \boldsymbol{R}_2)] \rangle_0}{\langle \exp[-\beta B_s(\boldsymbol{R}_1)] \rangle_0} \tag{7.139}$$

where $B_{ss}(\boldsymbol{R}_1, \boldsymbol{R}_2) = B_s(\boldsymbol{R}_1) + B_s(\boldsymbol{R}_2)$. Equation (7.139) can be rearranged into

$$\exp[-\beta \Delta A_s^*(\boldsymbol{R}_2/\boldsymbol{R}_1)] = g(\boldsymbol{R}_1, \boldsymbol{R}_2) \exp[-\beta \Delta A_s^*(\boldsymbol{R}_1)]. \tag{7.140}$$

Thus, the correlation function $g(\boldsymbol{R}_1, \boldsymbol{R}_2)$ "connects" the solvation Helmholtz energy on the first "site" and on the second "site." Equation (7.140) can also be rewritten as

$$W(\boldsymbol{R}_1, \boldsymbol{R}_2) = \Delta A_s^*(\boldsymbol{R}_2/\boldsymbol{R}_1)] - \Delta A_s^*(\boldsymbol{R}_2). \tag{7.141}$$

This means that the potential of mean force is the same as the solvation Helmholtz energy at $\boldsymbol{R}_2$ given a particle at $\boldsymbol{R}_1$, minus the solvation Helmholtz energy at $\boldsymbol{R}_2$. In other words the difference between the conditional and the unconditional quantities on the rhs of (7.141) is equal to the work required to bring the second s from infinity to position $\boldsymbol{R}_2$, given that another s is already placed at $\boldsymbol{R}_1$. If we are dealing with two different solutes, say s_1 and, s_2 then , instead of (7.141) we write

$$\begin{aligned} W_{s_1,s_2}(\boldsymbol{R}_1,\boldsymbol{R}_2) &= \Delta A^*_{s_2/s_1}(\boldsymbol{R}_2/\boldsymbol{R}_1) - \Delta A^*_{s_2}(\boldsymbol{R}_2) \\ &= \Delta A^*_{s_1/s_2}(\boldsymbol{R}_1/\boldsymbol{R}_2) - \Delta A^*_{s_1}(\boldsymbol{R}_1). \end{aligned} \tag{7.142}$$

In (7.142), we have written the work of bringing the two solutes from infinity to the positions $\boldsymbol{R}_1$ and $\boldsymbol{R}_2$ in two equivalent forms. The result of the two steps is the same work $W_{s_1,s_2}(\boldsymbol{R}_1, \boldsymbol{R}_2)$.

In the aforementioned examples, the conditional solvation Helmholtz energy includes the *direct* interaction between the two solute particles, as well as the effect of the solvent. In some applications it is found useful to exclude the direct interaction. This occurs whenever we want to estimate the contributions to the solvation Helmholtz energy of each part of a combined solute. In our definitions of both $\Delta A^*_s(\boldsymbol{R}_1)$ and $\Delta A^*_s(\boldsymbol{R}_2/\boldsymbol{R}_1)$, we transferred one solute s from a fixed position in an ideal gas into the liquid. Now suppose that we are given a pair of solutes at a distance $R = |\boldsymbol{R}_2 - \boldsymbol{R}_1|$ in an ideal gas. This pair of solutes can be viewed as a single molecule. We wish to know the contribution of each particle (1 and 2) to the Helmholtz energy of solvation of the pair. The latter is

$$\exp[-\beta A^*_{ss}(\boldsymbol{R}_1,\boldsymbol{R}_2)] = \langle \exp[-\beta B_{ss}(\boldsymbol{R}_1,\boldsymbol{R}_2)] \rangle_0. \tag{7.143}$$

Instead of transferring the pair as a single entity, we first transfer one particle. The solvation Helmholtz energy change is almost the same as in (7.134) but we must also add the energy required to break the *S–S* bond in the vacuum. In the second step, we transfer the second particle; the corresponding work is now exactly as in (7.139) where we now gain the *S–S* bond energy in the liquid. Therefore, in the entire process, the direct interaction, between the two solutes cancel out.

Thus, instead of equation (7.140), we now write the corresponding relation excluding the direct interaction. First, define

$$y(\boldsymbol{R}_1,\boldsymbol{R}_2) = g(\boldsymbol{R}_1,\boldsymbol{R}_2)\exp[\beta U(\boldsymbol{R}_1,\boldsymbol{R}_2)] \tag{7.144}$$

and rewrite the analog (7.140) as

$$\langle \exp[-\beta B_s(\boldsymbol{R}_2)] \rangle_{\boldsymbol{R}_1} = y(\boldsymbol{R}_1,\boldsymbol{R}_2)\langle \exp[\beta B_s(\boldsymbol{R}_1)] \rangle_0. \tag{7.145}$$

Note that the average on the rhs of (7.145) is the same as in (7.134); i.e., this is the same as the solvation Helmholtz energy of one s particle in a pure solvent. On the lhs of (7.145), we have the conditional solvation Helmholtz energy as in (7.137), but excluding the direct interaction between the two solutes. Note that since the solute at $\boldsymbol{R}_1$ affects the distribution of the solvent configurations, the average on the lhs of (7.145) is different from the average on the rhs of (7.145).

Another useful form of $y(\boldsymbol{R}_1, \boldsymbol{R}_2)$ is

$$y(\boldsymbol{R}_1, \boldsymbol{R}_2) = \frac{\langle \exp[-\beta B_s(\boldsymbol{R}_2)] \rangle_{\boldsymbol{R}_1}}{\langle \exp[-\beta B_s(\boldsymbol{R}_1)] \rangle_0} = \frac{\langle \exp[-\beta B_{ss}(\boldsymbol{R}_1, \boldsymbol{R}_2)] \rangle_0}{\langle \exp[-\beta B_s(\boldsymbol{R}_1)] \rangle_0^2}. \tag{7.146}$$

Note that both averages on the rhs of (7.146) are taken with the distribution function of the pure solvent (7.136). If $|\boldsymbol{R}_1 - \boldsymbol{R}_2| \to \infty$, the two solutes are uncorrelated and the average in the numerator of (7.146) can be factored into a product of two averages, i.e.,

$$\langle \exp[-\beta B_{ss}(\boldsymbol{R}_1, \boldsymbol{R}_2)] \rangle_0 = \langle \exp[-\beta B_s(\boldsymbol{R}_1)] \rangle_0^2. \tag{7.147}$$

In this case we have

$$y(\boldsymbol{R}_1, \boldsymbol{R}_2) = 1, \quad |\boldsymbol{R}_1 - \boldsymbol{R}_2| \to \infty \tag{7.148}$$

which means that there is no solvent-induced correlation.

7.8 Solvation of a molecule having internal rotational degrees of freedom

So far, in all of our discussions of the solvation phenomena, we assumed that the internal partition function is not affected by the solvent; i.e., q_s was assumed to be the same in the gas or in the liquid state. This assumption is approximately correct for the internal partition function of simple molecules in a simple solvent. There is one important exception where we must take into account the solvent effects even in simple solvents: the case in which the molecule can have different conformations, each with a different rotational partition function. Clearly, for large polymers, the rotational partition function of the extended polymer differs significantly from the rotational partition function of a compact conformer of the same molecule. Since these two conformations have different binding energies to the solvent, the relative weights given to each conformation will be different in the gas and in the liquid state. We shall demonstrate this effect for a small molecule such as butane (figure 7.6) and then generalize to larger polymers.

Consider a solute s with internal rotational degrees of freedom. We assume that the vibrational, electronic, and nuclear partition functions are separable and independent of the configuration of the molecules in the system. We define the pseudo-chemical potential of a molecule having a fixed conformation $\boldsymbol{P}_s$ as the change in the Helmholtz energy for the process of introducing s into the

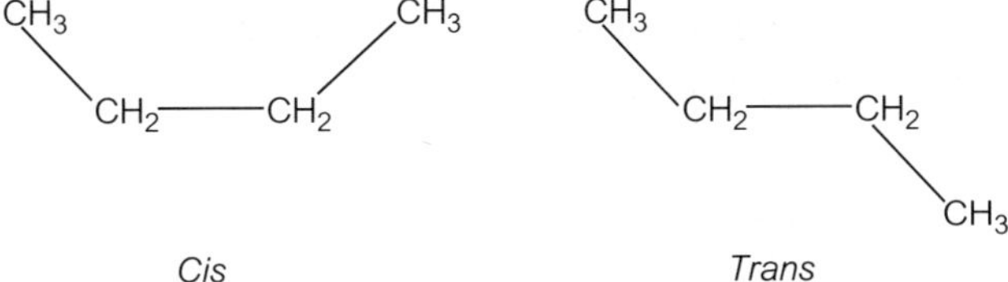

Figure 7.6 Schematic description of *n*-butane. Here we show only two conformers: the *cis* and *trans*.

system l (at fixed T, V) in such a way that its center of mass is at a fixed position R_s. If we release the constraint on the fixed position of the center of mass, we can define the chemical potential of the P_s conformer in the gas and liquid phases as follows:

$$\mu_s^g(\boldsymbol{P}_s) = \mu_s^{*g}(\boldsymbol{P}_s) + kT \ln \rho_s^g(\boldsymbol{P}_s)\Lambda_s^3 \tag{7.149}$$

$$\mu_s^l(\boldsymbol{P}_s) = \mu_s^{*l}(\boldsymbol{P}_s) + kT \ln \rho_s^l(\boldsymbol{P}_s)\Lambda_s^3. \tag{7.150}$$

Note that the rotational partition function of the entire molecule, as well as the internal partition functions of s, are included in the pseudo-chemical potential. In classical systems, the momentum partition function Λ_s^3 is independent of the environment, whether it is a gas or a liquid phase.

The solvation Helmholtz energy of the $\boldsymbol{P}_s$ conformer is defined as

$$\begin{aligned}\Delta\mu_s^*(\boldsymbol{P}_s) &= \mu_s^{*l}(\boldsymbol{P}_s) - \mu_s^{*g}(\boldsymbol{P}_s)\\ &= -kT \ln\langle \exp[-\beta B_s(\boldsymbol{P}_s)]\rangle_0\end{aligned} \tag{7.151}$$

i.e., this is the Helmholtz energy of transferring an s molecule, having a *fixed* conformation, from a fixed position in g to a fixed position in l. If we assume that all vibrational, electronic, and nuclear degrees of freedom are not affected by this transfer from g to l, we can write the second equality on the rhs of equation (7.151).

We now wish to find an expression for the Helmholtz energy of solvation of the molecule s, without specifying its conformation. We do this in two steps, and for convenience we use the T, V, N ensemble. Suppose first that s can attain only two conformations A and B, say the *cis* and *trans* conformations of a given molecule at equilibrium (figure 7.6). The pseudo-chemical potential of the conformer A is the change in the Helmholtz energy for placing an A molecule at a fixed position in l. The corresponding statistical-mechanical expression is

$$\exp(-\beta\mu_A^{*l}) = \frac{q_A \int d\boldsymbol{X}^N\, d\boldsymbol{\Omega}_A \exp[-\beta U_N(\boldsymbol{X}^N) - \beta B_A(\boldsymbol{X}^N) - \beta U^*(A)]}{(8\pi^2)\int d\boldsymbol{X}^N \exp[-\beta U_N(\boldsymbol{X}^N)]} \tag{7.152}$$

where $B_A(\boldsymbol{X}^N)$ is the binding energy of A to the rest of the system of N molecules at a specific configuration $\boldsymbol{X}^N$. $\boldsymbol{U}^*(A)$ denotes the intramolecular potential or the internal rotation potential function of s in the state A. We also assume that q_A has the form

$$q_A = q_v q_e q_{r,A} \tag{7.153}$$

where q_v and q_e are presumed to be independent of the conformation as well as of the solvent. $q_{r,A}$ is the rotational partition function of the entire molecule in state A. Clearly, since the two conformations have different moments of inertia, they will also have different rotational partition functions.

Equation (7.152) may be rewritten as

$$\exp(-\beta\mu_A^{*l}) = q_A \exp[-\beta U^*(A)]\langle\exp(-\beta B_A)\rangle_0 \tag{7.154}$$

where the average is over all the configurations of the N molecules in the system excluding the solvaton A. Likewise, in the gaseous phase, we have

$$\exp(-\beta\mu_A^{*g}) = q_A \exp[-\beta U^*(A)]. \tag{7.155}$$

Hence, the solvation Helmholtz energy of the conformer A is obtained from (7.154) and (7.155) i.e.,

$$\exp(-\beta\Delta\mu_A^{*l}) = \langle\exp(-\beta B_A)\rangle_0. \tag{7.156}$$

A similar expression holds for the conformer B.

To obtain the connection between $\Delta\mu_A^{*l}$, $\Delta\mu_B^{*l}$, and $\Delta\mu_s^{*l}$, we start with the equilibrium condition

$$\mu_s^l = \mu_A^l = \mu_B^l \tag{7.157}$$

or equivalently

$$\mu_s^{*l} + kT\ln\rho_s^l\Lambda_s^3 = \mu_A^{*l} + kT\ln\rho_A^l\Lambda_s^3 = \mu_B^{*l} + kT\ln\rho_B^l\Lambda_s^3 \tag{7.158}$$

where $\rho_s^l = \rho_A^l + \rho_B^l$ and ρ_A^l and ρ_B^l are the densities of A and B at equilibrium. Equation (7.158) may be rearranged to obtain

$$\exp(-\beta\mu_s^{*l}) = \frac{\rho_s^l}{\rho_A^l}\exp(-\beta\mu_A^{*l}) \tag{7.159}$$

and similarly

$$\exp(-\beta\mu_s^{*l}) = \frac{\rho_s^l}{\rho_B^l}\exp(-\beta\mu_B^{*l}). \tag{7.160}$$

On multiplying equation (7.159) by x_A^l, and equation (7.160) by x_B^l, and adding the resulting two equations (where $x_A^l = \rho_A^l/\rho_s^l$ and $x_B^l = 1 - x_A^l$), we get

$$\exp(-\beta\mu_s^{*l}) = \exp(-\beta\mu_A^{*l}) + \exp(\beta\mu_B^{*l}) \tag{7.161}$$

which is the required connection between the three pseudo-chemical potentials. Equation (7.161) is equivalent to the statement that the partition function of a system with one additional s particle at a fixed position is the sum of the partition function of the same system with one A particle at a fixed position and the partition function of the same system with one B particle at a fixed position.

We now write the corresponding expression for the ideal-gas phase, namely,

$$\exp(-\beta\mu_s^{*g}) = \exp(-\beta\mu_A^{*g}) + \exp(\beta\mu_B^{*g}). \tag{7.162}$$

Taking the ratio between expressions (7.161) and (7.162), with the aid of equations (7.154) and (7.156), we obtain

$$\exp(-\beta\Delta\mu_s^{*l}) = \frac{q_A \exp[-\beta U^*(A)] \exp(-\beta\Delta\mu_A^{*l}) + q_B \exp[-\beta U^*(B)] \exp(-\beta\Delta\mu_B^{*l})}{q_A \exp[-\beta U^*(A)] + q_B \exp[-\beta U^*(B)]} \tag{7.163}$$

or equivalently,

$$\exp(-\beta\Delta\mu_s^{*l}) = y_A^g \exp(-\beta\Delta\mu_A^{*l}) + y_B^g \exp(-\beta\Delta\mu_B^{*l}) \tag{7.164}$$

where y_A^g and y_B^g are the equilibrium mole fractions of A and B in the gaseous phase. These are defined by

$$y_A^g = \frac{q_A \exp[-\beta U^*(A)]}{q_A \exp[-\beta U^*(A)] + q_B \exp[-\beta U^*(B)]}, \quad y_B^g = 1 - y_A^g. \tag{7.165}$$

Equation (7.164) is the required relation between the solvation Helmholtz energy of the solute s and the solvation Helmholtz energies of the two conformers A and B. Note that if q_v and q_e are the same for the two conformations, they will cancel in (7.165) and (7.163). What remains is only the rotational partition function of the two conformers. Generalization to the case with n discrete conformations is straightforward: if there are n conformers, we have instead of (7.163) and (7.164)

$$\begin{aligned} \exp(-\beta\Delta\mu_s^{*l}) &= \frac{\sum_{i=1}^n q_i \exp[-\beta U^*(i)] \langle \exp(-\beta B_i) \rangle_0}{\sum_i q_i \exp[-\beta U^*(i)]} \\ &= \sum_{i=1}^n y_i^g \exp(-\beta\Delta\mu_i^{*l}) \end{aligned} \tag{7.166}$$

and for the continuous case , we have

$$\begin{aligned} \exp(-\beta\Delta\mu_s^{*l}) &= \frac{\int d\mathbf{P}_s\, q(\mathbf{P}_s) \exp[-\beta U^*(\mathbf{P}_s)] \langle \exp[-\beta B(\mathbf{P}_s) \rangle_0}{\int d\mathbf{P}_s\, q(\mathbf{P}_s) \exp[-\beta U^*(\mathbf{P}_s)]} \\ &= \int d\mathbf{P}_s\, y^g(\mathbf{P}_s) \exp[-\beta\Delta\mu^{*l}(\mathbf{P}_s)] = \langle \langle \exp[-\beta B(\mathbf{P}_s)] \rangle_0 \rangle. \end{aligned} \tag{7.167}$$

Here, $q(\boldsymbol{P}_s)$ denotes the rotational, vibrational, etc., partition function of a single s molecule at a *specific* conformation. $\boldsymbol{P}_s$, $y^g(\boldsymbol{P}_s)\, d\boldsymbol{P}_s$ is the mole fraction of s molecules at conformations between $\boldsymbol{P}_s$ and $\boldsymbol{P}_s + d\boldsymbol{P}_s$. The final expression on the rhs of equation (7.167) is a double average quantity; one, over all configurations of the N molecules (excluding the solvaton) and the second (the outer average), over all the conformations of the s molecule with distribution function $y^g(\boldsymbol{P}_s)$.

The chemical potential of s in an ideal gas phase can now be written as

$$\begin{aligned}\mu_s^{ig} &= \mu_s^{*g} + kT \ln \rho_s^g \Lambda_s^3 \\ &= -kT \ln\left\{ \int d\boldsymbol{P}_s\, q(\boldsymbol{P}_s) \exp[-\beta U^*(\boldsymbol{P}_s)] \right\} + kT \ln \rho_s^g \Lambda_s^3 \\ &= -kT \ln q_{\text{int}}^g + kT \ln \rho_s^g \Lambda_s^3 \end{aligned} \tag{7.168}$$

where the term in the curly brackets can be interpreted as the internal partition function of a single s in the gas phase. This is denoted by q_{int}^g.

In the liquid state we have the corresponding equation for the chemical potential of s

$$\begin{aligned}\mu_s^l &= \mu_s^{*l} + kT \ln \rho_s^l \Lambda_s^3 \\ &= -kT \ln\left\{ \int d\boldsymbol{P}_s\, q(\boldsymbol{P}_s) \exp[-\beta U^*(\boldsymbol{P}_s) \langle \exp[-\beta B(\boldsymbol{P}_s)] \rangle_0 \right\} + kT \ln \rho_s^l \Lambda_s^3. \end{aligned} \tag{7.169}$$

Here we cannot separate the internal partition function from the coupling work. The reason is that each conformation has a different binding energy to the solvent. In a formal way, we can use the definition of q_{int}^g from (7.168) to rewrite (7.169) as

$$\mu_s^l = -kT \ln q_{\text{int}}^g + \Delta\mu_s^{*l} + kT \ln \rho_s^l \Lambda_s^3. \tag{7.170}$$

In (7.170), the first term on the rhs is the same as in (7.168) but $\Delta\mu_s^{*l}$ includes both the coupling work of all the conformations, as well as the effect of the solvent on the internal degrees of freedom of the molecule.

In (7.165), we have written the mole fraction of the A and B conformers. These are also the probabilities of finding the solute s in A or B, respectively. When the conformation changes continuously, say in butane, as a function of the dihedral angle ϕ, (7.165) generalizes to

$$y^g(\phi) = \frac{q(\phi) \exp[-\beta U^*(\phi)]}{\int_0^{2\pi} d\phi\, q(\phi) \exp[-\beta U^*(\phi)]} \tag{7.171}$$

where $y^g(\phi)\,d\phi$ is the probability of finding the molecule at angle between ϕ and $\phi + d\phi$. Likewise, in the solution, the probability density is

$$y^l(\phi) = \frac{q(\phi)\exp[-\beta U^*(\phi) - \beta\Delta\mu^*(\phi)]}{\int_0^{2\pi} d\phi\, q(\phi)\exp[-\beta U^*(\phi) - \beta\Delta\mu^*(\phi)]}. \tag{7.172}$$

The difference between (7.171) and (7.172) is the addition of the solvation Gibbs energy $\Delta\mu^*(\phi)$. Note, that $y^g(\phi)$ and $y^l(\phi)$ are the actual distribution of the ϕ conformers for butanes in the ideal gas, and in the liquid phase. A related function, which has been calculated theoretically and by simulations [Rosenberg et al (1982), Jorgensen (1982), Jorgensen and Buckner (1987), Tobias and Brooks (1990), Zichi and Rossky (1986), Imai and Hirata (2003)], is the dihedral distribution of a molecule lacking translation and rotational degrees of freedom (i.e., as if we were holding the orientation of the $C_2{-}C_3$ bond fixed and measure the distribution of the angle of rotation about this bond). This function $s(\phi)$ is obtained from $y^g(\phi)$ or $y^l(\phi)$ by eliminating the rotational partition function $q(\phi)$ in (7.171) and (7.172).

Figure 7.7 shows the qualitative change in the distribution $s(\phi)$ when the molecule is transferred from an ideal gas into aqueous solution. Note that the *trans* conformer ($\phi = \pi$) dominates in the gaseous phase. In the liquid phase there is a shift of the distribution in favor of the *gauche* conformers ($\phi = \pi/3$) and ($\phi = 5\pi/3$).

The solvation Gibbs energy of the solute *s*, in this case, is given by

$$\exp\left[-\beta\Delta\mu_s^{*l}\right] = \int d\phi\, y^g(\phi)\exp[-\beta\Delta\mu^*(\phi)] \tag{7.173}$$

which is a particular case of (7.167), for a solute with one internal rotation.

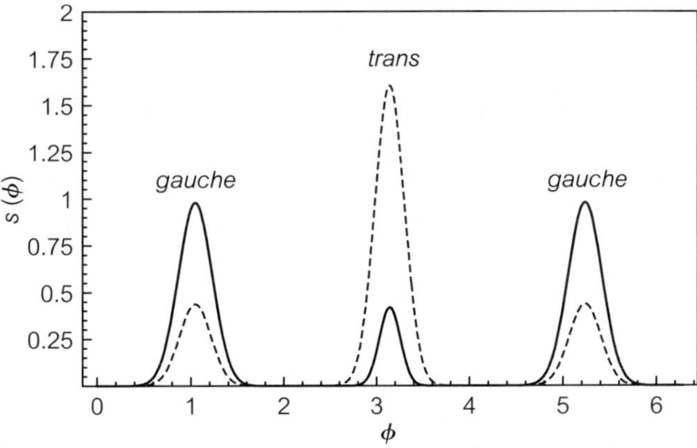

Figure 7.7 The angle distribution of butane, s(ϕ), in the gaseous phase (dashed line) and in the aqueous phase (solid line). The particular illustration here is an "average" result taken from various sources (see text).

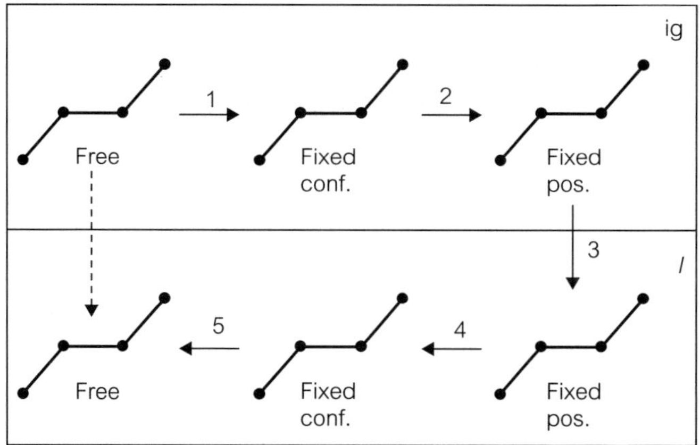

Figure 7.8 The five-step process of transferring butane from the gaseous to the liquid phase (*conf.* = conformation, *pos.* = position).

In this section, we have arrived at the expression for $\Delta\mu_s^{*l}$ by relying on the equilibrium conditions (7.157) among the different conformers.

It is instructive to re-derive this expression in a stepwise process, as depicted in figure 7.8. This stepwise process will also be helpful in understanding the process of solvation as carried out in the case of dissociable solute and will be discussed in the next section. The process of solvating s is carried out in five steps as depicted in figure 7.8. For convenience, we shall use the case of discrete species, i.e., the solute s is distributed among n isomers with mole fractions y_i^g. We can use the language of either a single-solvaton in the system distributed with probability y_i^g, or of a one-mole-of-solvatons distributed with mole fractions y_i^g. The latter is easier to visualize. We also assume that this mole of solvatons are far apart from each other hence, independent. For simplicity, we also assume that the solvent initially does not contain any solutes. We shall start with one mole s in an ideal gas phase. Each molecule has full translational and rotational freedom.

In the first step, we convert all the mole of s molecules into one specific isomer say $i=1$. The corresponding change in Gibbs energy (in the T, P, $\mathbf{N}$ ensemble) is

$$\Delta G_1 = [\mu_1^{*g} + kT\ln\rho_s^g\Lambda_s^3] - \sum y_i^g[\mu_i^{*g} + kT\ln\rho_i^g\Lambda_s^3]. \tag{7.174}$$

Since $\rho_i^g/\rho_s^g = y_i^g$ is the mole fraction of the isomer i in the gaseous phase, we can rewrite (7.174) as

$$\Delta G_1 = [\mu_1^{*g} - \sum y_i^g\mu_i^{*g}] - \sum kTy_i^g\ln y_i^g. \tag{7.175}$$

Note that there are two contributions to ΔG_1. One, in squared brackets, resulting from the changes in the energetics of all the species when transformed into the specific species $i=1$. The second is the increase in the Gibbs energy due to the *loss of identity* of the n distinguishable *species* (i.e., different conformers). This quantity is often referred to as the *demixing* free energy. It is not! In this process, we do not demix anything. Instead we have a process where we had initially n *different* species transformed into *one specific* species (i.e., the conformer $i=1$). We shall refer to this process as assimilation. Assimilation involves an increase in Gibbs energy (or decrease in entropy). See Appendix H and I for more details.

Next, we *freeze* the translational degrees of freedom of this mole of single species. The change in Gibbs energy is

$$\Delta G_2 = -kT \ln \rho_s^g \Lambda_s^3 \tag{7.176}$$

where ρ_s^g is the density of this particular single species in the gaseous phase.

In the third step, we solvate the species $i=1$. This involves the change

$$\Delta G_3 = \mu_1^{*l} - \mu_1^{*g}. \tag{7.177}$$

Next, we liberate the solvatons from their fixed positions in l. The corresponding change in Gibbs energy is

$$\Delta G_4 = kT \ln \rho_s^l \Lambda_s^3. \tag{7.178}$$

Finally, we release the constraint on the conformer $i=1$ and allow the solvaton to reach an equilibrium among the species but with a new distribution y_i^l in the liquid phase. The corresponding Gibbs energy change is

$$\Delta G_5 = \left[\sum_i y_i^l \mu_i^{*g} - \mu_1^{*l}\right] + \sum_i kT y_i^l \ln y_i^l. \tag{7.179}$$

Again, ΔG_5 consists of two parts: one due to the change in the internal energetics upon transferring the species $i=1$ into n species (n conformers); the second is due to the process of *acquiring* new identities, i.e., one species is split into n different species. We refer to this process as deassimilation. The deassimilation process always involves a decrease in Gibbs energy (or increase in entropy). Again, we note that the latter quantity is often referred to as the *mixing free energy*. Obviously, it is not. There is no process of *mixing* in this step, as there is no process of *demixing* in step one. We now combine the change in the Gibbs energies of the five steps to obtain

$$\begin{aligned}\sum_{k=1}^{5} \Delta G_k &= \sum_i y_i^l \mu_i^{*l} + kT \ln \sum_i y_i^l \ln \rho_i^l - \sum_i y_i^g \mu_i^{*g} - kT \ln \sum_i y_i^g \ln \rho_i^g \\ &= \sum_i y_i^l \mu_i^l - \sum_i y_i^g \mu_i^g. \end{aligned} \tag{7.180}$$

Note that (7.180) does not depend on the choice of the particular species $i=1$. We could have chosen any other species for carrying out the solvation step. Using the equilibrium conditions among all the species in both phases

$$\begin{aligned}\mu_s^l &= \mu_1^l = \mu_2^l = \cdots = \mu_n^l \\ \mu_s^g &= \mu_1^g = \mu_2^g = \cdots = \mu_n^g\end{aligned} \tag{7.181}$$

we can reach the conclusion that

$$\sum_{k=1}^{n} \Delta G_k = \Delta\mu_s^{*l} = \mu_s^{*l} - \mu_s^{*g} \tag{7.182}$$

which is the solvation Gibbs energy of the solvaton s (i.e., the solute at its equilibrium distribution with respect to all conformers). As we have seen in (7.166), $\Delta\mu_s^*$ can be expressed in terms of the solvation Gibbs energies of all the species as

$$\exp[-\beta\Delta\mu_s^{*l}] = \sum y_i^g \exp[-\beta\Delta\mu_i^{*l}]. \tag{7.183}$$

As a direct result of the equilibrium condition between all the species, we can write the mole fraction of the species i as

$$y_i^l = \frac{q_i \exp[-\beta U_i^* - \beta\Delta\mu_i^*]}{\sum q_i \exp[-\beta U_i^* - \beta\Delta\mu_i^*]}. \tag{7.184}$$

7.9 **Solvation of completely dissociable solutes**

In section 7.2, we introduced the process of solvation as the process of transferring a single molecule from a fixed position in an ideal-gas phase to a fixed position in the liquid. For solutes which do not dissociate into fragments, the solvation Helmholtz or Gibbs energy is related to experimental quantities by the equation

$$\Delta G_s^{*l} = kT \ln(\rho_s^{ig}/\rho_s^l)_{\text{eq}}. \tag{7.185}$$

In this section, we generalize this relation to solutes which dissociate in the liquid. The most important solutes of this kind are electrolytes, but the treatment is general and applies to any type of dissociable solutes.

We consider a molecule which is only in a state of a dimer D in the gaseous phase. When introduced into the liquid, it completely dissociates into two fragments A and B. Of foremost importance is the case of ionic solutes; e.g., D may be KCl, then A and B are K^+ and Cl^-, respectively. For simplicity, we assume that A and B do not have any internal degrees of freedom. The generalization in the case of multi-ionic solutes and polynuclear ions is quite straightforward.

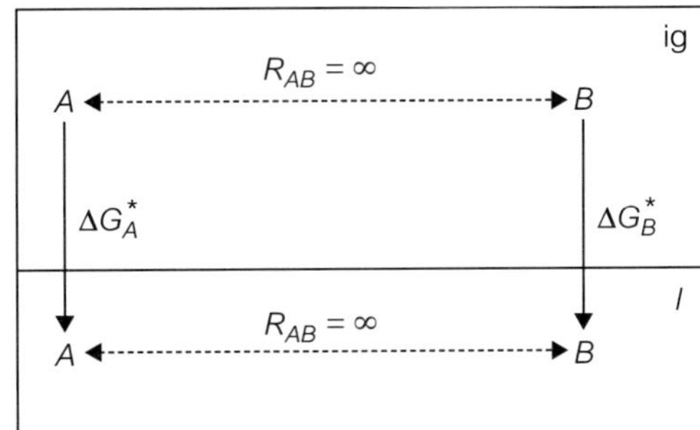

Figure 7.9 The solvation process of two fragments *A* and *B*. This is a straightforward generalization of the process of solvation of a solute *s* as depicted in figure 7.2.

The relevant solvation process is depicted in figure 7.9. Since the particles are presumed not to possess any internal degrees of freedom, the Gibbs energy of solvation of the pair of ions is simply the coupling work of *A* and *B* to the liquid phase *l*; thus

$$\Delta G_{AB}^* = W(A, B|l). \tag{7.186}$$

For simplicity, we shall use below the *T*, *V*, *N* ensemble. Hence, we shall derive an expression for ΔA_{AB}^*. However, it is easy to show that ΔA_{AB}^* at constant *T*, *V* is the same quantity as ΔG_{AB}^* at constant *T*, *P*, provided that the exact volume in the former is equal to the average volume in the latter.

Consider a system of N_W solvent molecules, say water, and N_D solute molecules contained in a volume *V* at temperature *T*. By "solute," we mean those molecules *D* for which we wish to evaluate the solvation thermodynamic quantities. The "solvent," which is usually water, may be any liquid or any mixture of liquids and could contain any number of other solutes besides *D*. In the most general case, N_W will be the total number of molecules in the system except those that are counted as "solutes" in N_D. However, for notational simplicity, we shall treat only two-component systems, *W* and *D*.

For a system of N_W solvent molecules and N_D solute molecules at given *V*, *T*, the corresponding partition function is

$$Q(T, V, N_W, N_D) = \frac{q_W^{N_w} q_A^{N_A} q_B^{N_B} \int d\mathbf{X}^{N_w} d\mathbf{X}^{N_A} d\mathbf{X}^{N_B} \exp[-\beta U(N_W, N_A, N_B)]}{\Lambda_w^{3N_w} \Lambda_A^{3N_A} \Lambda_B^{3N_B} N_A! N_B! N_w!}. \tag{7.187}$$

In writing (7.187), we have assumed that the solute molecules are completely dissociated into *A* and *B* in the liquid; $U(N_w, N_A, N_B)$ is the total potential energy of interaction among the N_w, N_A and N_B molecules at a specific configuration (with $N_D = N_A = N_B$).

We next add one dimer *D*, or equivalently one *A* and one *B*, to the liquid in two ways: first, without any further restrictions; and second, with the constraint that *A* and *B* be placed at two fixed positions $\mathbf{R}_A$ and $\mathbf{R}_B$, respectively. The corresponding partition functions are

$$Q(T, V, N_W, N_D + 1) = \frac{q_W^{N_w} q_A^{N_A+1} q_B^{N_B+1} \int d\mathbf{R}_A \, d\mathbf{R}_B \int d\mathbf{X}^{N_w+N_A+N_B} \exp[-\beta U(N_w, N_D + 1)]}{\Lambda_w^{3N_w} \Lambda_A^{3(N_A+1)} \Lambda_B^{3(N_B+1)} N_W!(N_A + 1)!(N_B + 1)!} \tag{7.188}$$

and

$$Q(T, V, N_W, N_D + 1; \mathbf{R}_A, \mathbf{R}_B) = \frac{q_W^{N_w} q_A^{N_A+1} q_B^{N_B+1} \int d\mathbf{X}^{N_w+N_A+N_B} \exp[-\beta U(N_w, N_D + 1)]}{\Lambda_w^{3N_w} \Lambda_A^{3N_A} \Lambda_B^{3N_B} N_w! \, N_A! \, N_B!}. \tag{7.189}$$

The differences in the two partition functions (7.188) and (7.189) should be noted carefully. In equation (7.188), we have two more integrations over $\mathbf{R}_A$ and $\mathbf{R}_B$, and also one more Λ_A^3 and one more Λ_B^3. Also, we have $(N_A + 1)!$ and $(N_B + 1)!$ in equation (7.188), but only $N_A!$ and $N_B!$ in (7.189). All these differences arise from the constraint we have imposed on the locations $\mathbf{R}_A$ and $\mathbf{R}_B$ in equation (7.189). The total potential energies of interaction in the two systems are related by the equation

$$U(N_W, N_D + 1) = U(N_w, N_D) + U_{AB}(\mathbf{R}_A, \mathbf{R}_B) + B_D(\mathbf{R}_A, \mathbf{R}_B) \tag{7.190}$$

where a shorthand notation is used for the configuration of the system, except for $\mathbf{R}_A$ and $\mathbf{R}_B$. The quantity $U_{AB}(\mathbf{R}_A, \mathbf{R}_B)$ is the direct interaction potential between A and B being at $\mathbf{R}_A$ and $\mathbf{R}_B$, respectively; $B_D(\mathbf{R}_A, \mathbf{R}_B)$ is the "binding energy," i.e., the total interaction energy between the solvaton, i.e., the pair A and B at $(\mathbf{R}_A, \mathbf{R}_B)$ and all the other particles in the system at some specific configuration.

The chemical potential of the solute D in the liquid l is obtained from the partition functions in equations (7.187) and (7.188):

$$\begin{aligned}
\mu_D^l &= -kT\ln\left[\frac{Q(T, V, N_w, \; N_D + 1}{Q(T, V, N_w, \; N_D)}\right] \\
&= -kT\ln\left\{\left[q_A q_B \int d\mathbf{R}_A \, d\mathbf{R}_B \exp[-\beta U_{AB}(\mathbf{R}_A, \mathbf{R}_B)]\right.\right. \\
&\quad \times \left.\int d\mathbf{X}^{N_w+N_A+N_B} \exp[-\beta U(N_w, N_D) - \beta B_D(\mathbf{R}_A, \mathbf{R}_B)]\right]. \\
&\quad \times \left.\left[\Lambda_A^3 \Lambda_B^3 (N_A + 1)(N_B + 1) \int d\mathbf{X}^{N_w+N_A+N_B} \exp[-\beta U(N_w, N_D)]\right]^{-1}\right\} \\
&= -\ln\left\{\frac{q_A q_B}{\Lambda_A^3 \Lambda_B^3 (N_A + 1)(N_B + 1)}\right. \\
&\quad \times \left.\int d\mathbf{R}_A \, d\mathbf{R}_B \exp[-\beta U_{AB}(\mathbf{R}_A, \mathbf{R}_B)] \, \langle \exp(-\beta B_D) \rangle_* \right\}
\end{aligned} \tag{7.191}$$

Here the symbol $\langle\ \rangle_*$ stands for an average over all configurations of the $N_w + N_A + N_B$ particles excluding only the solvatons A and B at $\mathbf{R}_A$ and $\mathbf{R}_B$.

Equation (7.191) may be transformed into a simpler form as follows. The pair distribution function for the species A and B in the liquid is defined by

$$\rho_{AB}^{(2)}(\mathbf{R}_A, \mathbf{R}_B) = (N_A + 1)(N_B + 1) \times \frac{\int d\mathbf{X}^{N_w+N_A+N_B} \exp[-\beta U(N_w, N_D + 1)]}{\int d\mathbf{X}^{N_w+(N_A+1)+(N_B+1)} \exp[-\beta U(N_w, N_D + 1)]}. \tag{7.192}$$

The pair distribution function at infinite separation is denoted by $\rho_{AB}^{(2)}(\infty)$. We now write the following ratio:

$$\frac{\rho_{AB}^{(2)}(\mathbf{R}_A, \mathbf{R}_B)}{\rho_{AB}^{(2)}(\infty)} = \frac{\int d\mathbf{X}^{N_w+N_A+N_B} \exp[-\beta U(N_w, N_D) - \beta B_D(\mathbf{R}_A, \mathbf{R}_B) - \beta U_{AB}(\mathbf{R}_A, \mathbf{R}_B)]}{\int d\mathbf{X}^{N_w+N_A+N_B} \exp[-\beta U(N_w, N_D) - \beta B_D(\infty) - \beta U_{AB}(\infty)]} \tag{7.193}$$

where we have used equation (7.190) and also introduced the notation $B_D(\infty)$ and $U_{AB}(\infty)$ for the binding energy and interaction energy at infinite separation, respectively. Using now the same probability distribution as was used in relation (7.191), we rewrite equation (7.193) in the form

$$\frac{\rho_{AB}^{(2)}(\mathbf{R}_A, \mathbf{R}_B)}{\rho_{AB}^{(2)}(\infty)} = \frac{\exp[-\beta U_{AB}(\mathbf{R}_A, \mathbf{R}_B)]\langle \exp[-\beta B_D(\mathbf{R}_A, \mathbf{R}_D])\rangle_*}{\exp[-\beta U_{AB}(\infty)]\langle \exp[-\beta B_D(\infty)]\rangle_*}. \tag{7.194}$$

Equation (7.194) is now introduced into relation (7.191) to obtain

$$\mu_D^l = -kT \ln\left(\frac{q_A q_B \int d\mathbf{R}_A d\mathbf{R}_B \rho_{AB}^{(2)}(\mathbf{R}_A, \mathbf{R}_B) \exp[-\beta U_{AB}(\infty)]\langle \exp[-\beta B_D(\infty)]\rangle_*}{\Lambda_A^3 \Lambda_B^3 (N_A + 1)(N_B + 1)\rho_{AB}^{(2)}(\infty)}\right). \tag{7.195}$$

We can further simplify (7.195) as follows.

Since we define our zero potential energy at $R_{AB} = \infty$, we have $U_{AB}(\infty) = 0$. Also at $R_{AB} = \infty$, we have $\rho_{AB}^{(2)}(\infty) = \rho_A \rho_B$. The normalization condition for $\rho_{AB}^{(2)}$ in a closed system is

$$\int d\mathbf{R}_A\, d\mathbf{R}_B \rho_{AB}^{(2)}(\mathbf{R}_A, \mathbf{R}_B) = (N_A + 1)(N_B + 1). \tag{7.196}$$

Hence, expression (7.195) can be rewritten in the final form:

$$\mu_D^l = -kT \ln\langle \exp[-\beta B_D(\infty)]\rangle_* + kT \ln\left(\frac{\rho_A \rho_B \Lambda_A^3 \Lambda_B^3}{q_A q_B}\right). \tag{7.197}$$

Note that in relation (7.191), we wrote μ_D^l as an average over all possible locations of A and B; the simplification achieved in equation (7.197) was rendered possible because there is equilibrium among all pairs of A and B at any specific configuration $\mathbf{R}_A$, $\mathbf{R}_B$. This fact allows us to choose one configuration $R_{AB} = \infty$ and use it to obtain the simplified form of the chemical potential μ_D^l.

Next, we identify the liberation Helmholtz energy of the pair A and B. For any specific configuration, this may be obtained from the ratio of the partition functions (7.188) and (7.189), i.e.,

$$\begin{aligned}
&\Delta A(\text{Lib}) \\
&= -kT \ln \frac{Q(T, V, N_w, N_D + 1)}{Q(T, V, N_w, N_D + 1; \mathbf{R}_A, \mathbf{R}_A)} \\
&= -kT \ln\left(\frac{\int d\mathbf{R}_A d\mathbf{R}_B \int d\mathbf{X}^{N_w+N_A+N_B} \exp[-\beta U(N_w, N_D + 1)]}{\Lambda_A^3 \Lambda_B^3 (N_A + 1)(N_B + 1) \int d\mathbf{X}^{N_w+N_A+N_B} \exp[-\beta U(N_w, N_D + 1)])}\right) \\
&= kT \ln[\Lambda_A^3 \Lambda_B^3 \rho^{(2)}(\mathbf{R}_A, \mathbf{R}_B)]
\end{aligned} \tag{7.198}$$

which is a generalization of the expression for the liberation Helmholtz energy of one particle. Here, the liberation Helmholtz energy depends on the particular configuration of A and B from which these particles are being released. For our application, we only need the liberation energy at infinite separation namely,

$$\Delta A(Lib, \infty) = kT \ln(\Lambda_A^3 \Lambda_B^3 \rho_A \rho_B). \tag{7.199}$$

Extracting the liberation Helmholtz energy (7.199) from (7.197), we can identify the pseudo-chemical potential of the pair A and B, i.e.,

$$\begin{aligned}
\mu_D^* &= \Delta A_{AB}^* - kT \ln q_A q_B \\
&= -kT \ln\langle \exp[-\beta B_D(\infty)]\rangle_* - kT \ln q_A q_B.
\end{aligned} \tag{7.200}$$

The chemical potential of D in the gaseous phase, where it is assumed to exist only in dimeric form, is simply

$$\mu_D^g = kT \ln \rho_D^g \Lambda_D^3 q_D^{-1} \tag{7.201}$$

where ρ_D^g is the number density and Λ_D^3 the momentum partition function of D; q_D includes the rotational, vibrational, and electronic partition functions of D. For all practical purposes, we may assume that the dimer D at room temperature is in its electronic and vibrational ground states. We also assume that the rotation may be treated classically; hence, we write

$$q_D = q_{\text{rot}} \exp(-\tfrac{1}{2}\beta h\nu) \exp(-\beta\varepsilon_{\text{el}}) = q_{\text{rot}} \exp(\beta D_0) \tag{7.202}$$

where $\varepsilon_{\text{el}} = U_{AB}(\sigma) - U_{AB}(\infty)$ is the energy required to bring A and B from infinite separation to the equilibrium distance $R_{AB} = \sigma$. The experimental dissociation energy, as measured relative to the vibrational zero-point energy, is

$$D_0 = -\tfrac{1}{2}h\nu - \varepsilon_{\text{el}} = -\tfrac{1}{2}h\nu - [U_{AB}(\sigma) - U_{AB}(\infty)] \tag{7.203}$$

where $\frac{1}{2}h\nu$ is the zero-point energy for the vibration D. Note that in the case of an ionic solution we can either separate the two *ions* from σ to ∞, or first separate the neutral *atoms* and then add the ionization energy and electron affinity of the cation and anion, respectively. Using the equilibrium condition for D in the two phases, and assuming for simplicity that $q_A = q_B = 1$, we have

$$\begin{aligned} 0 &= \mu_D^l - \mu_D^g \\ &= [-kT \ln(\rho_D^g \Lambda_D^3 q_{\text{rot}}^{-1}) - \tfrac{1}{2}h\nu] + [U_{AB}(\infty) - U_{AB}(\sigma)] \\ &\quad - kT \ln\langle \exp[-\beta B_D(\infty)]\rangle_* + [kT \ln \rho_A \rho_B \Lambda_A^3 \Lambda_B^3] \end{aligned} \tag{7.204}$$

From Equation (7.204), we can eliminate the required Helmholtz energy of solvation of the pair AB to obtain the final relation

$$\begin{aligned} \Delta A_{AB}^* &= -kT \ln\langle \exp[-\beta B_D(\infty)]\rangle_* \\ &= -D_0 + kT \ln\left(\frac{\Lambda_D^3}{\Lambda_A^3 \Lambda_B^3 q_{\text{rot}}}\right)\left(\frac{\rho_D^g}{\rho_A^l \rho_B^l}\right)_{\text{eq}}. \end{aligned} \tag{7.205}$$

We see that in this case we need not only the number densities of D and of A and B at equilibrium, but also the molecular quantities Λ_A^3, Λ_B^3, Λ_D^3, q_{rot}, and D_0. Some numerical values of ΔA_{AB}^* are given in Ben-Naim (1987).

Equation (7.204) corresponds to the following stepwise process. At equilibrium $\mu_D^l = \mu_D^g$. We first "freeze in" the translational and rotational degrees of freedom of D, then separate the two fragments A and B from their equilibrium distance to two fixed positions but at infinite separation. Next, solvate the two fragments A and B when they are at infinite separation from each other, finally release the two fragments to attain their liberation free energy.

In deriving equation (7.200) for the pseudo-chemical potential μ_D^*, we have selected one specific distance between the two particles A and B. This was rendered possible by using (7.194). There is an analogy between the procedure used in section 7.8, figure 7.8, and the procedure that we took in this case.

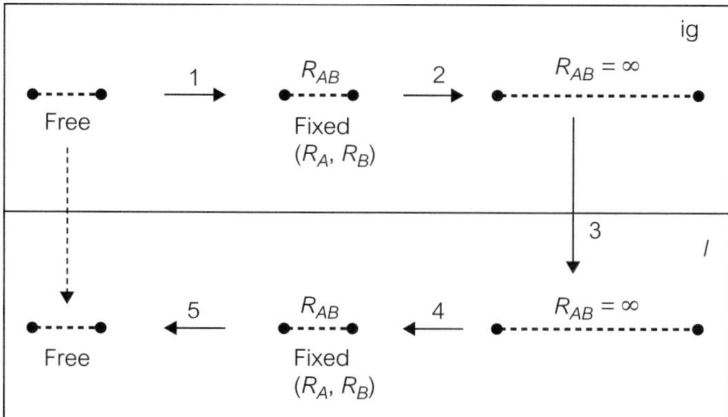

Figure 7.10 An analogous five-step process to transfer a dissociated solute from the gaseous phase to the liquid phase. Here the various "conformers" are the various "dimers" defined by distance R_{AB} (instead of the angle ϕ in figure 7.8). The analog of the species 1 in figure 7.8 is here the "species" at $R_{AB}=\infty$.

To see that suppose we view the molecule D in the liquid l as a mixture of "species," each characterized by the distance R_{AB}. These are the analogs of the different conformers of the solvaton in the previous section. As in section 7.8, since there exists chemical equilibrium between all the species, we could write in analogy with (7.184), the ratio between the two mole fractions of the two "species" in (7.194). In contrast to the case of section 7.8, we have here one particular "conformer," the one for which $\mathbf{R}_{AB}=\infty$, which is convenient because of $U_{AA}(\infty)=0$. This rendered the simplification of (7.191) into (7.195), and hence (7.200) too. A process analogous to the one carried out in section 7.8, figure 7.8, is depicted in figure 7.10. In this process, we transfer the two fragments from the gaseous to the liquid phase. This is slightly different from the process discussed in this section where we assumed that A and B form a dimmer D in the gaseous phase.

7.10 Solvation in water: Probing into the structure of water

In this section, we demonstrate how one can use solvation Gibbs energies to extract structural information on liquid water. We present here only a tiny part of a very large and important field of research. For more details, see Ben-Naim (1992).

The idea that liquid water retains much of the structure of ice upon melting is very old. It has been used in numerous ways to explain the peculiar properties of liquid water and aqueous solutions. For many years, the concept of the structure of water (structure of water) has been defined in terms of some mixture-model description of liquid water. The simplest model of this kind views water as a mixture of two components; one having structure similar to ice, and a second as a random packed fluid. The structure of water in this model is simply the concentration, or the mole fraction of the icy component of the liquid. In the next section, we shall define the concept of structure of water based on a particular form assumed for the water–water pair potential. With this definition, we shall show that from the isotope effect on the solvation Gibbs energy of water in liquid water, we can extract a measure of the structure of the liquid. Adding a solute to pure water will in general change the structure of water. Again, the isotope effect of the solvation Gibbs energy of the solute provide information on the extent of change in the structure of water induced by the solute.

7.10.1 **Definition of the structure of water**

For convenience, let us start with the assumption that the water–water pair potential has the general form (more details can be found in chapter 6 of Ben-Naim 1992)

$$U(\boldsymbol{X}_i, \boldsymbol{X}_j) = U(R_{ij}) + U_{\text{el}}(\boldsymbol{X}_i, \boldsymbol{X}_j) + \varepsilon_{\text{HB}} G(\boldsymbol{X}_i, \boldsymbol{X}_j) \tag{7.206}$$

where $U(R_{ij})$ is a spherically symmetric contribution to the total interaction between two water molecules. This part may conveniently be chosen to have the Lennard-Jones form with the appropriate parameters of neon, an atom isoelectronic with a water molecule. Its main function is to account for the very short-range interaction. The electrostatic interaction part $U_{\text{el}}(\boldsymbol{X}_i, \boldsymbol{X}_j)$ may include the interaction between a few electric multipoles, such as dipole and quadrupoles. This part is to account for the long-range interaction between two water molecules. The third term on the rhs of equation (7.206) is to account for the interaction energy at intermediate distances around 2.8Å. There exists no information on the analytical form of the potential function in this intermediate range of distances. However, recognizing the known fact that two water molecules form a hydrogen bond (HB) at some quite well-defined configuration $(\boldsymbol{X}_i, \boldsymbol{X}_j)$, we shall refer to this part of the potential as the HB part. There have been several suggestions for an explicit form for this part of the potential. However, for our purposes we shall not need to describe this function in any detail. Instead, it will be sufficient to describe the most important feature of this function that is essential for our definition of the structure of water. This part should be present in any reasonable pair potential used to study liquid water†.

We assume that the HB part consists of an energy parameter ε_{HB}, which we refer to as the HB energy, and a geometric function $G(\boldsymbol{X}_i, \boldsymbol{X}_j)$, which is essentially a stipulation on the relative configuration of the pair potential of molecules i and j. Namely, this function attains a maximum value of unity whenever the configuration of the two molecules is most favorable to form a HB. Its value drops sharply to zero when the configuration deviates considerably from the one required for the formation of a HB.

With this qualitative description, we define the function $G(\boldsymbol{X}_i, \boldsymbol{X}_j)$ by

$$G(\boldsymbol{X}_i, \boldsymbol{X}_j) = \begin{cases} 1 & \text{if } i \text{ and } j \text{ are in configurations} \\ & \text{favorable for a HB formation} \\ 0 & \text{for all other configurations.} \end{cases} \tag{7.207}$$

Although we did not specify which configurations are favorable for a HB formation, we have in mind configurations such that the O–O distance is about 2.76 Å, and that one of the O–H bonds on one molecule is directed toward one of the lone pair electrons on the second molecule. Also, in equation (7.207), we let the value of $G(\boldsymbol{X}_i, \boldsymbol{X}_j)$ drop sharply to zero. One way of achieving such a drop in a continuous manner is by using Gaussian functions or similar functions. For more details, see chapter 6 in Ben-Naim (1992).

Clearly, with the aforementioned qualitative description of $G(\boldsymbol{X}_i, \boldsymbol{X}_j)$, we leave a great deal of freedom as to the manner in which we split the potential (7.206) into its various contributions, and as to the specific form we wish to choose for this function. The important feature that we need for our definition of the structure of water is the following: let $\boldsymbol{X}^N = \boldsymbol{X}_1, \ldots, \boldsymbol{X}_N$ be a specific configuration of all the N water molecules in the system at some T and P. We select one molecule, say the ith one, and define the quantity

$$\psi_i(\boldsymbol{X}^N) = \sum_{\substack{j=1 \\ j \neq i}}^{N} G(\boldsymbol{X}_i, \boldsymbol{X}_j). \tag{7.208}$$

Since $G(\boldsymbol{X}_i, \boldsymbol{X}_j)$ by definition (7.207) contributes unity to the sum on the rhs of equation (7.208) whenever the jth molecule is hydrogen bonded to the ith molecule, $\psi_i(\boldsymbol{X}^N)$ measures the number of HBs in which the ith molecule participates when the entire system is at the configuration $\boldsymbol{X}^N$. Based on what

† It should be stressed that (7.206) is not, and should not be, the actual pair potential for water molecules in vacuum. The latter, even if exactly known, will not be useful in the study of the properties of liquid water. The reason is that the structure of water is determined not only by the pair potential but also by higher order potentials. Therefore, (7.206) should be viewed as an effective pair potential designed for the study of the properties of liquid water.

we know about the behavior of water molecules, $\psi_i(\boldsymbol{X}^N)$ may attain values between zero and four.[†]

When $\psi_i(\boldsymbol{X}_N) = 4$, we may say that the local structure around the *i*th molecule, at the configuration, $\boldsymbol{X}^N$ is similar to that of ice. When $\psi_i(\boldsymbol{X}^N) = 0$, the local structure around the *i*th molecule bears no resemblance to that of ice. Hence, we can use $\psi_i(\boldsymbol{X}^N)$ to serve as a measure of the *local* structure around the *i*th molecule at the given configuration of the whole system.

Next, we define the average value of $\psi_i(\boldsymbol{X}^N)$ in the *T*, *P*, *N* ensemble:

$$\langle \psi \rangle_0 = \int dV \int d\boldsymbol{X}^N P(\boldsymbol{X}^N, V) \psi_i(\boldsymbol{X}^N) \tag{7.209}$$

where $P(\boldsymbol{X}^N, V)$ is the fundamental distribution function in the *T*, *P*, *N* ensemble. Since all molecules in the system are equivalent, the average in (7.209) is independent of the subscript *i*; hence, we denote this quantity $\langle \psi \rangle_0$ and not $\langle \psi_i \rangle_0$. The subscript 0 indicates that the average is taken with the distribution $P(\boldsymbol{X}^N, V)$ for *pure* water; this should be distinguished from a conditional average discussed in the following sections.

Thus, $\langle \psi \rangle_0$ is the average number of the HBs formed by (any) single water molecule in pure water at a given *T* and *P*. This quantity may serve as a definition of the local structure around a given molecule in the system. The total average of the HBs in the entire system is evidently

$$\langle HB \rangle_0 = \frac{N}{2} \langle \psi \rangle_0. \tag{7.210}$$

The division by 2 is required since in $N\langle \psi \rangle_0$ we count each HB twice.

From the definition of $\psi_i(\boldsymbol{X}^N)$, we may also rewrite equation (7.210) in the form

$$\begin{aligned}
\langle \mathrm{HB} \rangle_0 &= \frac{N}{2} \int dV \int d\boldsymbol{X}^N P(\boldsymbol{X}^N, V) \sum_{\substack{j=1 \\ j \neq i}}^{N} G(\boldsymbol{X}_i, \boldsymbol{X}_j) \\
&= \frac{1}{2} \int dV \int d\boldsymbol{X}^N P(\boldsymbol{X}^N, V) \sum_{i=1}^{N} \sum_{\substack{j=1 \\ j \neq i}}^{N} G(\boldsymbol{X}_i, \boldsymbol{X}_j) \\
&= \int dV \int d\boldsymbol{X}^N P(\boldsymbol{X}^N, V) \sum_{i=1}^{N} \underset{i<j}{} \sum_{j=1}^{N} G(\boldsymbol{X}_i, \boldsymbol{X}_j).
\end{aligned} \tag{7.211}$$

[†] Basically, ψ_i is a counting function, as defined in section 2.7. Because of the "on and off" nature of the HB, the distribution of ψ_i is discrete, i.e., the *i*th molecule can have 0, 1, 2, 3, 4, hydrogen-bonded neighbors.

In subsequent sections, we shall use the last form of the rhs of relation (7.211) as our definition of the structure of water. Clearly, either $\langle\psi\rangle_0$ or $\langle \mathrm{HB}\rangle_0$ can be used for that purpose. Also, as we shall see in the following sections, we need not assume a full pairwise additivity of the total potential energy of the system. Instead, we require only that

$$U(\mathbf{X}^N) = U'(\mathbf{X}^N) + \varepsilon_{\mathrm{HB}} \sum_{i=1}^{N} \sum_{\substack{j=1 \\ i<j}}^{N} G(\mathbf{X}_i, \mathbf{X}_j) \tag{7.212}$$

where in U' we lumped together both the Lennard-Jones and the electrostatic interactions among all the N water molecules. The pairwise additivity is required only from the total HB interactions, as is explicitly written on the rhs of equation (7.212).

7.10.2 **General relations between solvation thermodynamics and the structure of water**

We first consider the case of a simple spherical solute s in a very dilute solution in water w. By very dilute, we mean that all solute–solute interactions may be neglected. Formally, this is equivalent to a system containing just one s solute and N water molecules. Also, for convenience, we assume that the system is at a given temperature T and volume V. The Helmholtz energy of solvation of s in this system is

$$\Delta A_s^* = -kT \ln\langle \exp(-\beta B_s)\rangle_0 \tag{7.213}$$

where

$$\begin{aligned}\langle \exp(-\beta B_s)\rangle_0 &= \frac{\int d\mathbf{X}^N \exp[-\beta U_N(\mathbf{X}^N) - \beta B_s(\mathbf{R}_s, \mathbf{X}^N)]}{\int d\mathbf{X}^N \exp[-\beta U_N(\mathbf{X}^N)]} \\ &= \int d\mathbf{X}^N P(\mathbf{X}^N) \exp[-\beta B_s(\mathbf{R}_s, \mathbf{X}^N)].\end{aligned} \tag{7.214}$$

We have denoted by $B_s(\mathbf{R}_s, \mathbf{X}^N)$ the total binding energy of s to all the N water molecules at the specific configuration $\mathbf{X}^N$. The solute is presumed to be at some fixed position $\mathbf{R}_s$. However, since the choice of $\mathbf{R}_s$ is irrelevant to the value of the solvation Helmholtz energy, it does not appear in the notation on the lhs of equation (7.214). We also note that the average $\langle\,\rangle_0$ is taken with the probability distribution of the N water molecules of *pure* liquid water, i.e., in the absence of a solute s at $\mathbf{R}_s$. The subscript zero serves to distinguish this average from a conditional average introduced below.

The solvation entropy of s is obtained by taking the derivative of equation (7.214) with respect to T, i.e.,

$$\Delta S_s^* = -(\partial \Delta A_s^*/\partial T)_{V,N}$$
$$= k\ln\langle \exp(-\beta B_s)\rangle_0 + \frac{1}{T}(\langle B_s\rangle_s + \langle U_N\rangle_s - \langle U_N\rangle_0) \qquad (7.215)$$

where the symbol $\langle\,\rangle_s$ stands for a *conditional* average over all configurations of the solvent molecules, given a solute s at a fixed position $\mathbf{R}_s$. This conditional distribution is defined by

$$P(\mathbf{X}^N/\mathbf{R}_s) = \frac{\exp[-\beta U_N(\mathbf{X}^N) - \beta B_s(\mathbf{R}_s, \mathbf{X}^N)]}{\int d\mathbf{X}^N \exp[-\beta U_N(\mathbf{X}^N) - \beta B_s(\mathbf{R}_s, \mathbf{X}^N)]}. \qquad (7.216)$$

The solvation energy is obtained from equations (7.213) and (7.215) i.e.,

$$\Delta E_s^* = \Delta A_s^* + T\Delta S_s^* = \langle B_s\rangle_s + \langle U_N\rangle_s - \langle U_N\rangle_0. \qquad (7.217)$$

Had we used the solvation process at constant T, P, we should have obtained the same formal expression for ΔG_s^* in (7.213), for ΔS_s^* in (7.215) and for ΔH_s^* in (7.217), but with the interpretation of all the averages as averages in the T, P, N ensemble.

The solvation energy as presented in equation (7.217) has a very simple interpretation. It consists of an average binding energy of s to the system, and a change in the average total interaction energy among all the N water molecules, induced by the solvation process. For any solvent, this change in the total potential energy may be reinterpreted as a *structural change* induced by s on the solvent. This aspect is dealt with in great detail in chapter 5 of Ben-Naim (1992). For the special case of liquid water, we use the split of the total potential energy as in equation (7.212) to rewrite equation (7.217) as

$$\Delta E_s^* = \langle B_s\rangle_s + \langle U_N'\rangle_s - \langle U_N'\rangle_0 + \varepsilon_{\mathrm{HB}}(\langle HB\rangle_s - \langle HB\rangle_0). \qquad (7.218)$$

Hence, we see that ΔE_s^* explicitly contains a term which originates from the change in the average number of HBs in the system, induced by the solvation process. We also note that the same term also appears in the solvation entropy as written in equation (7.215). Furthermore, when we form the combination of $\Delta E_s^* - T\Delta S_s^*$, this term cancels out. The conclusion is that any structural change in the solvent induced by the solute might affect ΔE_s^* and ΔS_s^*, but will have no effect on the solvation Helmholtz energy. This conclusion has been derived for more general processes and for more general notions of structural changes in the solvent (Ben-Naim 1992).

The second case of interest is the solvation of a water molecule in pure liquid water[†]. Again, we assume that we have a T, V, N system and we add one water molecule to a fixed position, say $\boldsymbol{R}_W$. The solvation Helmholtz energy is

$$\Delta A_W^* = -kT \ln\langle \exp(-\beta B_W)\rangle_0 \tag{7.219}$$

where $\langle\,\rangle_0$ indicates an average over all the configurations of the N water molecules excluding the solvaton. The corresponding entropy and energy of solvation of a water molecule are obtained by standard relationships. The results are

$$\Delta S_W^* = k \ln\langle \exp(-\beta B_W)\rangle_0 + \frac{1}{T}(\langle B_W\rangle_W + \langle U_N\rangle_W - \langle U_N\rangle_0) \tag{7.220}$$

and

$$\Delta E_W^* = \langle B_W\rangle_W + \langle U_N\rangle_W - \langle U_N\rangle_0 \tag{7.221}$$

where here the conditional average is taken with the probability distribution

$$P(\mathbf{X}^N/\mathbf{X}_W) = \frac{\exp[-\beta U_{N+1}(\mathbf{X}^{N+1})}{\int d\mathbf{X}^N \exp[-\beta U_{N+1}(\mathbf{X}^{N+1})]} \tag{7.222}$$

which is the probability density of finding a configuration $\boldsymbol{X}^N$ given one water molecule at a specific configuration $\boldsymbol{X}_W$. Hence,

$$P(\mathbf{X}^N/\mathbf{X}_W) = \frac{P(\mathbf{X}^N, \mathbf{X}_W)}{P(\mathbf{X}_W)} \tag{7.223}$$

Formally, equations (7.220) and (7.221) are similar to the corresponding equations (7.215) and (7.217). It follows that one can also give a formal interpretation to the various terms as we had done before. However, since we have here the case of *pure* liquid water, it is clear that the average binding energy of a water molecule is the same as the *conditional* average binding energy

[†] We note that this quantity could note have been defined within the traditional definitions of solvation where only a solute in dilute solutions in a solvent could be investigated. In this sense, the new measure of solvation can be viewed as a "generalization," from two to one component systems.

of a water molecule, i.e.,

$$\begin{aligned}\langle B_W \rangle_W &= \int d\boldsymbol{X}^N P(\boldsymbol{X}^N/\boldsymbol{X}_W) B_W \\ &= \frac{\int d\boldsymbol{X}^N P(\boldsymbol{X}^N, \boldsymbol{X}_W) B_W}{P(\boldsymbol{X}_W)} \\ &= \frac{\int d\boldsymbol{X}^N P(\boldsymbol{X}^N, \boldsymbol{X}_W) B_W}{8\pi^2/V} \\ &= \int d\boldsymbol{X}_W d\boldsymbol{X}^N P(\boldsymbol{X}^N, \boldsymbol{X}_W) B_W \\ &= \langle B_W \rangle_0 \end{aligned} \tag{7.224}$$

where $\langle B_w \rangle_0$ is the average binding energy of a water molecule in a system of pure water (of either N or $N+1$ molecules). Furthermore,

$$\begin{aligned}\langle U_N \rangle_W &= \frac{\int d\boldsymbol{X}^N P(\boldsymbol{X}^N, \boldsymbol{X}_W) U_N}{P(\boldsymbol{X}_W)} \\ &= \frac{\int d\boldsymbol{X}^N P(\boldsymbol{X}^N, \boldsymbol{X}_W) U_N}{8\pi^2/V} \\ &= \int d\boldsymbol{X}_W d\boldsymbol{X}^N P(\boldsymbol{X}^N/\boldsymbol{X}_W) U_N \\ &= \int d\boldsymbol{X}_W d\boldsymbol{X}^N P(\boldsymbol{X}^N, \boldsymbol{X}_W)(U_{N+1} - B_W) \\ &= \langle U_{N+1} \rangle_0 - \langle B_W \rangle_0. \end{aligned} \tag{7.225}$$

Using the latter equation, we can rewrite the expression (7.221) for ΔE^*_W in the simpler form

$$\Delta E^*_W = \langle B_W \rangle_0 + \frac{N+1}{2} \langle B_W \rangle_0 - \langle B_W \rangle_0 - \frac{N}{2} \langle B_W \rangle_0 = \frac{1}{2} \langle B_W \rangle_0. \tag{7.226}$$

Thus, the solvation energy of a water molecule in pure liquid water is simply half the average binding energy of a single water molecule. This result is, of course, more general and applies to any pure liquid (Ben-Naim 1992).

Finally, we note that for convenience, we have used the T, V, N ensemble in this section. Similar results may be obtained in any other ensemble as well.

7.10.3 Isotope effect on solvation Helmholtz energy and structural aspects of aqueous solutions

In the previous section, we derived a general and formal relationship between thermodynamics of solvation and structural changes induced in the water. Now, we present an approximate relationship between the structure of water, and the isotope effect on the solvation Helmholtz energy of a water molecule. Similarly, from the isotope effect on the solvation Helmholtz energy of an inert solute, we can estimate the extent of structural change induced by the solute on the solvent. The presentation in this section is quite brief. A more detailed treatment may be found in chapter 6 of Ben-Naim (1992).

Pure liquid water

As in section (7.10.1), we assume in this section that water molecules interact according to a potential function of the form (7.206). We also assume that H_2O and D_2O have essentially the same pair potential function, except for the HB energy parameter. It is assumed that $|\varepsilon_{HB}|$ is slightly larger for D_2O than for H_2O. We denote by ε_D and ε_H the energy parameters for D_2O and H_2O respectively.

Viewing ΔA_W^* as a function of the HB energy parameter, we expand $\Delta A_{D_20}^*$ about $\Delta A_{H_2O}^*$ to first order in ε_D-ε_H and obtain

$$\Delta A_{D_2O}^* = \Delta A_{H_2O}^* + \frac{\partial \Delta A_W^*}{\partial \varepsilon_{HB}}(\varepsilon_D - \varepsilon_H) + \cdots \tag{7.227}$$

The derivative of ΔA_W^* with respect to ε_{HB} may be obtained from the general expression

$$\begin{aligned}\Delta A_W^* &= -kT \ln\langle \exp(-\beta B_W)\rangle_0 \\ &= -kT \ln\left[\int d\mathbf{X}^N \exp(-\beta U_{N+1}) \Big/ \int d\mathbf{X}^N \exp(-\beta U_N)\right]\end{aligned}$$

Next, we use the general form for the total potential energy (7.212) of the systems of N and $N+1$ particles to perform the differentiation of ΔA_W^* with respect to ε_{HB}. The result is

$$\begin{aligned}\frac{\partial \Delta A_W^*}{\partial \varepsilon_{HB}} &= \frac{\int d\mathbf{X}^N \exp(-\beta U_{N+1}) \sum_{i<j}^{N+1} G(i,j)}{\int d\mathbf{X}^N \exp(-\beta U_{N+1})} \\ &\quad - \frac{\int d\mathbf{X}^N \exp(-\beta U_{N+1}) \sum_{j<j}^{N} G(i,j)}{\int d\mathbf{X}^N \exp(-\beta U_N)} \\ &= \int d\mathbf{X}^N P(\mathbf{X}^N/\mathbf{X}_0) \sum_{i<j}^{N+1} G(i,j) - \int d\mathbf{X}^N P(\mathbf{X}^N) \sum_{i<j}^{N} G(i,j)\end{aligned} \tag{7.229}$$

where the first integral on the rhs of (7.229) is identified as a conditional average, i.e., an average over all the configurations of the N water molecules, given one water molecule at $\boldsymbol{X}_0$. Thus, using equation (7.211), we rewrite (7.229) as

$$\frac{\partial \Delta A_W^*}{\partial \varepsilon_{HB}} = \langle HB \rangle_W^{N+1} - \langle HB \rangle_0^N. \tag{7.230}$$

We note that the conditional average number of HBs in a system of $N+1$ water molecules, given that one of them is at a fixed configuration $\boldsymbol{X}_0$, will not be changed if we release the condition. In other words, the average number of HBs in the system of $N+1$ water molecules is invariant to fixing the location and orientation of *one* of its molecules. Hence, we may write

$$\langle HB \rangle_W^{N+1} = \langle HB \rangle_0^{N+1} = \frac{N+1}{2} \langle \psi \rangle_0 \tag{7.231}$$

and

$$\langle HB \rangle_0^N = \frac{N}{2} \langle \psi \rangle_0. \tag{7.232}$$

Hence, the derivative of ΔA_W^* may be simplified to

$$\frac{\partial \Delta A_W^*}{\partial \varepsilon_{HB}} = \frac{N+1}{2} \langle \psi \rangle_0 - \frac{N}{2} \langle \psi \rangle_0 = \frac{1}{2} \langle \psi \rangle_0 \tag{7.233}$$

and the approximate relation (7.227) is now rewritten as

$$\Delta A_{D_2O}^* - \Delta A_{H_2O}^* \cong \tfrac{1}{2} \langle \psi \rangle_0 (\varepsilon_D - \varepsilon_H). \tag{7.234}$$

On the lhs we have a measurable quantity, the isotope effect on the solvation Helmholtz energy of a water molecule in pure liquid water. On the rhs, we have a measure of structure of pure water $\langle \psi \rangle_0$ (see section 7.10.1). This quantity may be evaluated if we can estimate the difference $\varepsilon_D - \varepsilon_H$. Relation (7.234) has been used to estimate the structure of pure water (Marcus and Ben-Naim 1985; Ben-Naim 1987 Ben-Naim 1992).

Dilute solution of inert solutes

We now extend the treatment for pure liquid water to dilute aqueous solutions. We assume that the solute s does not form any HBs with the solvent molecules. Furthermore, we assume that the solute–solvent interaction is the same between s and H_2o and between s and D_2o. The solvation Gibbs energy of s is

$$\Delta A_s^* = -kT \langle \exp(-\beta B_s) \rangle_0 \tag{7.235}$$

where the average is over all the configurations of the solvent molecules, in the T, V, N ensemble.

Again, viewing H_2O and D_2O as essentially the same liquid, except for a small difference in the HB energy parameter, we write the following expansion similar to equation (7.227) namely

$$\Delta A_s^*(D_2O) = \Delta A_s^*(H_2O) + \frac{\partial \Delta A_W^*}{\partial \varepsilon_{HB}}(\varepsilon_D - \varepsilon_H). \tag{7.236}$$

where now the solvaton is the solute *s*, rather than a water molecule as in (7.227).

The derivative of ΔA_S^* with respect to ε_{HB} is

$$\begin{aligned}\frac{\partial \Delta A_S^*}{\partial \varepsilon_{HB}} &= \frac{\int d\mathbf{X}^N \exp(-\beta U_N - \beta B_S) \sum_{i<j}^N G(i,j)}{\int d\mathbf{X}^N \exp(-\beta U_N - \beta B_S)} \\ &\quad - \frac{\int d\mathbf{X}^N \exp(-\beta U_N) \sum_{i<j}^N G(i,j)}{\int d\mathbf{X}^N \exp(-\beta U_N)} \\ &= \langle \mathrm{HB} \rangle_s^N - \langle \mathrm{HB} \rangle_0^N. \end{aligned} \tag{7.237}$$

Therefore, the expansion to first order in first ε_D-ε_H becomes

$$\Delta A_s^*(D_2O) - \Delta A_s^*(H_2O) \cong \left(\langle \mathrm{HB} \rangle_s^N - \langle \mathrm{HB} \rangle_0^N\right)(\varepsilon_D - \varepsilon_H). \tag{7.238}$$

On the lhs of equation (7.238), we have measurable quantities – the solvation Helmholtz (or Gibbs) energy of a solute in H_2O and in D_2O. On the rhs, we have the quantity $\left(\langle \mathrm{HB} \rangle_s^N - \langle \mathrm{HB} \rangle_0^N\right)$, which measures the change in the average number of HBs in a system of N water molecules, induced by the solvation process. This may be estimated, provided we have an estimated value for the difference $\varepsilon_D - \varepsilon_H$. Thus, whereas in (7.234) we obtained an estimate of the *structure* of *pure* water, here we got an estimate of the structural changes induced by a simple solute. More details can be found in Ben-Naim (1992).

7.11 **Solvation and solubility of globular proteins**

We end this long chapter with a brief discussion of a very important subject of intensive research. We present here only a few aspects of protein solvation. Since proteins do not have any measurable vapor pressure, their solvation Gibbs energy cannot be measured. It is also extremely difficult to compute it either theoretically or by simulation methods. However, owing to the utmost importance in understanding biochemical processes involving proteins (such as protein folding, binding of small molecules to proteins or binding of proteins to DNA), it is worthwhile to invest time and effort, if only to gain a feeling for the degree of complexity of the solvation of these molecules.

As we have noted in section 7.1, the solvation Gibbs energy of molecules is needed whenever we are interested in the standard Gibbs energy of a reaction carried out in a solvent. We have seen that if we know the solvation Gibbs energies of all the molecules involved in a chemical reaction, we can calculate the equilibrium constant of the reaction in the liquid, from the knowledge of the equilibrium constant of the same reaction in the gaseous phase. Protein folding is an example of such a reaction, which we write as

$$U \rightarrow F \tag{7.239}$$

where U and F are the unfolded and the folded forms of the protein. Clearly, knowing the solvation Gibbs energies of both U and F will tell us quantitatively why and to what extent the folded form is much more stable in aqueous solutions than the unfolded form. Studying the dependence of the solvation Gibbs energy on temperature, pressure and solvent composition will also tell us when we can expect to observe destabilization of the folded form leading to denaturation of the protein.

Clearly, the solvation Gibbs energy of proteins cannot be studied experimentally. This would require measurable vapor of the protein. Therefore, the only option of studying the solvation of proteins is by theoretical means. As we shall see below, in spite of the enormous complexity of the problem, theory *does* suggest some guidance as to how to dissect the problem into relatively small and manageable problems.

Before describing this, it should be noted that the solvation Gibbs energy of a globular protein is in itself an important quantity since it determines the *solubility* of the protein; the solubility of proteins is a marvel in itself. The problem is this: suppose we consider a medium size protein of about 150 amino acid residues; there are some 20^{150} possible sequences of polypeptides of such a length. In the "beginning," when polypeptides of random sequences formed spontaneously, they were probably very insoluble in water (see below). Evolution has probably not acted on this immense number of polypeptides but only on a tiny fraction of these which were soluble in aqueous solutions. Today, we have both soluble and insoluble (in water) proteins in living systems. We shall only be interested in the former: those which can be carried by water – the main component of the blood – from one place to another within the living system.

But what makes this special class of protein soluble? As we shall see, the answer is not fully known. However, some indications as to the molecular reasons for the solubility were recently revealed (Wang and Ben-Naim 1996, 1997).

The solubility is determined relative to the solid phase. At saturation we have the equilibrium condition[†]

$$\mu_s^s = \mu_s^l = \mu_s^{*l} + kT \ln \rho_s^l \Lambda_s^3. \tag{7.240}$$

Thus, the solubility is determined by the quantity

$$(\rho_s^l)_{\text{eq}} = \Lambda_s^{-3} \exp[-\beta(\mu_s^{*l} - \mu_s^s)]. \tag{7.241}$$

Since μ_s^s, the chemical potential of the pure solid, is constant we can estimate the relative solubilities of s in two liquids, say and l_1 and l_2 by

$$\left(\frac{\rho_s^{l_1}}{\rho_s^{l_2}}\right)_{\text{eq}} = \exp[-\beta(\mu_s^{*l_1} - \mu_s^{*l_2})] = \exp[-\beta(\Delta\mu_s^{*l_1} - \Delta\mu_s^{*l_2})] \tag{7.242}$$

where $\Delta\mu_s^{*l_1}$ is the solvation Gibbs energy of s in liquid l_1. Since this cannot be measured, we must resort to some theoretical estimates of these quantities. Here we encounter two serious difficulties. One is the large number of contributions to $\Delta\mu_s^{*l}$. The second is more subtle. If the solubility is known to be small, then $\Delta\mu_s^{*l}$ or (μ_s^{*l}) is independent of ρ_s^l. Therefore, having an estimate of $\mu_s^{*l} - \mu_s^s$ (or $\Delta\mu_s^{*l}$) can give us the solubility of s. However, if the solubility is large (beyond the limit of DI behavior), then both μ_s^{*l} and $\Delta\mu_s^{*l}$ depend on ρ_s^l. The problem is that one must compute $\Delta\mu_s^{*l}$ in the liquid l, having a solute with an *unknown* concentration ρ_s^l. In other words, $(\rho_s^l)_{\text{eq}}$ is *determined* by $\Delta\mu_s^{*l}$. But the computation of $\Delta\mu_s^{*l}$ depends on the knowledge of $(\rho_s^l)_{\text{eq}}$, which depends on the knowledge of $\Delta\mu_s^{*l}$ and so on. Of course, had we had an analytical expression for the dependence of μ_s^{*l} on ρ_s^l, we could have tried to solve the implicit equation

$$(\rho_s^l)_{\text{eq}} = \Lambda_s^{-3} \exp[-\beta(\mu_s^{*l}(\rho_s^l) - \mu_s^s]. \tag{7.243}$$

Such a relation unknown, even for simple solutes. We note that this problem exists for any solute, the solubility of which is beyond the region of DI behavior.

We shall leave this problem for a while. We next discuss some aspects of the solvation Gibbs energy of protein, and at the end of this section present some tentative conclusions regarding the molecular reasons for the solubility of proteins.

[†] The subscript s is for solute. The superscript s is for solid phase.

We now turn to the theoretical consideration of the solvation Gibbs energy of a globular protein which is very diluted in water. Assume for simplicity that the protein is a rigid molecule (it is certainly not), and that all the internal degrees of freedom are unaffected by the insertion process – again, there is such an effect, and probably large. However, for the qualitative discussion below, we ignore these effects, in which case, the solvation Gibbs energy is simply the coupling work of the protein s to the solvent l:

$$\Delta\mu_s^{*l} = W\langle s|l\rangle = -kT\ln\langle\exp[-\beta B_s\rangle_0 \tag{7.244}$$

where B_s is the total binding energy of the protein to all solvent molecules at a specific configuration, $\boldsymbol{X}^N$, and the average in (7.244) is over all the configurations of the solvent molecules, in the absence of s. Clearly, since the protein molecule is very large (having many groups on its surface, each of which interacts differently with the solvent molecules), a calculation of $W(s\,|\,l)$ is beyond the reach of our computational means.

The procedure we undertake here is to dissect the solvation Gibbs energy into small, more manageable quantities. Here, only a brief description is presented. For more details, see Ben-Naim (1992). First, one assumes a pairwise additivity for the solute–solvent interactions, i.e., we write the binding energy of the protein α† to the solvent at a specific configuration $\boldsymbol{X}_1, \ldots, \boldsymbol{X}_N$ as

$$B_\alpha = \sum_{i=1}^{N} U(\boldsymbol{X}_\alpha, \boldsymbol{X}_i). \tag{7.245}$$

For simple solute α, such as argon, one can separate each of the solute–solvent pair potentials into two contributions (see section 7.7)

$$U(\boldsymbol{X}_\alpha, \boldsymbol{X}_i) = U^H + U^S \tag{7.246}$$

where U^H is the "hard," or the repulsive part of the interaction potential function, and U^S is the "soft," or the van der Waals part of the interaction. For such a solute, the solvation Gibbs energy may be written as a sum of two terms

$$\Delta G_\alpha^* = \Delta G_\alpha^{*H} + \Delta G_\alpha^{*S/H}. \tag{7.247}$$

The first term corresponds to the solvation of the hard core of the solute. This is equivalent to the work required to create a cavity of a suitable size and shape to accommodate the solute. The second term is the *conditional* solvation Gibbs energy of the soft part, given that the hard part has already been turned on (see section 7.7).

† We change notation for the solute form s to α since we shall need s to denote the "soft" interaction below.

For proteins, the simple split of the solute–solvent pair potential as in (7.246) is not appropriate; one needs a more elaborate description of the ingredients of this pair potential. There are several ways of performing such a split into a sum of contributions. One way is to recognize that in addition to the hard and soft parts of the potential, there are also specific functional groups such as charged or polar groups which interact with water in a way different from a simple nonpolar group.

To account for these specific interactions, we write the solute–solvent pair interaction for protein α, in generalization of (7.246), as

$$U(\boldsymbol{X}_\alpha, \boldsymbol{X}_i) = U^H + U^S + \sum_k U_k \tag{7.248}$$

where U_k is the contribution of the kth functional group to the solute–solvent interaction. Note that U_k includes both hard and soft, as well as any specific interaction with the solvent. Figure 7.11 shows a schematic example of such a split of the total solute–solvent interaction. When the solvation Gibbs energy is computed through (7.244) using (7.248) we obtain

$$\Delta G_\alpha^* = \Delta G_\alpha^{*H} + \Delta G_\alpha^{*S/H} + \sum_k \Delta G_\alpha^{*k/H,S} + \sum_{k,j} \Delta G^{*k,j/H,S} + \sum_{i,j,k} \Delta G_\alpha^{*i,j,k/H,S} + \cdots \tag{7.249}$$

The first two terms on the rhs of (7.249) are the same as in (7.247). The third term is the coupling work of all the independently solvated groups. The fourth term includes all the pair correlated groups, and so forth.

Thus, even when the additivity assumption in (7.248) is a good approximation, or even exact, the solvation Gibbs energy of α is, in general, not additive with respect to the contributions of all the functional groups. The reason is that the solvation of a group of two or more functional groups might

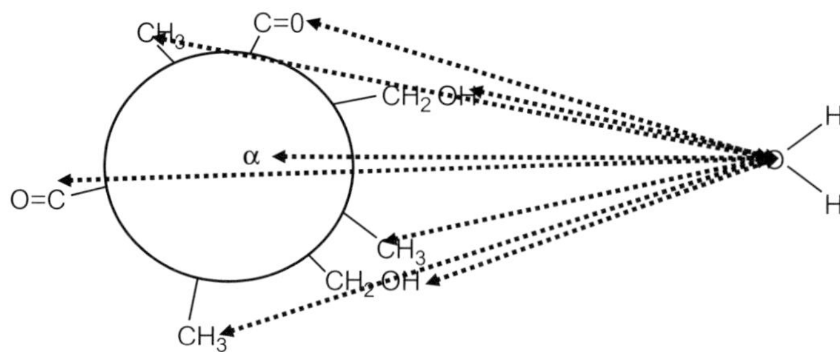

Figure 7.11 A schematic split of the protein–water interaction into a sum of "group" interactions, similar to the split depicted in figure 7.4.

be correlated, and therefore, depending on the extent of correlation, one must account for independent functional groups, pair correlated groups, triplet correlated groups, etc.

The expansion of the solvation Gibbs energy as done in (7.249) is useful in the study of the various contributions to the solubility of a globular protein. Of course, the specific implementation of such an expansion depends upon many, some arbitrary, decisions on how to split the potential function into its ingredients. For instance, one must decide where to make the cut-off between the hard and soft part of the interaction (a problem which exists even for the simplest solute such as argon), how to select the criterion which distinguishes between functional groups that are exposed or unexposed to the solvent, etc. These details are not described here; see Wang and Ben-Naim (1996, 1997). The qualitative meaning of the expansion (7.249) is very simple. Instead of inserting the whole protein to some fixed position in the liquid, we can first insert the hard part H, i.e., creating a cavity of radius $R_{c\alpha v}$. Then, we turn on the soft interaction S; the contribution is the conditional solvation Gibbs energy $\Delta G_{\alpha}^{*S/H}$. Next, we turn on each of the functional groups on the surface of the protein. It is here that one should be careful to turn on first all the independently solvated functional groups, then all the pair correlated groups, and so on. The corresponding sums are on the rhs of (7.249). The expansion (7.249) allow us to estimate the solvation Gibbs energy of the protein from estimates of each of the contributions to ΔG_{α}^{*}, from either theoretical or experimental source.

We next describe very briefly the procedure for estimating some of these contributions.

(1) *The hard part.* The first term in (7.249), ΔG^{*H}, corresponds to turning on the hard part of the protein–solvent interaction potential. This is the same as the free energy of creating a cavity of suitable size at some fixed position in the solvent. We assume that the globular protein is spherical with an effective diameter d_p, so we can calculate ΔG^{*H} using the scaled particle theory (see section and Appendix N). If we choose $d_w = 2.8$ Å as the diameter of a water molecule, then the cavity suitable to accommodate the protein has a radius of

$$R_{c\alpha v} = (d_p + d_w)/2. \tag{7.250}$$

According to SPT, the Gibbs energy of cavity formation is calculated by (see appendix N).

$$\Delta G^{*H} = K_0 + K_1 R_{\mathrm{cav}} + K_2 R_{\mathrm{cav}}^2 + K_3 R_{\mathrm{cav}}^3 \tag{7.251}$$

where the coefficients K_i are

$$\begin{aligned}K_0 &= kT(-\ln(1-y) + 4.5z^2) - \pi P d_w^3/6\\ K_1 &= -kT(6z + 18z^2)/d_w + \pi P d_w^2\\ K_2 &= kT(12z + 18z^2)/d_w^2 - 2\pi P d_w\\ K_3 &= 4\pi P/3\end{aligned}$$

$$y = \pi \rho_w d_w^3/6 \qquad z = y(1-y) \tag{7.252}$$

where ρ_w is the solvent density and P is the pressure. Taking the density of water at 298.15 K, $\rho_w = 3.344 \times 10^{22}$ molecules/cm^3 and P as 1 atmosphere, we can calculate the solvation Gibbs energy of the hard part as a function of protein diameter d_p. The result is that the solvation Gibbs energy is a steeply increasing function of d_p, corresponding to a very steep decrease of the solubility (see figure N.2 in appendix N).

(2) *The soft part.* The second term in (7.249) corresponds to the soft part of the protein–water interaction potential. This is given by (see section 7.7)

$$\Delta G^{*S/H} = -kT \ln\langle \exp[-\beta B_\alpha^S]\rangle_H \tag{7.253}$$

where B^S is the total van der Waals interaction of the protein with the surrounding water molecules.

The soft part $\Delta G^{*S/H}$ is calculated by assuming that the surface of the globular protein consists of methane-like groups, so that the total soft interaction between α and a water molecule is the sum of the interactions between these methane-like molecule and a water molecule. These pair interactions are assumed to be of a Lennard-Jones type. As expected, this part of the solvation Gibbs energy is positive. However, it is about an order of magnitude smaller than the values of the cavity work. Hence, a solute α of the size of a typical globular protein having only hydrophobic groups (methane-like) on its surface will be extremely insoluble in water.

(3) *The specific hydrogen-bonding interactions.* The next and the most critical step is to add the interaction between all hydrophilic groups which are exposed to the solvent and the water molecules. As expected and as is well known, the addition of the hydrophilic groups to the surface of the protein reduces dramatically the solvation Gibbs energy, i.e., making the protein *less insoluble* than the "hard and soft" protein. What was less expected, and to a large extent a surprising finding, was that when one adds all the hydrophilic groups of a specific protein, and assuming that each of these is solvated *independently*, the reduction in ΔG_α^*, though dramatic, is not enough to make ΔG_α^* change its sign

from positive to negative. It was found that the addition of pair and triplet *correlations* between the solvation of the hydrophilic groups is crucial in changing the sign of ΔG_{α}^{*} making it negative, i.e., making α soluble, or very soluble in aqueous solutions. In other words, if we "turn-on" all the hydrophilic groups as being *independent*, each contributing some negative free energy to the total ΔG_{α}^{*}, we shall find this hypothetical protein to be only slightly soluble in water. What makes it very soluble is the fact that some of these hydrophilic groups on the surface are *correlated*, giving an additional negative Gibbs energy to the solvation Gibbs energy of the protein. It has been estimated that each pair correlated hydrophilic groups (at the right distance and orientation) can contribute about $-2.5\,\text{kcal/mol}$ to ΔG_{α}^{*}, which is translated to a factor of about 100 to the solubility of the protein. For details, see Wang and Ben-Naim (1997).

EIGHT

Local composition and preferential solvation

In this chapter, we discuss an important application of the Kirkwood–Buff integrals. We first define the local composition around any molecule in the mixture. Comparison of the local with the global composition leads to the concept of preferential solvation (PS).

After defining the local composition and preferential solvation, we turn to discuss these quantities in more detail: first, in three-component systems and later in two-component systems. This "order" of systems is not accidental. The concept of PS was first defined and studied only in three-component systems: a solute s diluted in a two-component solvent. It is only in such systems that the concept of PS could have been defined within the traditional approach to solvation[†]. However, with the new concept of solvation, as defined in section 7.2, one can define and study the PS in the entire range of compositions of *two-component* systems. In the last section of this chapter, we present a few representative examples of systems for which a complete local characterization is available. These examples should convince the reader that *local* characterization of mixture is not only equivalent to its *global* characterization, but also offers an alternative and more informative view of the mixture in terms of the local properties around each species in the mixture. We also present here a brief discussion of two difficult but important systems: electrolyte and protein solutions. It is hoped that these brief comments will encourage newcomers into the field to further study these topics of vital importance.

[†] There are other methods of studying PS which do not depend on the concepts of solvation thermodynamics. Perhaps the earliest treatment of PS of ions in a two-component solvent was presented by Grunwald et al. (1960). This was followed by Covington and Newman (1976, 1988). For a review see Engberts (1979).

8.1 **Introduction**

The problem of preferential solvation (PS) arises naturally in many studies of the physical-chemical properties of a solute in mixed solvents. Suppose we are interested in a property δ which could be chemical reactivity, spectroscopy, diffusion coefficient, etc., of a solute s in mixed solvents of A and B. The question is how to relate the value of the measured property δ_{AB} of the solute s, in the mixture of composition x_A, to the values δ_A and δ_B in the pure solvents A and B, respectively. We shall first discuss the simplest case of one solute s which is very diluted in a solvent composed of two components A and B.

The most naive assumption would be to write δ_{AB} as an average of δ_A and δ_B of the form

$$\delta_{AB} = x_A \delta_A + x_B \delta_B. \tag{8.1}$$

Such an average would be reasonable for an ideal-gas mixture. In most cases, however, the property δ is affected by the interactions between the solute s and the surrounding solvent molecules. Unless these interactions are very weak, one cannot expect that a relationship of the form (8.1) will hold[†]. The reason is simple: the solute–solvent interactions have limited range, normally of a few molecular diameters, far smaller than the size of the macroscopic system. Therefore, we should expect that all the effects of the solvent molecules on the property δ of the solute arise from the solvent molecules in the neighborhood of the solute, i.e., solvent molecules that are in some *local* "sphere" around the solute s.

In general, in such a local sphere around the solute, the composition of the solvent might be different from the bulk composition x_A. Let x_A^L (and $x_B^L = 1 - x_A^L$) be the composition of the solvent in this local sphere. We shall refer to x_A^L as the *local* composition of the solvent around the solute. We shall discuss further the meaning of the word *local* in the next section, but in the meantime we assume that we can choose some sphere of radius R_a centered at the center of s, where R_a is on the order of a few molecular diameters.

Since only the solvent molecules in *this* sphere are presumed to affect the property δ, a better approximation for δ_{AB} might be

$$\delta_{AB} = x_A^L \delta_A + x_B^L \delta_B. \tag{8.2}$$

Unlike relation (8.1), this new relationship cannot be tested. In general, we do not know what is the local composition x_A^L. However, since the quantities

[†] Actually, since all the quantities in (8.1) are measurable, one can easily test the validity of (8.1) experimentally.

δ_A, δ_B, and δ_{AB} are measurable, one can *define* operationally the *local* composition by assuming the validity of (8.2). In other words, one *defines* x_A^L as

$$x_A^L = \frac{\delta_{AB} - \delta_B}{\delta_A - \delta_B}. \tag{8.3}$$

Clearly, if x_A^L is *defined* in terms of the property δ, we should expect to obtain different values of the local compositions for different properties δ. This is certainly an undesirable feature of a quantity which is supposed to describe the *local* composition of the solvent around s.

In the following sections, we shall define a measure of the local composition around a molecule (not necessarily a solute in dilute solutions) in various solvents. This measure is independent of the property δ of the molecule. In principle, one can use this measure of the local composition to test the validity of an equation of the type (8.2).

Once we have obtained a measure of the *local* composition, we can define the preferential solvation (PS) of a "solute" s with respect to a solvent molecule of species i simply by the difference

$$\mathrm{PS}(i|s) = x_i^L(s) - x_i \tag{8.4}$$

where $x_i^L(s)$ is the mole fraction of the species i in some correlation sphere of radius R_a and of volume $V_a = 4\pi R_\alpha^3/3$ around s.

We shall say that the PS of s with respect to species i is *positive* (or negative) whenever $\mathrm{PS}(i|s)$ is positive (or negative). Some possible cases of PS are shown schematically in figure 8.1.

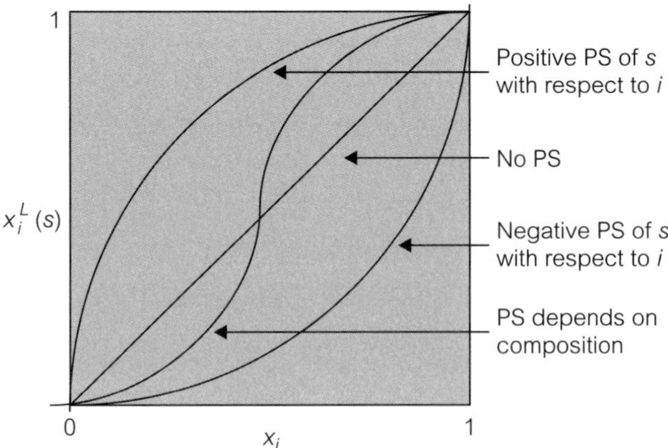

Figure 8.1 Possible signs of preferential solvation of s with respect to i.

8.2 Definitions of the local composition and the preferential solvation

Consider the general case of a multicomponent system of composition $(x_1, \ldots, x_c)$, (with $\sum x_i = 1$) at some temperature T and pressure P. We select an arbitrary molecule of type s in the system. It is convenient to refer to this particularly selected molecule as the solvaton. The solvaton s is identical to all the N_s molecules of the same type. We have selected this particular molecule, referred to as the solvaton, from the center of which we want to examine its local environment. It goes without saying that whatever we find for the solvation s will be equally true for any other s molecule in the system. Also, take note that we do not impose any restriction on the relative concentration of s in the system.

Consider now the volume $V_a = 4\pi R_\alpha^3/3$ of a sphere of radius R_a, centered at the center of solvaton s. At the moment, R_a is any arbitrarily chosen radius. Let $\overline{N}_i(s, R_a)$ be the average number of molecules of type i, in the volume V_a around the solvaton s. The *local* mole fraction of i in V_a is *defined* by

$$x_i^L(s, R_a) = \frac{\overline{N}_i(s, R_a)}{\sum_{j=1}^{c} \overline{N}_j(s, R_a)}. \tag{8.5}$$

We recall that $\rho_i\, g_{is}(R)dR$ is the average number of i particles in the element of "volume" dR, relative to an s molecule at the center of our coordinate system. It is assumed that we have already integrated over all orientations of both i and s to obtain the radial (or the spatial) distribution function, i.e., a function depending on the scalar R only. Hence, $\rho_i\, g_{is}(R)4\pi R^2\, dR$ is the average number of i particles in a spherical shell of radius R and width dR centered at s. We can now write the *local* composition for each i as

$$x_i^L(s, R_a) = \frac{\rho_i \int_0^{R_a} g_{is}(R)4\pi R^2\, dR}{\sum_{j=1}^{c} \rho_j \int_0^{R_a} g_{js}(R)4\pi R^2\, dR}. \tag{8.6}$$

This is valid for any R_a. However, since in general we do not know the various radial distribution functions, this relation is not useful. Before transforming (8.6) into a more useful and computable form, we recall the following two characteristic features of the radial distribution functions $g_{is}(R)$.

First, the correlation between i and s originates from two sources: from the interactions between i and s; and from the closure of the system with respect to the number of particles. We have seen in section 2.5 that if there are no intermolecular interactions in the system (the theoretical ideal gas), there still exists correlation. Placing a particle, say of species s, at a fixed position in the system, changes the average density of s at any other point of the system from N_s/V to $(N_s - 1)/V$. Hence, there is always a residual correlation, which is proportional to N_s^{-1}, which is due to the closure of the system with respect to this species. This kind of correlation does not feature in an open system (see Appendix G).

Second, the correlation due to the intermolecular interactions is usually (except for perfect solids or near the critical point) of short range (normally, a few molecular diameters). For the present discussion, "diameter" could be defined as an average diameter of the species in the system. We do not need any more precise definition of this quantity here. All we need to assume is that the chosen molecular diameter is much smaller than the size of the macroscopic system. Hence, we assume that there exists a correlation distance R_C such that for $R > R_C$ there are no correlations due to intermolecular interactions. The existence of such a correlation distance is supported both by experiments and by theoretical considerations (for more details see Appendix G).

If we now assume that we have taken the thermodynamic limit, i.e., all $N_i \to \infty$, $V \to \infty$ but $\rho_i = N_i/V$ constants, or equivalently if we take all the pair correlations in an open system with respect to all species, we can safely assume that there exists a distance R_C, such that for all i

$$g_{is}(R) \approx 1 \quad \text{for } R > R_C. \tag{8.7}$$

For the specific choice of $R_a = R_C$, beyond which (8.7) is valid, we can write

$$\begin{aligned}\overline{N}_i(s, R_C) &= \rho_i \int_0^{R_C} [g_{is}(R) - 1]4\pi R^2\, dR + \int_0^{R_C} \rho_i 4\pi R^2\, dR \\ &= \rho_i \int_0^{\infty} [g_{is}(R) - 1]4\pi R^2\, dR + \rho_i V_C \\ &= \rho_i G_{is} + \rho_i V_C.\end{aligned} \tag{8.8}$$

We may refer to R_C as the *correlation* radius and to V_C as the correlation volume[†]. In general, R_C would be dependent on the species i and s. Therefore, in (8.8) we shall take the largest R_C which fulfills relation (8.7) for all i. It should be noted that R_C has the true meaning of the correlation distance in the sense that, beyond R_C, there exists *no correlations* due to intermolecular forces.

[†] It should be noted that our definition of R_C is in terms of the pair correlation function as in (8.7). Mansoori and Ely (1985) defined the correlation radius (or the "radius of the sphere of influence") as the distance R_C, for which the *integral* $\int_{R_C}^{\infty} [g_{ij}(R) - 1]4\pi R^2\, dR$ is zero for all pairs of species i and j. This is an unacceptable definition of a correlation radius. Because of the oscillatory nature of $g_{ij}(R)$, one can have more than one R_C for which this integral is zero. Therefore, such a definition of R_C does not confer the meaning of a correlation radius.

Although we have started with an arbitrarily chosen R_a in (8.6), once we have reached the last equality on the rhs of (8.8), specifically the replacement of R_C in the upper limit of the integration by infinity, we are committed to a choice of $R_a \geq R_C$ in (8.8).

With this commitment we have undertaken, we can identify the quantities G_{is} in (8.8) as the Kirkwood–Buff integrals. As we have seen in chapter 4, the KB integrals may be obtained from experimental data for any multicomponent system. The *local* composition in the volume V_a is now written as

$$x_i^L(s, R_a) = \frac{\overline{N}_i(s, R_a)}{\sum_{j=1}^c \overline{N}_j(s, R_a)} = \frac{\rho_i G_{is} + \rho_i V_a}{\sum_{j=1}^c (\rho_j G_{js} + \rho_j V_a)}. \tag{8.9}$$

Equation (8.9) is valid for any $R_a > R_C$ i.e., when R_a is at least as large as the correlation radius of the system. Thus, having all the G_{is} and a choice of a volume V_a, we can calculate the local composition for any mixture. Clearly, if R_a is very large then $x_i^L(s, R_a)$ will approach the bulk composition and we shall miss the local character of x_i^L. Therefore, we have to choose R_a large enough to take into account all the effects of s on its environment, but not too large that the local composition is washed out.

We next define the *preferential solvation* (PS) of s with respect to i as the difference[†]

$$\mathrm{PS}(i|s) = x_i^L(s, R_a) - x_i = \frac{x_i(G_{is} - \sum_{j=1}^c x_j G_{js})}{\sum_{j=1}^c x_j G_{js} + V_a} = \frac{x_i \sum_{j \neq i} x_j (G_{is} - G_{js})}{\sum_{j=1}^c x_j G_{js} + V_a}. \tag{8.10}$$

Note that G_{is} are independent of V_C, provided we have chosen R_C large enough so that the replacement of the upper limit of the integral in (8.8) by infinity is valid.

Note also that the first equality in (8.10) can be defined for any R_a provided we use R_a as the upper limit of the integral in (8.6). The second and third equalities hold true only for $R_a > R_C$ where R_C is the correlation distance in the system. Since we can obtain all the G_{ij} from the inversion of the KB theory, we can also compute $\mathrm{PS}(i|s)$ for each i and s. Clearly, for very large R_a, we have $\mathrm{PS}(i|s) = 0$. This makes sense, since for very large volume V_a, the "local" composition must approach the bulk composition, hence the PS of s with respect to all i will tend to zero. As with the local composition,

[†] It should be noted that a quasi-lattice, quasi-chemical theory of preferential solvation has been developed by Marcus (1983, 1988, 1989, 2002). In the author's opinion, this approach is not adequate to describe PS in liquid mixtures, especially when the different species have widely different sizes.

the PS of s with respect to i may be calculated for any chosen V_a provided we have chosen $R_a > R_C$.

Because of the abovementioned property of PS, and since in general we do not know the value of V_C we expand PS$(i|s)$ in power series about V_a^{-1} to obtain the first order of the PS in V_a^{-1}. The result is

$$\mathrm{PS}(i|s) = 0 + \frac{x_i \sum_{j \neq i} x_j [G_{is} - G_{js}]}{V_a} + O\left(\frac{1}{V_a^2}\right). \tag{8.11}$$

Thus, the first-order *coefficient* of this expansion is

$$\mathrm{PS}^0(i|s) = x_i \sum_{j \neq i} x_j [G_{is} - G_{js}]. \tag{8.12}$$

Clearly, this quantity is independent of V_a. Operationally, PS^0 is the limiting slope of the PS$(i|s)$ drawn as a function of $\varepsilon = V_a^{-1}$. At $\varepsilon = 0$, $PS(i|s) = 0$, hence PS$^0(i|s)$ gives the *direction* in which the PS$(i|s)$ is changing when we increase ε (or decrease V_a from infinity).

Note that the sign of the PS is determined by the numerator of (8.10). Although each of the G_{ij} can be either positive or negative, the entire denominator is always positive. This follows from the fact that the denominator of (8.10) is proportional to the average number of particles as in (8.9). Therefore, the sign of the first-order coefficient in (8.11) gives the correct sign of the PS at any $R_a > R_C$.

Extreme care must be exercised in interpreting both PS$(i|s)$ and PS$^0(i|s)$. First, equation (8.9), when G_{ij} are used as the KB integrals, is valid only for any $R_a > R_C$. Therefore, one cannot compute values of the PS in first, second, etc., coordination spheres where the radius of the coordination sphere is smaller than R_C.†

Clearly, one can compute the PS for any R_a provided that all the G_{ij}'s in (8.9) and (8.10) are defined as in (8.6), i.e., with a *finite* upper limit of the integrals. This requires a detailed knowledge of all $g_{ij}(R)$ as a function of R. If, however, we use G_{ij} from the KB theory, then we must commit ourselves to R_a *beyond* the correlation radius R_C. It is meaningless to compute the PS$(i|s)$ from (8.10) with G_{ij} from the KB theory, for the first or the second coordination spheres.

The quantity PS$^0(i|s)$ in (8.12) is not the PS at any specific volume, not even in a volume larger than the correlation volume V_C. The correct interpretation of this quantity is the following.

† This error has been committed by the author in the original publication of this definition of PS. It has been followed by others, calculating the PS for radii smaller than R_C.

Suppose we use the definition of PS as in (8.10) but with finite radius R_a. In this case, equation (8.10) is a function of R_a of the form

$$\mathrm{PS}(R_a) = \frac{A(R_a)}{B(R_a) + 4\pi R_a^3/3} \tag{8.13}$$

where $A(R_a)$ and $B(R_a)$ depend on R_a through the upper limit of integration.

Plotting PS(R_a) as a function of R_a would give us a very complicated function PS(R_a) (equation 8.13), but once we cross $R_a = R_C$, $A(R_a)$, and $B(R_a)$ become constants and hence, we are guaranteed that this function behaves as

$$\mathrm{PS}(R_a) = \frac{A(R_C)}{B(R_C) + V_a} \tag{8.14}$$

with $A(R_C)$ and $B(R_C)$ constants; the dependence of PS on R_a is only through V_a. This property is illustrated in figure 8.2. The figures here correspond to a two-component system of A and B as discussed in section 8.4. They are shown here only to illustrate the behavior of PS at large V_a.

Taylor expansion of the PS as a function of $\varepsilon = V_a^{-1}$ has the form

$$\mathrm{PS}(\varepsilon) = 0 + \varepsilon \mathrm{PS}^0(i|s) + O(\varepsilon^2). \tag{8.15}$$

Thus, $\mathrm{PS}^0(i|s)$ measures the slope of the function PS(ε) at $\varepsilon = 0$. As we increase ε from zero to $\varepsilon = V_C^{-1}$, we are guaranteed that PS(ε) is a *monotonic* function of ε of the form

$$\mathrm{PS}(\varepsilon) = \frac{\varepsilon A(R_C)}{\varepsilon B(R_C) + 1} \tag{8.16}$$

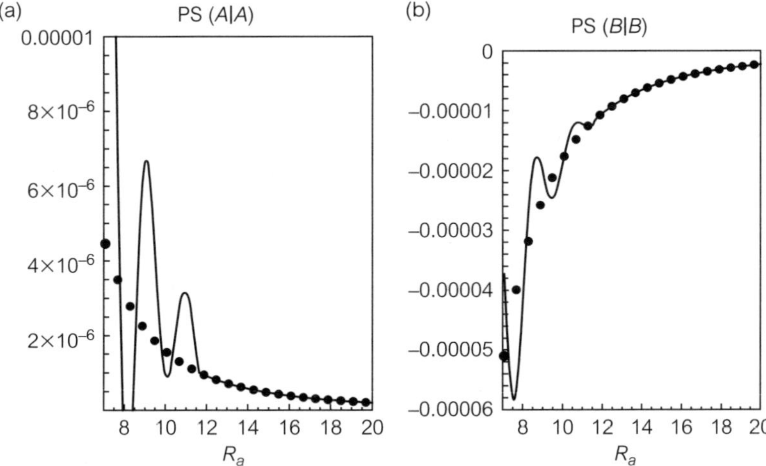

Figure 8.2 The behavior of the function PS($A \mid A$) and PS($B \mid B$) as a function of R_a for large values of R_a (equation (8.13) but applied to a two-component system). The dotted curve is the limiting behavior of the *PS* as $R_a \to \infty$ (equation 8.14) but applied to a two-component system of A and B.

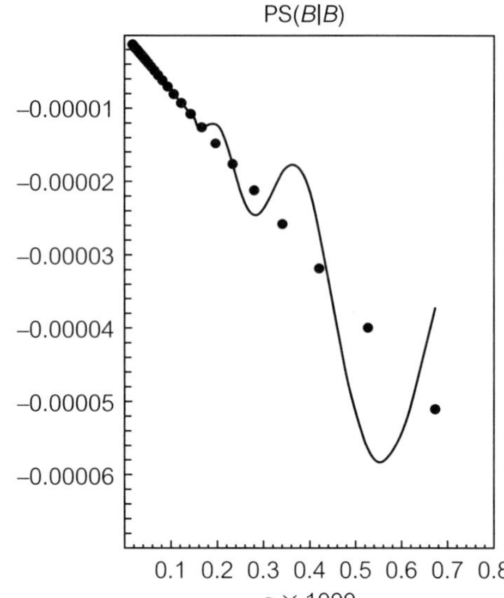

Figure 8.3 Illustration of the behavior of the preferential solvation in the limit $\varepsilon = V_a^{-1} \rightarrow 0$. The full curve was calculated for the same system as in figure 8.2. The dotted curve shows the limiting behavior as $\varepsilon \rightarrow 0$.

with constants $A\,(R_C)$ and $B\,(R_C)$. By taking the limiting slope of this function at $\varepsilon = 0$, we obtain the *direction* of the change of PS(ε) up to $\varepsilon = V_C^{-1}$. Hence, PS^0 gives the *correct sign* of the PS beyond the correlation volume V_C. Figure 8.3 shows the behavior of the PS as a function of ε. The linear behavior (8.15) is the dotted line.

8.3 **Preferential solvation in three-component systems**

In the previous section, we have defined the concepts of the *local* composition and the preferential solvation for any mixture. In a three-component system of *s*, *A*, and *B*, we can define three local compositions around each of the molecules, e.g., $x_A^L(s), x_B^L(s)$ and $x_S^L(s)$.

However, since the sum of these is unity (provided we have chosen the same correlation volume for the three cases), we have only two independent quantities for each type of solvaton. Altogether, we have six independent local compositions $x_i^L(j)$ in the system. Likewise there are three PS(i/s):

$$\begin{aligned}\text{PS}(s|s) &= x_S^L(s) - x_S \\ \text{PS}(A|s) &= x_A^L(s) - x_A \\ \text{PS}(B|s) &= x_B^L(s) - x_B.\end{aligned} \tag{8.17}$$

Since these are related by the equation

$$\mathrm{PS}(s|s) + \mathrm{PS}(A|s) + \mathrm{PS}(B|s) = 0, \tag{8.18}$$

only two of these are independent. Hence, in our system we have six independent PS quantities, two for each species selected as the solvaton.

As in the general case discussed in section 8.2, for any choice of a volume V_a which is at least the size of correlation volume V_C, one can obtain all the KB integrals from the inversion of the KB theory. Hence, we can compute all the local compositions as well as the preferential solvation around any species in the system. To the best of our knowledge, such a complete computation has not been undertaken for any three-component system. However, there exists abundant information, both experimental and theoretical, on a three-component system where one *solute* say, *s*, is very dilute in the mixed *solvents* of *A* and *B*.[†] Although one can define the local composition and *PS* around *s*, *A* and *B*, only one of these has been studied, the component *s* which is diluted in the mixed solvent. It is worthwhile noting that in the traditional approach to solvation thermodynamics, only very dilute solutions could be studied, i.e., a dilute solution of *s* in a mixed solvent of *two* components was a minimal requirement for studying PS. We shall see in the next section that PS can be studied in a two-component system as well.

From now on we focus on the solute s and define the local composition around it. Since $x_s \to 0$, we have from (8.9)

$$x_A^L(s) = \frac{x_A G_{As} + x_A V_a}{x_A G_{As} + x_B G_{Bs} + V_a}. \tag{8.19}$$

Since $x_A^L(s) + x_B^L(s) = 1$, only one local composition around *s* is defined and the corresponding PS is

$$\mathrm{PS}(A|s) = x_A^L(s) - x_A = \frac{x_A x_B (G_{As} - G_{Bs})}{x_A G_{As} + x_B G_{Bs} + V_a}. \tag{8.20}$$

PS($A|s$) is the preferential solvation of *s* with respect to *A*. A positive value of PS($A|s$) means that the *local* mole fraction $x_A^L(s)$ is larger than the bulk mole fraction x_A. In this system, PS($A|s$) = − PS($B|s$), i.e., a positive PS($A|s$) implies a negative PS($B|s$). Hence, in this system, we can also say that a positive PS($A|s$) implies a *preference* of *s* to have *A*'s in its surroundings more than *B*'s. This is also clear from the quantity $G_{As} - G_{Bs}$ in the numerator of (8.20). A positive PS($A|s$) is equivalent to the statement that the affinity between *A* and *s* is larger than the affinity between *B* and *s*. This statement is not valid in

[†] See, for example, Zielkiewicz (1995a, b, 1998, 2000, 2003).

general when s (or any other solvaton) has more than two species in its surroundings, as is clear from the general expression (8.10). In our case, since there are only *two* species in the surroundings of s (as also will be the case discussed in the next section), PS($A|s$) has a true meaning of preferences for one species over the other.

From (8.20), we see that when $G_{As} = G_{Bs}$ (equal affinities), then there is no PS. Furthermore, when either of the components A or B is very dilute, say when $x_A \to 0$, there is only one component in the surroundings of s and PS($A|s$) = 0. On the other hand, if PS($A|s$) = 0 for all compositions x_A, then it follows that $G_{As} = G_{Bs}$.

Note again that the sign of the PS is determined by the numerator of (8.20), i.e., by $G_{As} - G_{Bs}$ of (8.20). The denominator is always positive (though each of the G_{is} could either be positive or negative), and is a monotonically increasing function of R_a as R_a^3 provided that $R_a > R_C$.

The first-order behavior of PS($A|s$) as a function of $\varepsilon = V_a^{-1}$ is

$$\mathrm{PS}(A|s) = \varepsilon x_A x_B (G_{As} - G_{Bs}). \tag{8.21}$$

In figure 8.3, we illustrate the behavior of the PS($A|s$) as a function of ε. The function starts at PS($A|s$) = 0 for $\varepsilon = 0$, and the slope at $\varepsilon = 0$ is $x_A x_B$ $(G_{As} - G_{Bs})$.

All the G_{is}'s in equation (8.20) or (8.21) may be computed from KB theory. For our special case, when s is highly diluted in a solvent mixture of s and B of composition x_A, the chemical potential of the solute is

$$\mu_s^l(T, P, x_A, \rho_s) = \mu_s^{*l}(T, P, x_A) + kT \ln \rho_s \Lambda_s^3. \tag{8.22}$$

The derivative of (8.22) with respect to N_A is

$$\left(\frac{\partial \mu_s^l}{\partial N_A}\right)_{T,P,N_B,N_S} = \left(\frac{\partial \mu_s^{*l}}{\partial N_A}\right)_{T,P,N_B,N_S} - \frac{kT\overline{V}_A}{V} \tag{8.23}$$

where $\overline{V}_A$ is the partial molar volume of A in the mixture. This derivative, as well as $\overline{V}_A$, may be expressed in terms of the Kirkwood–Buff integrals. The algebra is quite lengthy; therefore, we present here the final result (some further details are discussed in Appendix K)

$$\lim_{\rho_S \to 0} \left(\frac{\partial \mu_s^{*l}}{\partial x_A}\right)_{T,P} = \frac{kT(\rho_A + \rho_B)^2}{\eta}(G_{Bs} - G_{As}) \tag{8.24}$$

where η is defined as in chapter 4

$$\eta = \rho_A + \rho_B + \rho_A \rho_B (G_{AA} + G_{BB} - 2G_{AB}). \tag{8.25}$$

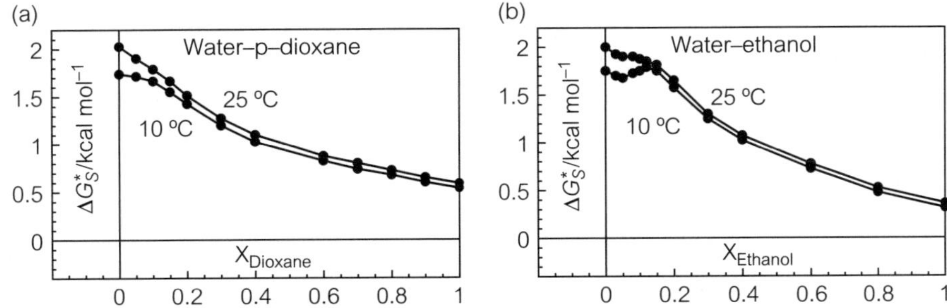

Figure 8.4 Solvation Gibbs energy of methane in mixtures of (a) water-p-dioxane, and in (b) water ethanol at two temperatures.

Let μ_s^{*g} be the pseudo-chemical potential of s in an ideal-gas phase. Clearly, μ_s^{*g} is independent of x_A. Therefore, we may rewrite (8.24) as

$$\lim_{\rho_S \to 0} \left(\frac{\partial \Delta G_s^*}{\partial x_A} \right)_{P,T} = \frac{kT(\rho_A + \rho_B)^2}{\eta} (G_{Bs} - G_{As}) \tag{8.26}$$

where $\Delta G_s^* = \mu_s^{*l} - \mu_s^{*g}$ is the solvation Gibbs energy of s in our system. Thus, from the slope of the solvation Gibbs energy as a function of x_A, we can extract the required difference $G_{Bs} - G_{As}$.

Recall that η is a measurable quantity through the inversion of the Kirkwood–Buff theory. Since $\eta > 0$, the entire quantity $kT(\rho_A + \rho_B)^2/\eta$ is always positive. Therefore, the sign of the derivative on the lhs of (8.25) is the same as the sign of $G_{Bs} - G_{As}$.

Figure 8.4 shows the solvation Gibbs energy of methane in mixtures of water (A) and p-dioxane (B)(figure 8.4a) and water (A), and ethanol (B),(figure 8.4b) as a function of the mole fraction of the organic component (B) throughout the entire range compositions. Regions in which the slope is positive correspond to $G_{As} - G_{Bs} > 0$,[†] which in the case of water–ethanol means that methane is preferentially solvated by water. Note that this occurs only in a very small region, say $0.1 \leq x_{\text{ethanol}} \leq 0.15$. In most of the composition range, methane is preferentially solvated by ethanol. The same is true for the PS in the water–dioxane system.

Relation (8.26) is useful for solute s, the solvation Gibbs energy of which can be determined. If we are interested in proteins as solutes, then (8.26) is impractical. However, we can still measure the solvation free energy of s in a

[†] Note that a positive slope in the curve of ΔG_S^* as a function of x_B corresponds to negative slope of ΔG_S^* as a function of x_A (A being water in this example).

mixture of l relative to, say, pure A (e.g., solvation of protein in a solution relative to the solvation in pure water). The relevant relation is

$$\Delta\Delta G_s^* = \Delta G_s^{*l} - \Delta G_s^{*A} = kT \ln\left(\frac{\rho_s^A}{\rho_s^l}\right)_{\text{eq}} . \tag{8.27}$$

Thus, by measuring the density of s in l, and in pure A at equilibrium with respect to the pure solid s, we can determine $\Delta\Delta G_s^*$ from (8.27). Since ΔG_s^{*A} is independent of x_A, we can apply (8.27) in (8.26) to obtain

$$\lim_{\rho_S \to 0} \frac{\partial(\Delta\Delta G_s^*)}{\partial x_A} = \frac{kT(\rho_A + \rho_B)^2}{\eta}(G_{Bs} - G_{As}) \tag{8.28}$$

which is a useful relation for solutes, the solvation Gibbs energy of which are not measurable. Note also that the Kirkwood–Buff theory allows us to express both G_{Bs} and G_{As} in terms of measurable quantities. Again, the algebra involved is quite lengthy. We therefore present the final result only (more details are in Appendix K). First, we express the partial molar volume of s in the limit of very dilute solution in terms of the Kirkwood–Buff integrals. This relation is

$$\overline{V}_s^0 = \lim_{\rho_S \to 0} \overline{V}_s = kT\kappa_T - \rho_A \overline{V}_A G_{As} - \rho_B \overline{V}_B G_{Bs}. \tag{8.29}$$

In equations (8.26) and (8.29), all the quantities κ_T, η, ρ_A, ρ_B, $\overline{V}_A$, $\overline{V}_B$, $\overline{V}_s^0$, and $\partial\Delta G_s^*/\partial x_A$ are experimentally determinable. Hence, these two equations may be used to eliminate the required quantities G_{Bs} and G_{As}. Thus, denoting

$$a = \lim_{\rho_s \to 0}\left(\frac{\partial \Delta G_s^*}{\partial x_A}\right)_{P,T}, \quad b = \frac{kT(\rho_A + \rho_B)^2}{\eta}, \quad c = kT\kappa_T \tag{8.30}$$

we may solve for G_{AS} and G_{BS}. The results are

$$G_{Bs} = c - \overline{V}_S^0 + \frac{a}{b}\rho_A \overline{V}_A \tag{8.31}$$

$$G_{As} = c - \overline{V}_S^0 - \frac{a}{b}\rho_B \overline{V}_B \tag{8.32}$$

which are the required quantities.

It is interesting to note that if the mixed solvent of A and B forms a symmetrical ideal (SI) solutions, i.e., when

$$G_{AA} + G_{BB} - 2G_{AB} = 0 \tag{8.33}$$

then, equation (8.26) reduces to

$$\lim_{\rho_S \to 0}\left(\frac{\partial \Delta G_s^*}{\partial x_A}\right)_{P,T} = kT(\rho_A + \rho_B)(G_{Bs} - G_{As}). \tag{8.34}$$

Thus, even when A and B are "similar" in the sense of (8.33), they can still have *different* affinities towards a third component. This was pointed out in both the original publication on the *PS* [Ben-Naim (1989, 1990b)] in two-component systems (see next section) as well as in Ben-Naim (1992). It was stressed there that "similarity" does not imply lack of PS. These are two different phenomena. Failing to understand that has led some authors to express their astonishment in finding out that symmetrical ideal solutions manifest preferential solvation. As we have seen above, SI behavior of the mixed solvents of A and B does not imply anything on the PS of s. This can have any value. In the next section, we shall see that the PS in two-component mixtures is *related* to the condition (8.33). However, the PS is not *determined* by the condition of SI solutions. In a three-component system, even when we assume the stronger condition of SI for the whole system, not only on the solvent mixture, i.e., when in addition to (8.33) we also have

$$G_{ss} + G_{AA} - 2G_{As} = 0$$

$$G_{ss} + G_{BB} - 2G_{Bs} = 0 \tag{8.35}$$

then we get

$$G_{AA} - G_{BB} = 2(G_{As} - G_{Bs}). \tag{8.36}$$

Thus, complete ideality [in the sense of (8.33) and (8.35)] does not imply lack of PS. Conversely, lack of PS (in the sense of $G_{As} = G_{Bs}$) does not imply ideality. We shall further discuss this point in relation to PS in *two*-component systems in the next section.

Equations (8.26) and (8.28) are also important in connection with the problem of the effect of added co-solvent on the solubility of s. When the co-solvent is a salt, this effect is well-known as the "salting-out" or the "salting-in" effect.

Let s be any solute diluted in a mixed solvent. Its solubility relative to the solubility in pure A is determined by equation (8.27). If B is a salt and A is water, then for dilute solutions of s we have

$$\left(\frac{\partial \Delta\Delta G_s^*}{\partial x_B}\right)_{P,T} = -\left(\frac{\partial (kT \ln \rho_s^l)_{\mathrm{eq}}}{\partial x_B}\right)_{P,T}. \tag{8.37}$$

If we start from pure A (say water) and add B (say electrolyte), an increase in the solvation Gibbs energy of s is equivalent to a *decrease* in solubility of s. This may be referred as a "salting-out" effect. A decrease in ΔG_s^* upon adding A is equivalent to an increase in solubility of s, hence a "salting-in" effect. From

(8.26) or (8.28), we see that the "salting-out" effect is equivalent to positive PS($A|s$) and the "salting-in" effect is equivalent to negative PS($A|s$). Of course, these conclusions are valid for any solution not necessarily containing electrolytes. It should be noted also that the derivative in (8.37) is related to the Sechenov coefficient – see for example Ruckenstein and Shulgin (2002). The latter is usually expressed in terms of Henry's constant, but in dilute solutions, the Henry constant is equivalent to the solvation Gibbs energy of the solute.

8.4 Local composition and preferential solvation in two-component systems

In the previous section, we discussed the theory of preferential solvation of a solute s in a two-component system. In the traditional concept of solvation thermodynamics, only very dilute solutions could be treated. Therefore, the minimum number of components required for such a study are three: a solute and a two-component solvent. However, the question of PS can also be asked in a two-component system, say of A and B. At any composition x_A, we may focus on an A solvaton and ask what is the PS of A with respect to the two components A and B. Likewise, we may focus on a B solvaton and ask the same, but independent question of the PS of B with respect to the two components A and B. In this sense, the treatment of the two-component system is a "generalization" of the corresponding three-component system, as discussed in the previous section.

Let R_a be any arbitrary radius, and let $V_a = 4\pi R_a^3/3$ be the corresponding sphere, the center of which coincides with the center of an A solvaton. The local composition in the volume V_a is defined as (see section 8.2)

$$x_A^L(A, R_a) = \frac{x_A G_{AA} + x_A V_a}{x_A G_{AA} + x_B G_{AB} + V_a}. \tag{8.38}$$

Clearly, in a two-component system, there is only one *local* composition around A, x_A^L (and $x_B^L = 1 - x_A^L$). Likewise, there is one local composition around B which we choose to define as

$$x_B^L(B, R_a) = \frac{x_B G_{AB} + x_B V_a}{x_A G_{AB} + x_B G_{BB} + V_a}. \tag{8.39}$$

Since there are only two components around either an A solvaton or a B solvaton, we have only one PS of A and one PS of B. For reasons of symmetry, we choose these as

$$\mathrm{PS}(A|A) = x_A^L(A, R_a) - x_A = \frac{x_A x_B (G_{AA} - G_{AB})}{x_A G_{AA} + x_B G_{AB} + V_a} \tag{8.40}$$

and

$$\mathrm{PS}(B|B) = x_B^L(B, R_a) - x_B = \frac{x_A x_B (G_{BB} - G_{AB})}{x_B G_{BB} + x_A G_{AB} + V_a}. \tag{8.41}$$

Note that PS($A|A$) and PS($B|B$) are two independent quantities, but

$$\mathrm{PS}(A|A) = -\mathrm{PS}(B|A)$$

$$\mathrm{PS}(B|B) = -\mathrm{PS}(A|B). \tag{8.42}$$

Again, we stress that all the quantities in (8.38)–(8.41) can be defined for *any* R_a, provided that we choose R_a as the upper limit of the KB integrals as in (8.6). However, if we want to use the G_{ij} from the KB theory, we must take the infinite size limiting behavior of $g_{ij}(R)$ so that only correlations due to intermolecular interactions are captured in G_{ij}, and choose R_a to be at least as large as the correlation distance $R_a \geq R_C$ for all the pairs of species. The PS($A|A$) can be correctly assigned the meaning of *preferential solvation*, in the sense that a positive PS($A|A$) means that an A solvaton *prefers* to be solvated by A compared to B. This is also reflected by the numerator of (8.40), the denominator being always positive. A positive PS($A|A$) is equivalent to a larger affinity between the pair AA, relative to affinity between the pair AB. A similar interpretation applies to PS($B|B$). As in the more general cases discussed in sections 8.2 and 8.3, the PS will necessarily tend to zero as we increase R_a up to the macroscopic size of the system. Hence, it is useful to take the first-order expansion of the PS with respect to $\varepsilon = V_a^{-1}$ to obtain

$$\mathrm{PS}(A|A) = \varepsilon \mathrm{PS}^0(A|A) + \cdots = \varepsilon x_A x_B (G_{AA} - G_{AB}) + \cdots \tag{8.43}$$

$$\mathrm{PS}(B|B) = \varepsilon \mathrm{PS}^0(B|B) + \cdots = \varepsilon x_A x_B (G_{BB} - G_{AB}) + \cdots \tag{8.44}$$

Again, we note that if A and B form a SI solution, in the sense of (8.33), we can write it as

$$G_{AA} - G_{AB} = -(G_{BB} - G_{AB}). \tag{8.45}$$

This does not imply anything regarding neither the sign nor the magnitude of the PS. This fact was stressed and explained in the original publication of the concept of PS in a two-component system (Ben-Naim 1989). The two properties of SI solutions and of PS arise from the two different attributes of the molecules in the system. One arises from the "similarity" of the two species, as defined in (8.45), and the other arises from the difference in the affinities between the species. Hence, in general, one behavior does not imply anything about the other.

Although it is very clear that SI behavior does not necessarily imply lack of PS, some authors expressed puzzlement at finding that PS exists in an SI solution (e.g. Marcus 2002). In the author's opinion, the puzzlement is a result of confusing the condition of *similarity*, as defined in chapter 4, which is a necessary and sufficient condition for SI behavior, with "indistinguishable" intermolecular interactions. In fact, Matteoli (1997), in referring to SI solutions, writes "this reference mixture, for which, by definition, all interactions between species are indistinguishable." Indeed if all interactions between species are indistinguishable, then all of G_{ij} will necessarily be *equal*, hence, no PS could occur and the mixture would be SI. Matteoli has also suggested to "correct" these integrals by subtracting a quantity G_{ij}^{id} which pertains to a hypothetical SI solution. In the author's opinion, these new quantities do not have a clear-cut meaning.

Note, however, that if both $PS(A|A)$ and $PS(B|B)$ are zero, then both sides of (8.45) are zero, hence we have a special case of an SI solution. However, the reverse of this is, in general, not true. In an SI solution (8.45) holds, but nothing is implied regarding either the sign or the magnitude of the PS. It does imply that $PS(A|A)$ and $PS(B|B)$ have opposite signs, or $PS(A|A)$ and $PS(A\,|\,B)$ have the same signs.

Again, we stress that both the KB integrals and preferential solvation are well-defined, well-interpreted, and meaningful quantities. There is no need to "correct" these quantities and replace them with *less* meaningful quantities.[†]

Unfortunately, several authors have already computed these "corrected" quantity,[††] a practice that the present author believes should be abandoned. Furthermore, there is no need to seek a reference state for the proper interpretation of the PS. The PS has a "built-in" reference state in its very definition.

In chapter 4, we have derived a general relationship between the KB integrals and experimentally measurable quantities. Hence, in principle we can compute all the KB integrals for the two-component system, and then obtain the required quantities for the local composition and the preferential solvation.

[†] The present author specifically objects to the usage of the adjective "corrected," in conjunction with the modified KB integrals, as first suggested by Matteoli (1997), and later followed by Marcus (2001), and others. Even if the modified quantities do convey any new information, it does not mean that the original quantities are somehow defective or subject to "correction." This is a fortiori true when the original quantities convey significant information that is not conveyed by the new quantities.

[††] See Marcus (2001) and shulgin and Ruckenstein (2005a, b, c).

To obtain the limiting coefficient of the PS in (8.43) and (8.44), we can use equations (4.49), (4.50) and (4.51) to obtain the quantities

$$G_{AA} - G_{AB} = \frac{kT\rho_T\overline{V}_B}{\rho_B\mu_{BB}} - \frac{1}{\rho_A}$$

$$G_{BB} - G_{AB} = \frac{kT\rho_T\overline{V}_A}{\rho_A\mu_{AA}} - \frac{1}{\rho_B} \qquad (8.46)$$

where $\mu_{ij} = (\partial\mu_i/\partial N_j)_{P,T,N'_j}$. It is seen that the sign of PS in the system is determined by the derivatives of the chemical potentials and the partial molar volumes; no need for the isothermal compressibility of the mixture. One can also use excess Gibbs energies and excess volume to calculate the two PS's.

8.5 **Local composition and preferential solvation in electrolyte solutions**

All the definitions of the local composition and preferential solvation are applicable for any mixture, including electrolyte solutions. Suppose for simplicity we have a solution of an electrolyte D in water W. Viewing this system as a mixture of a two-component system, we can apply all the equations of the previous section.

However, we could also view this system as a three-component system of W, the anion A, and the cation C. If the system is open with respect to W and D, but not to the individual ions, then we must have the following conditions:

$$\rho_D = \rho_A = \rho_C = \rho. \qquad (8.47)$$

The total number of A's around W must be equal to the total number of C's around W, hence

$$\rho_A \int_0^\infty [g_{WA}(R) - 1]4\pi R^2\, dR - \rho_C \int_0^\infty [g_{WC}(R) - 1]4\pi R^2\, dR = 0 \qquad (8.48)$$

or equivalently

$$G_{WA} = G_{WC}. \qquad (8.49)$$

The upper limit of the integrals in (8.48) should be understood to be very large compared with the correlation distance. Likewise, the total number of C around A must be equal to the total number of C around C, hence

$$\rho_C \int_0^\infty [g_{AC}(R) - 1]4\pi R^2\, dR = \rho_C \int_0^\infty [g_{CC}(R) - 1]4\pi R^2\, dR + 1 \qquad (8.50)$$

or equivalently

$$G_{CC} - G_{AC} = -1/\rho. \tag{8.51}$$

Similarly, the conservation of the total number of A's in the system leads to

$$\rho_A \int_0^\infty [g_{AC}(R) - 1] 4\pi R^2\, dR = \rho_A \int_0^\infty [g_{AA}(R) - 1] 4\pi R^2\, dR + 1 \tag{8.52}$$

or equivalently

$$G_{AA} - G_{AC} = -1/\rho. \tag{8.53}$$

The conditions (8.49), (8.51), and (8.53), are often referred to as the electroneutrality conditions since in most cases they apply to electrolyte solutions where the total charge in the system is fixed. However, it is clear that the same conditions apply to any solute D that dissociates into two fragments

$$D \rightarrow A + C \tag{8.54}$$

where A and C could be either charged or neutral. As can be easily checked, inserting the conditions (8.49), (8.51), and (8.53) into the matrix $\boldsymbol{B}$ of the KB theory (see equation 4.23), leads to singular matrix that cannot be inverted as in (4.27).

Clearly, the conditions above hold because the system, though open with respect to D, is not open with respect to A and C *individually*. The KB theory applies for any three-component system of W, A, and C without any restriction on the concentrations of A and C, i.e., when the system is open to each of its components*. If this happens for an electrolyte solution, clearly the conservation of the total charge in the system will not hold, and fluctuations in A and C would lead to fluctuations in the net charge of the system. One should not interpret equations (8.49), (8.51), or (8.53), as implying anything on the preferential solvation of W, A, or C. The reason is that the condition of the conservation of the total number of A and C must impose a long-range behavior on the various pair correlation functions. This is similar to a two-component system of A and B in a *closed* system, where we have (see section 4.2) the conservation relations

$$\rho_A G_{AA}^{\text{closed}} = -1 \tag{8.55}$$

$$\rho_A G_{AB}^{\text{closed}} = 0 \tag{8.56}$$

hence

$$G_{AA}^{\text{closed}} - G_{AB}^{\text{closed}} = -1/\rho_A. \tag{8.57}$$

* Here, as in section 4.7, the so-called electro-neutrality conditions are not relevant to the KB theory

This cannot be interpreted in terms of PS. Because of the long-range correlations imposed by the closure condition, these $G_{\alpha\beta}$ are not the Kirkwood–Buff integrals. In the same sense, the relations (8.49), (8.51), and (8.53) hold true because of the closure condition with respect to the individual particles A and C. Clearly, one cannot conclude from (8.49) that the PS of W is zero. The sign of PS of W is determined by the difference G_{WA} and G_{WC}, provided that G_{WA} and G_{WC} are evaluated in a system open with respect to the three components.

Of course, we can define the local quantities

$$G_{ij}^{\text{local}} = \int_0^{R_a} [g_{ij}(R) - 1] 4\pi R^2 \, dR \tag{8.58}$$

for all the species in the three-component system W, A, and C, and even compute these for any given R_a to obtain the local excess densities and composition. However, one cannot obtain these *local* quantities from the inversion of the KB theory.

8.6 **Preferential solvation of biomolecules**

One of the most important applications of the theory of PS is to biomolecules. There have been numerous studies on the effect of various solutes (which may be viewed as constituting a part of a solvent mixture) on the stability of proteins, conformational changes, aggregation processes, etc., (Arakawa and Timasheff 1985; Timasheff 1998; Shulgin and Ruckenstein 2005; Shimizu 2004). In all of these, the central quantity that is affected is the Gibbs energy of solvation of the biomolecule s. Formally, equation (8.26) or equivalently (8.28), applies to a biomolecule s in dilute solution in the solvent mixture A and B. However, in contrast to the case of simple, spherical solutes, the pair correlation functions g_{AS} and g_{BS} depend in this case on both the location and the relative orientation of the two species involved (figure 8.5). Therefore, we write equation (8.26) in an equivalent form as:

$$\begin{aligned} \lim_{\rho_s \to 0} \left(\frac{\partial \Delta G_s^*}{\partial x_A} \right)_{T,P} &= \frac{kT\rho_T^2}{\eta} (G_{Bs} - G_{As}) \\ &= \frac{kT\rho_T^2}{\eta} \left\{ \int_V [g_{Bs}(\mathbf{X}_B) - 1] \, d\mathbf{X}_B - \int_V [g_{As}(\mathbf{X}_A) - 1] \, d\mathbf{X}_A \right\} \end{aligned} \tag{8.59}$$

where the integration is over all the locations and orientations of A and B, relative to a fixed position and orientation of s. However, the Gibbs energy of

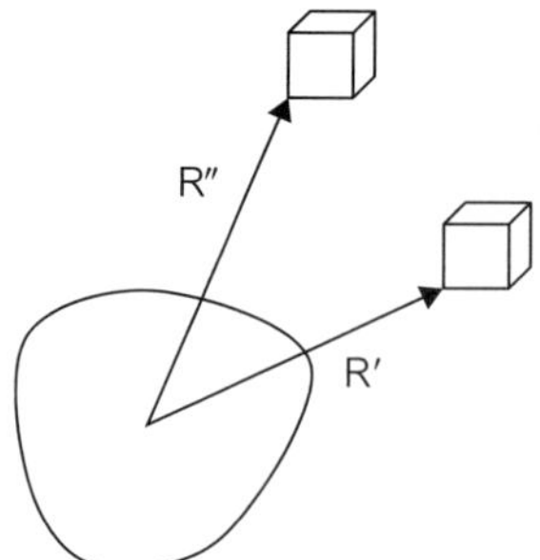

Figure 8.5 Schematic description of the dependence of the correlation on the orientation. The solvent densities at two points R' and R'' are different although the scalar distances $R = |R'| = |R''|$ are equal.

solvation of a biomolecule may be considered to be composed of several ingredients, each of which contributes differently to the overall PS. In order to study these individual contributions, we write the Gibbs energy of solvation of s as [see section 7.11 and Ben-Naim (1992)]

$$\Delta G_s^* = \Delta G_s^{*H} + \Delta G_s^{*S/H} + \sum_{i=1}^{m} \Delta G_s^{*i/H,S} + \sum_{i,j} \Delta G^{*i,j/H,S} + \cdots \tag{8.60}$$

where the first two terms on the rhs of (8.60) correspond to the work required to introduce the "hard" and "soft" parts of the interaction of s with the solvent. The third term is the sum of all the contributions due to specific functional groups, or side chains protruding from s that are independently solvated. The fourth term takes into account pair-correlated functional groups, etc. [for more details, see section 7.11 and Ben-Naim (1992)].

Corresponding to each of the contributions to the solvation Gibbs energy of s in (8.60), we can write the affinities G_{AS} and G_{BS} as a sum of terms, i.e.,

$$G_{As} = G_{As}^{H} + G_{As}^{S/H} + \sum_{i} G_{As}^{i/H,S} + \sum_{i,j} G_{As}^{i,j/H,S} + \cdots \tag{8.61}$$

and similar expansion for G_{Bs}. The meaning of each of the terms in (8.61) is as follows. The first term is the affinity between the hard part of the interaction between the solute s and A. Next, we "turn on" the soft part of the interaction; the corresponding change in the affinity is $G_{As}^{S/H}$. Next, we turn on each of the independently solvated functional groups. Each functional group i contributes a change in the affinity denoted by $G_{As}^{i/H,S}$. The sum over i corresponds to integration over the correlation regions of all the independently solvated functional groups. Thus, formally, the integration over $\mathbf{X}_A$ is extended only over each of the regions that are affected by the functional group i. Similarly, we have contributions due to pair-correlated, and higher correlated functional groups that are exposed to the solvent. Altogether, the integration over $\boldsymbol{X}_A$ (and similarly over $\boldsymbol{X}_B$) will cover the entire correlation region affected by all the groups on the surface of *s*. Clearly, this is a very complicated expression and cannot be studied at present for any real protein. However, one can study each contribution to the solvation Gibbs energy by using simple model compounds. The methodology of such an approach has been discussed in great detail in Ben-Naim (1992). The same methodology can be applied in the study of the PS of each functional group, using simple model compounds. Once this information is available, we can estimate the PS of the entire protein from the sum of the contributions of all the functional groups.

It is well-known that the solvent-induced driving force for protein folding, or protein denaturation, can be expressed as the difference in the solvation Gibbs energy of the protein in the folded, and in the unfolded forms (or any other two conformers, e.g., in the case of conformational change of hemoglobin. See Ben-Naim (2001); see also Shimizu and Boon (2004) and Shimizu and Smith (2004). Therefore, the study of the PS of a protein in any two conformations can tell us the direction of the change in the equilibrium constant induced by changes in the solvent composition.

8.7 Some illustrative examples

Traditionally, mixtures of two (or more) components were characterized and studied by examining the excess thermodynamic functions. These offer a kind of *global* view of the system. In this book, we have developed the *local* point of view of mixtures. This view consists of the KB integrals, the local composition, the PS, and the solvation thermodynamic quantities. All of these convey *local* information of the system. Therefore, the study of these quantities provides new and complementary information on the local behavior around each molecule in the mixture. We present, in this last section, a few illustrative examples of systems for which such a complete set of local quantities is available.

8.7.1 Lennard-Jones particles with the same ε but different diameter σ

The first example is a mixture of two kinds of Lennard-Jones (LJ) particles. These particles are defined by the pair potentials

$$U_{ij}(R) = 4\varepsilon_{ij}\left[\left(\frac{\sigma_{ij}}{R}\right)^{12} - \left(\frac{\sigma_{ij}}{R}\right)^{6}\right] \tag{8.62}$$

with parameters

$$\frac{\varepsilon_{AA}}{kT} = \frac{\varepsilon_{BB}}{kT} = \frac{\varepsilon_{AB}}{kT} = 0.5$$

$$\sigma_{AA} = 1, \quad \sigma_{BB} = 1.5, \quad \sigma_{AB} = \tfrac{1}{2}(\sigma_{AA} + \sigma_{BB}). \tag{8.63}$$

The pair correlation functions for this system were calculated by solving the Percus–Yevick equations as described in section 2.9, and Appendices D and E.

The calculations were done for mixtures at eleven mole fractions $x_A = 0.01$, 0.1, 0.2, 0.3, 0.4, 0.5, 0.6, 0.7, 0.8, 0.9, 0.99, and a fixed volume density $\eta = 0.4$. The functions $g_{ij}(R)$ were calculated at 100 points in the range of $(0, R_{\max})$ where $R_{\max} = 12\sigma_{AA}$. The KB integrals were calculated as

$$G_{ij}(R) = \int_0^{R_{\max}} [g_{ij}(R) - 1] 4\pi R^2 \, dR \tag{8.64}$$

and it is assumed that $R_{\max}$ is larger than the correlation length in the system.

Figure 8.6a shows the variation of G_{ij} as a function of composition in units of cm^3/mole. The first calculations of this type were done over 30 years ago (Ben-Naim 1977). Similar calculations were also performed by Kojima et al. (1984). The general behavior of these quantities as recorded here is qualitatively similar to the results in previous publications. In all of these composition ranges, G_{AA} is negative and almost independent of x_A. G_{AB} and G_{BB} start with a small negative slope at small values of x_A, then continue with larger slope at higher values of x_A. In figure 8.6b, we plot the derived quantities

$$\begin{aligned} \Delta_{AB} &= G_{AA} + G_{BB} - 2G_{AB} = \delta_A + \delta_B \\ \delta_A &= G_{AA} - G_{AB} \\ \delta_B &= G_{BB} - G_{AB}. \end{aligned} \tag{8.65}$$

The first quantity measures the extent of deviations from SI solution. The second and the third quantities are the two contributions to Δ_{AB} that reflect the

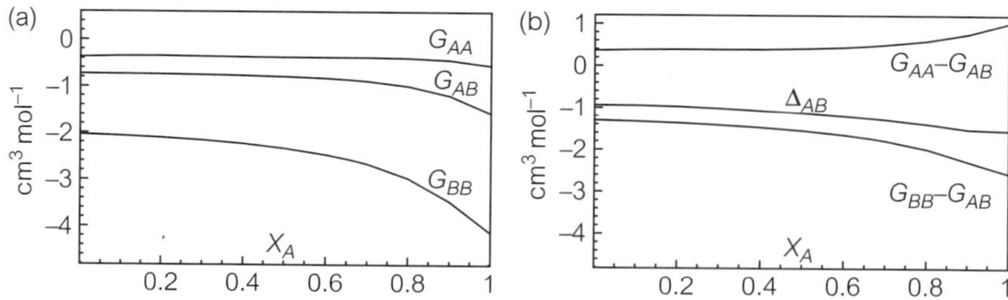

Figure 8.6 Computed results for a mixture of LJ particles with parameters as described in equation (8.63): (a) values of G_{ij} in cm^3/mol; (b) values of Δ_{AB}, δ_A, and δ_B as defined in equation (8.65).

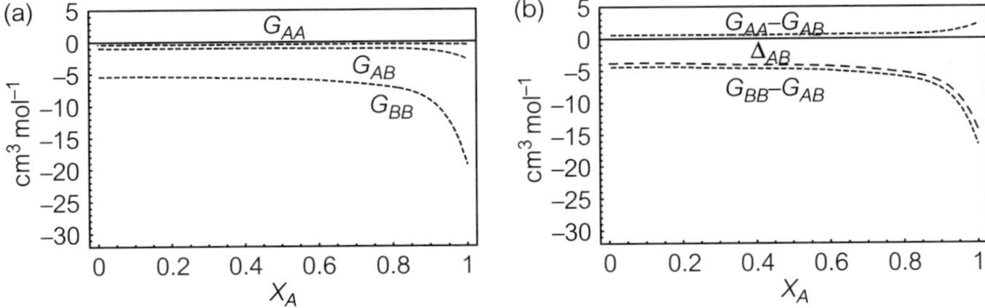

Figure 8.7 Same as in figure 8.6 but with parameters as described in equations (8.66).

signs of the limiting PS about an A particle and B particle, respectively. We see that in this system, the deviation from SI is negative and quite small (relative to the system discussed below), and the PS about A is positive, while the PS about B is negative in the entire range of compositions. Similar results were obtained for LJ particles with diameter $\sigma_{AA}=1$ and $\sigma_{BB}=2$, and also $\sigma_{AA}=1$, and $\sigma_{BB}=3$ (Ben-Naim 1977).

Having numerical values of Δ_{AB} as a function of x_A, we can integrate equation (6.1) to obtain the excess Gibbs energy for this system†. From the excess Gibbs energies, one can also compute the excess solvation Gibbs energies. See the examples below.

8.7.2 Lennard-Jones particles with the same σ but with different ε

Figure 8.7 shows the same set of G_{ij} for LJ particles with parameters

$$\sigma_{AA} = \sigma_{BB} = \sigma_{AB} = 1$$

$$\frac{\varepsilon_{AA}}{kT} = 0.5, \quad \frac{\varepsilon_{BB}}{kT} = 0.25, \quad \varepsilon_{AB} = \sqrt{\varepsilon_{AA}\varepsilon_{BB}}. \tag{8.66}$$

The calculations were done at the same set of composition as in subsection (8.62) and for $\eta = 0.4$.

The results are quite similar to the case presented in figure 8.6. Again, G_{AA} is very small and almost independent of x_A. G_{AB} is almost constant except for a small region near $x_A \approx 1$, and G_{BB} drops more sharply as $x_A \to 1$. The signs of the PS about A and about B are positive and negative, respectively, as in the previous example (figure 8.6b), and the deviations from SI are negative throughout the entire concentration range.

† Note that the calculations were carried out at constant volume density $\eta = \pi\rho_T (x_A\sigma_{AA}^3 + x_B\sigma_{BB}^3)/6$. From this relation, we can derive the total density of the mixtures at each composition x_A.

For spherical particles, a convenient way of analyzing the variation of $G_{\alpha\beta}$ with composition, is to write

$$\begin{aligned} G_{\alpha\beta} &= \int_0^\infty [g_{\alpha\beta}(R) - 1]4\pi R^2\, dR \\ &= \int_0^{\sigma_{\alpha\beta}} -4\pi R^2\, dR + \int_{\sigma_{\alpha\beta}}^\infty [g_{\alpha\beta}(R) - 1]4\pi R^2\, dR \\ &= -V_{\alpha\beta} + I_{\alpha\beta}. \end{aligned} \tag{8.67}$$

$V_{\alpha\beta}$ is essentially the volume of a sphere of radius $\sigma_{\alpha\beta}$, where $\sigma_{\alpha\beta}$ is the distance below which the pair correlation function is practically zero. $I_{\alpha\beta}$ is the overall *total* correlation between α and β in the range between $\sigma_{\alpha\beta}$, and the correlation length R_C. Whereas the first term is negative and almost composition independent, the second term may be either positive or negative. The negative values observed for $G_{\alpha\beta}$ for the LJ particles are probably due to the dominance of the volume term $-V_{\alpha\beta}$.

8.7.3 **The systems of argon–krypton and krypton–xenon**

Figure 8.8 shows the excess Gibbs energy (g^{EX}/kT), and the excess volume for the argon–krypton and krypton–xenon systems. Figure 8.9 shows the values of G_{ij}, and the quantities δ_A, δ_B, and Δ_{AB} for the argon–krypton system at 115.77 K, based on data from Davies et al. (1967), and Chui and Canfield (1971), and Calado and Staveley (1971). Here, the values of G_{ij} are negative in most of the composition range except at the very edges (i.e., $x_A \approx 0.0$ and $x_A \approx 1.0$). The deviations from SI solution are positive in the entire composition range. Also, the two components of Δ_{AB}; δ_A, and δ_B, are positive and show strong dependence on the composition. We present in figure 8.9 also the

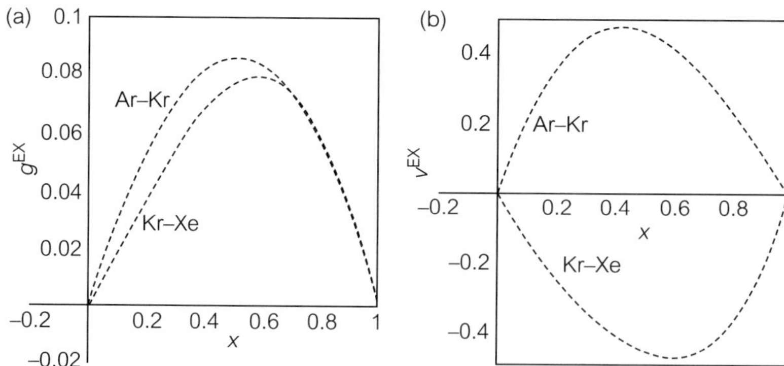

Figure 8.8 Excess Gibbs energy and excess volume for the systems of argon–krypton and krypton–xenon. Based on data from Davies et al. (1967), Chui and Canfield (1971), Calado and Staveley (1971).

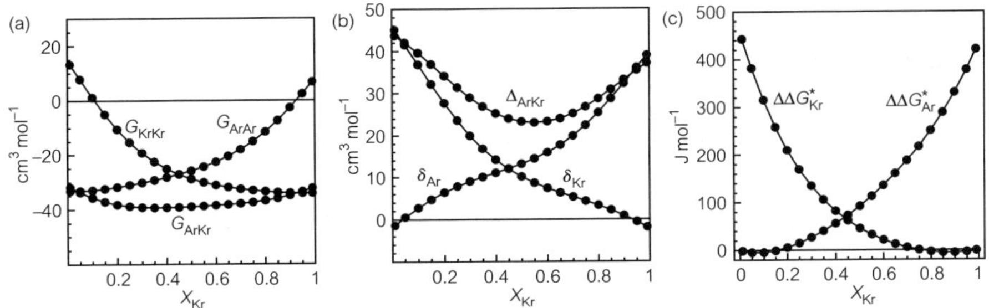

Figure 8.9 Calculated values of G_{ij} and the derived quantities Δ_{AB}, δ_A, δ_B, and $\Delta\Delta G^*_A$ for the argon–krypton system.

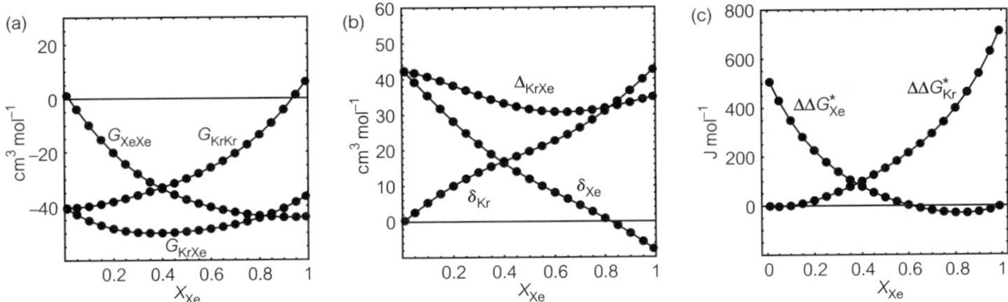

Figure 8.10 Same as in figure 8.9 but for the krypton–xenon system.

values of the excess of the solvation Gibbs energies $\Delta\Delta G^*$ for argon and krypton. These were calculated from data on excess Gibbs energies and excess volumes. Details of the calculations are reported by Ben-Naim (1987):

$$\Delta\Delta G^*_\alpha = \Delta G^*_\alpha(\text{in the mixture}) - \Delta G^*_\alpha(\text{in pure } \alpha). \qquad (8.68)$$

The set of G_{ij}'s along with the derived quantities Δ_{AB}, δ_A, δ_B, and $\Delta\Delta G^*_\alpha$, provide detailed *local* information on these mixtures. The first calculations of $\Delta\Delta G^*_\alpha$ for this system were reported by Ben-Naim (1987). In figure 8.9, we have recalculated these data for completeness. We note that the solvation Gibbs energy of argon in pure argon is negative (Ben-Naim 1987). From figure 8.9, we see that as we add krypton to pure argon, $\Delta\Delta G^*_A$ is *positive*, which means that the absolute magnitude of the solvation Gibbs energy *decreases* upon the addition of krypton. This finding was interpreted as arising from the change in the coordination number around the argon solvaton. Similarly, ΔG^*_K of krypton in pure krypton is negative. Adding argon to pure krypton causes a *decrease* in the solvation Gibbs energy of krypton. This was interpreted as a result of the weaker interaction between argon and krypton, compared with the krypton–krypton interaction (Ben-Naim 1987).

Figure 8.10 presents the same set of results on the krypton–xenon system. The general form of the curves is similar to the case of argon–krypton. The deviations from SI are still positive, somewhat larger than the case of argon–krypton. Also, the excess solvation Gibbs energies are somewhat larger than the values for argon and krypton, but the general form of the curves is similar to figure 8.9. The interpretation of the form of the curves of $\Delta\Delta G^*$ of both Kr and Xe is similar to the interpretation given for the system of Ar and Kr.

8.7.4 **Mixtures of water and alcohols**

Here, we present some examples of a non-simple system. Before discussing mixtures, we first show in figure 8.11 the values of G^0_{WW} for pure water as a function of temperature. These were calculated from the compressibility equation (Ben-Naim 1977). G^0_{WW} is always negative and decreases with temperature.

In figure 8.11, we also plot the values of G^0_{AA} for a series of linear alcohols $CH_3(CH_2)_{n-1}OH$ as a function of n at $t=0\,°C$. We also show the value of G^0_{WW} at the same temperature. Note that the value G^0_{WW} is quite near the value that can be extrapolated from the linear plot of $G^0_{AA}(n)$ at $n=0$ ($n=1$ corresponds to methanol, $n=2$ to ethanol, and $n=0$ correspond to an extrapolated "alcohol" with no methyl group). This is an important observation and it might indicate that the value of G^0_{WW} is not very sensitive to the extent of the structure of water. This is in sharp contrast to the behavior of the entropy of solvation of inert gases in water, and in a series of alcohols. It is known (Ben-Naim 1987) that the value of ΔS^*_A of say, argon or methane in water, is far more negative than the value that one can extrapolate from the solvation entropy in a series of alcohols.

Figure 8.12 presents some "global" information on the systems of water and a few alcohols. The corresponding local information on water–methanol is

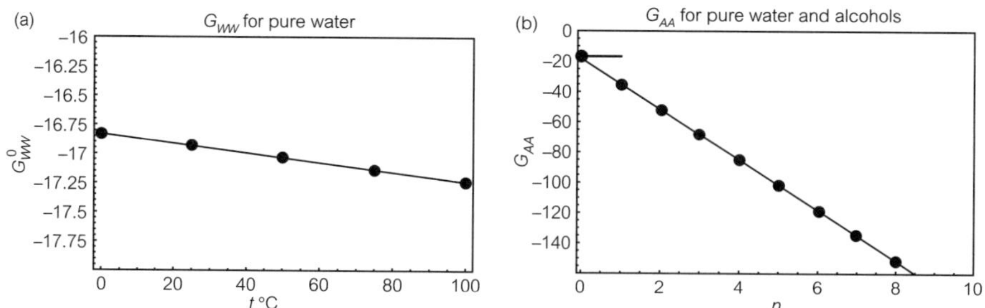

Figure 8.11 (a) Values of G^0_{WW} for pure water as a function of temperature, and (b) values of G^0_{AA} for pure alcohols as a function of the number of carbon atoms (n).

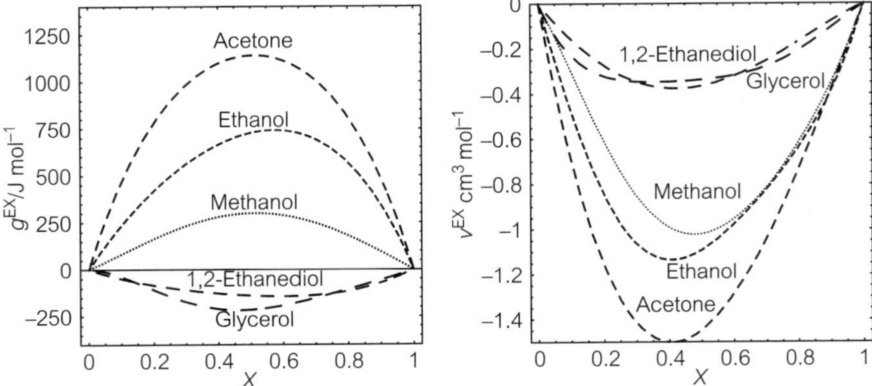

Figure 8.12 Excess Gibbs energy (per mole of mixture) and excess volume (per mole of mixture) for various aqueous mixtures at 25 °C. Based on data from Marcus (2002).

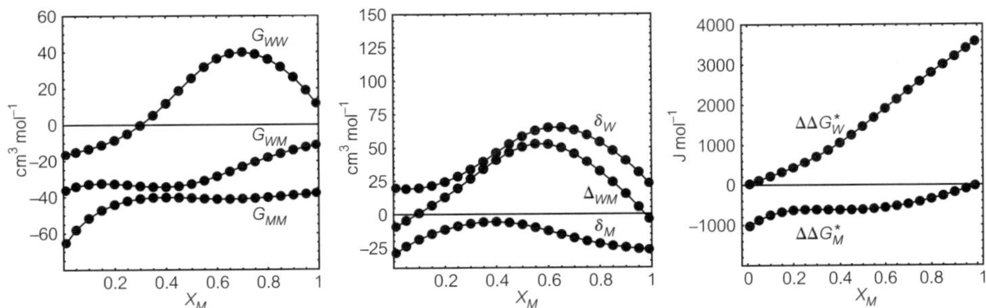

Figure 8.13 Local quantities for the water–methanol system at 25 °C. [Based on data from Marcus (1999, 2002)]

shown in figure 8.13. It is seen that while G_{WM} and G_{MM} are negative and relatively independent of composition, the values of G_{WW} start from a negative value of about $-17\,\text{cm}^3/\text{mol}$ at pure water (at 25 °C), and increases to large positive values as we add methanol to the system. Thus, the water–water affinity increases up to mole fraction of about 0.7, then decreases when more methanol is added. This behavior indicates that G_{WW} does not have a simple interpretation in terms of the extent of structure of water in these mixtures.

The slight increase in the methanol–methanol affinity in the water-rich region might be interpreted as resulting from hydrophobic interaction which peaks at about $x_{\text{methanol}} \approx 0.2$, and then stays almost constant for the entire range of compositions.

Figure 8.13 also shows the quantities δ_W, δ_M, and Δ_{WM} in the entire range of compositions. Note that in almost the entire range of compositions, Δ_{WM} is positive. In the water-rich, and in the methanol-rich, regions, the value of Δ_{WM} is nearly zero. It should be noted that the near-zero value of Δ_{WM} at $x_M \approx 0$ and

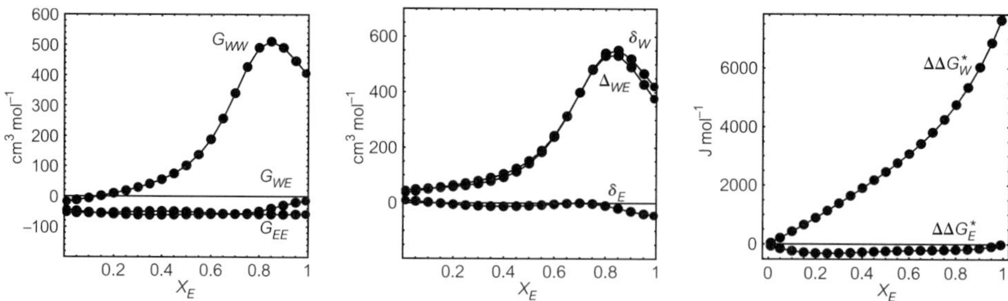

Figure 8.14 Local quantities for the water–ethanol system at 25 °C. [Based on data from Marcus (1999, 2002)]

at $x_M \approx 1$ is a result of the opposite signs of the PS about M and about W. This clearly shows that the magnitudes of the PS of the two components could be quite large, yet the combinations of the two produce nearly SI behavior. It is also interesting to note that the extent of deviations from SI behavior goes through a maximum as the composition changes. This information is not so obvious from the curves of the global properties.

The excess solvation Gibbs energies of water and methanol are also shown in figure 8.13. The Value of ΔG_W^* in pure water at $t = 25\,°C$ is about -27 kJ / mol. We see from figure 8.13 the solvation Gibbs energy of water in the mixture, reaching the value of about -23 kJ/mol in pure methanol. On the other hand, the variation of the solvation Gibbs energy of methanol is far less significant in the entire range of compositions.

Figure 8.14 shows similar local results for the water–ethanol system. Qualitatively, the results are similar to the case of water–methanol. In all cases, G_{WE} and G_{EE} are negative and relatively small, whereas G_{WW} climbs to large values of about 50 cm^3/mol and reaches a maximum at about $x_E \approx 0.8$. Note also that the deviations from SI are much larger than in the case of water–methanol. The quantity Δ_{WE} is almost identical to δ_W in the entire range of compositions, whereas δ_E is nearly zero. Also, we note that $\Delta\Delta G_W^*$ increases more dramatically when we add ethanol compared to the addition of methanol. This means that the solvation Gibbs energy of water in pure ethanol is much smaller (in absolute magnitude) than in pure methanol.

8.7.5 **Mixtures of water: 1,2-ethanediol and of water–glycerol**

The global information of these two systems is shown in figure 8.12. The excess Gibbs energies are negative for these two systems, in contrast to the systems discussed in subsection 8.7.4. Also, the excess volumes are negative but relatively small in the entire range of composition. Figures 8.15 and 8.16 show the relevant local quantities for these two systems. It in interesting to note that the deviations from SI behavior are negative but relatively small. In water–glycerol, Δ_{WG} is almost zero at $x_G = 0.7$. This indicates that the solution is almost an SI solution at this composition. However, δ_W and δ_G are quite large and have opposite signs. The behavior of $\Delta\Delta G_W^*$ and $\Delta\Delta G_G^*$, and of $\Delta\Delta G_D^*$, are very similar in the two systems.

8.7.6 **Mixtures of water and acetone**

The global quantities for the water–acetone system are also included in figure 7.12. It is seen that both the excess Gibbs energies, and the excess volumes for this system are quite large relative to all the systems discussed in the previous subsections.

The relevant local information is shown in figure 8.17. The results are qualitatively similar to the results of water–ethanol, only the magnitudes of the quantities such as G_{WW}, Δ_{WA}, and $\Delta\Delta G_W^*$ are much larger in this case. Recently, Perera et al. (2004, 2005) have done an extensive examination of the data for this system obtained by different methods and by different authors. They found large discrepancies in the data from the different sources.

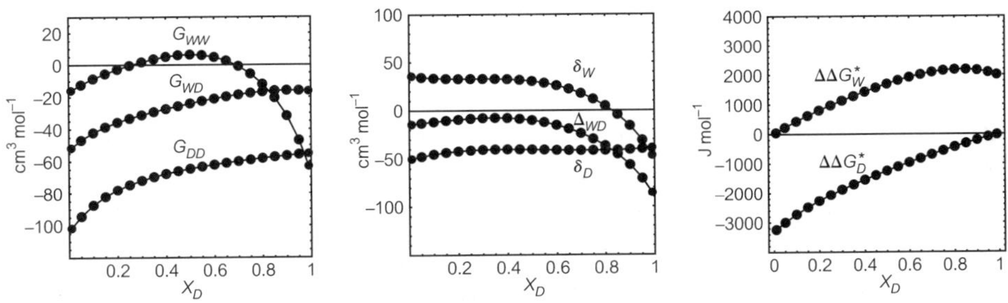

Figure 8.15 Local quantities for the system of water and 1-4 ethanediol at 25 °C. [Based on data from Marcus (2002)]

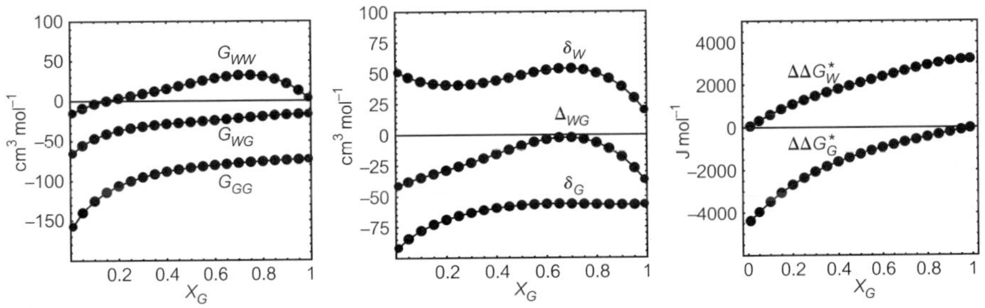

Figure 8.16 Local quantities for the water–glycerol system at 25 °C. [Based on data from Marcus (2002)]

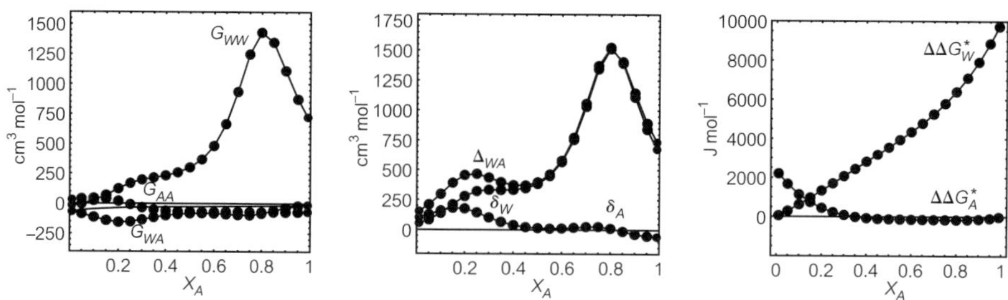

Figure 8.17 Local quantities for the water–acetone system at 25 °C. [Based on data from Marcus (2002)]

8.7.7 **Aqueous mixtures of 1-propanol and 2-propanol**

The last examples presented here are the two systems of water-1-propanol and water-2-propanol. These two systems illustrate quite dramatically the difference between the *global* and the *local* views of the mixtures. The two isomers of propanol are quite simple molecules and similar in structure. Only one group, the hydroxyl group, moves from one carbon atom to another. The curves for g^{EX} and V^{EX} for these two systems are shown in figure 8.18. The excess Gibbs energies of two systems are nearly the same. The maximum for the water-1-propanol is slightly shifted to the left, whereas for water-2-propanol, it is shifted to the right. The values at the maximum are nearly the same, differing by at most 20%. Similarly, the excess volume, though differing considerably in values, both have a shift of the minimum to the left.

On the other hand, the information provided by the *local* quantities in figures 8.19 and 8.20 is much richer, sharper, and accentuate the differences in these two systems. We note some of these features.

(1) The excess Gibbs energy of solvation of water is always positive. This means *decreasing* the absolute magnitude of the solvation Gibbs energy relative to solvation in pure water. The solvation of water in pure-1-propanol is considerably larger (in absolute values) than in pure-2-propanol. This could be a result of the relative ease of 1-propanol to form hydrogen bonds with water as compared with 2-propanol. If 1-propanol molecules can form more hydrogen bonds with a water molecule, compared with the 2-propanol isomer, then this can account for the larger reduction of the solvation Gibbs energy of water in pure-2-propanol. The excess solvation Gibbs energy of 2-propanol in the entire range of composition is negative, very small and nearly independent of composition. Adding water to 2-propanol almost does not change the solvation Gibbs energy of 2-propanol, indicating again that water molecule do not form more hydrogen bonds with the 2-propanol molecule, compared with propanol-propanol hydrogen bonding. On the other hand, for 1-propanol, the picture is

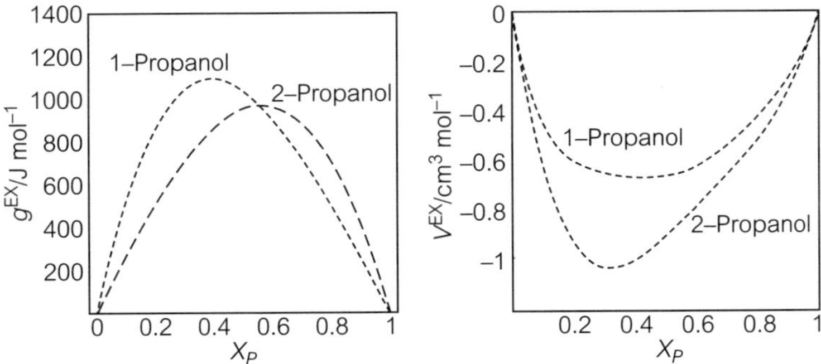

Figure 8.18 Excess Gibbs energies and excess volumes for the system of water-1-propanol and water-2-propanol at 25 °C. Based on data from Nakagawa (2002) and Marcus (2002) and Benson et al (1980).

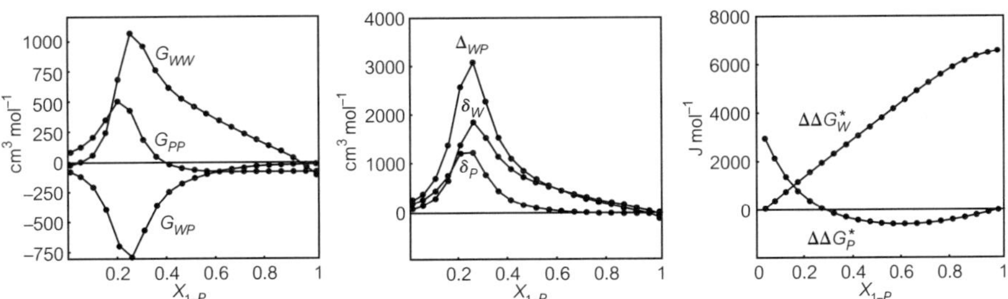

Figure 8.19 Local quantities for the system of water-1-propanol at 25 °C. Based on data from Benson and Kiyohara (1980) and Nakagawa (2002)

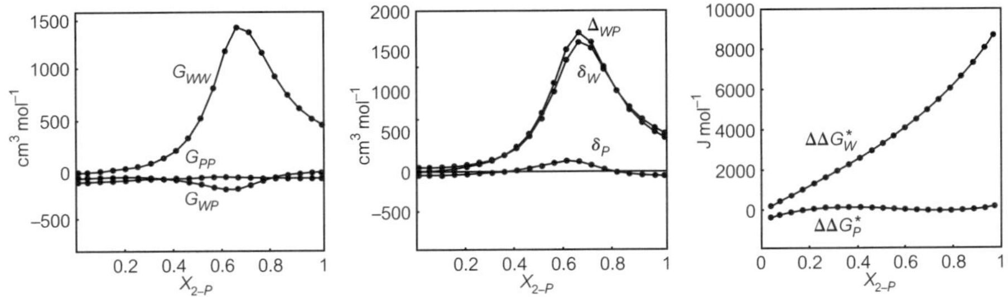

Figure 8.20 Local quantities for the system of water-2-propanol at 25 °C. Based on data from Marcus (2002)

quite different. Adding water to pure 1-propanol has a negative effect initially, probably because water forms more hydrogen bonds with the 1-propanol. However, as we add more and more water, the excess Gibbs energy of 1-propanol becomes positive. The interpretation of this phenomena on a molecular level is unclear.

(2) The extent of deviations from SI solutions as measured by Δ_{WP} has a sharp maximum at about $x_P \approx 0.2$ for 1-propanol, whereas at about $x_P \approx 0.7$ for 2-propanol. This is similar to the information on the Gibbs excess functions in figure 8.18. However, the effect here is much sharper, and also the magnitude of the deviation from SI is considerably larger in 2-propanol compared with the 1-propanol solutions. Here, in contrast to the *global* information, we can also tell which of the two "ingredients" δ_W or δ_P is responsible for the deviations from SI. In 2-propanol, δ_P is almost zero, indicating almost the same affinities between 2-propanol-2-propanol and water-2-propanol. The corresponding values of δ_P are much larger in solutions of 1-propanol, indicating that the 1-propanol-1-propanol affinity is larger than the 1-propanol-water affinity.

(3) The KBIs for the two systems also show quite different magnitude and composition dependence in these two systems. The values of G_{PP} and G_{WP} are much larger and strongly dependent on the composition.

To conclude, we have demonstrated that a host of *local* information may be obtained from the KBI, the derived quantities Δ_{AB}, δ_A, δ_B and from the solvation Gibbs energies. Such information cannot be obtained directly from the *global* quantities of these systems.

More results on local properties of both binary and tertiary systems can be found in Matteoli and Lepori (1984, 1990a, b, 1995), Ruckenstein and Shulgin (2001a, b, c), Rubio et al (1987), Marcus (2002) and Zielkiewicz (1995a, b, 1998, 2000, 2003), Zielkiewicz and Mazerski (2002).

Appendices

APPENDIX A

A brief summary of some useful thermodynamic relations

Statistical mechanics provides relations between macroscopic thermodynamic quantities and microscopic molecular properties. Thermodynamics, on the other hand, provides only relationships between various thermodynamic quantities. The multitude of these relationships arise from the freedom we have in choosing the independent variables to describe a thermodynamic system. For instance, we can choose the variables T, V, N to describe the system. Hence, all the other variables such as energy, entropy, pressure, etc., are viewed as functions of these independent variables; or we could choose T, P, N to describe the system and view all other variables, such as energy, entropy, volume, etc., as functions of T, P, N.

The most fundamental relationships are[†]

$$dE = T\,dS - P\,dV + \sum \mu_i\,dN_i \tag{A.1}$$

$$dH = T\,dS + V\,dP + \sum \mu_i\,dN_i \tag{A.2}$$

$$dA = -S\,dT - P\,dV + \sum \mu_i\,dN_i \tag{A.3}$$

$$dG = -S\,dT + V\,dP + \sum \mu_i\,dN_i. \tag{A.4}$$

In equation (A.1), we view the energy E as a function of the entropy S, the volume V, and the number of particles of each species N_i. In equation (A.2), we view the enthalpy H as a function of S, P, and $\mathbf{N} = N_1, \ldots, N_c$. In equation (A.3), we view the Helmholtz energy A as a function of T, V, $\mathbf{N}$ and in equation (A.4), we view the Gibbs energy G as a function of the most commonly used independent variables T, P, $\mathbf{N}$.

[†] Note that in thermodynamics, U is used for the total internal energy of the system. In this book, we use E instead of U and reserve the symbols U for the total potential energy of the system.

From equations (A.1)–(A.4), one can obtain immediately

$$P = -\left(\frac{\partial E}{\partial V}\right)_{S,N} = -\left(\frac{\partial A}{\partial V}\right)_{T,N} \tag{A.5}$$

$$V = \left(\frac{\partial H}{\partial P}\right)_{S,N} = \left(\frac{\partial G}{\partial P}\right)_{T,N} \tag{A.6}$$

$$S = -\left(\frac{\partial A}{\partial T}\right)_{V,N} = -\left(\frac{\partial G}{\partial T}\right)_{P,N}. \tag{A.7}$$

The most important relation is for the chemical potential[†]

$$\mu_i = \left(\frac{\partial E}{\partial N_i}\right)_{S,V,N_i'} = \left(\frac{\partial H}{\partial N_i}\right)_{S,P,N_i'} = \left(\frac{\partial A}{\partial N_i}\right)_{T,V,N_i'} = \left(\frac{\partial G}{\partial N_i}\right)_{T,P,N_i'} \tag{A.8}$$

where N_i' is the vector comprising all the components of $N = N_1, \ldots, N_c$ except the component N_i i.e., $N_i' = (N_1, \ldots, N_{i-1}, N_{i+1}, \ldots, N_c)$.

Many more relationships can be obtained from equations (A.1)–(A.4). Some of the more useful are the heat capacity at constant volume and at constant pressure:

$$C_V = \left(\frac{\partial E}{\partial T}\right)_{V,N} = -T\left(\frac{\partial^2 A}{\partial T^2}\right)_{V,N} \tag{A.9}$$

$$C_P = \left(\frac{\partial H}{\partial T}\right)_{P,N} = -T\left(\frac{\partial^2 G}{\partial T^2}\right)_{P,N}; \tag{A.10}$$

the isothermal compressibility

$$\kappa_T = -\frac{1}{V}\left(\frac{\partial V}{\partial P}\right)_{T,N} = -\frac{1}{V}\left(\frac{\partial^2 G}{\partial P^2}\right)_{T,N} = -\frac{\left(\frac{\partial^2 G}{\partial P^2}\right)_{T,N}}{\left(\frac{\partial G}{\partial P}\right)_{T,N}}; \tag{A.11}$$

and the coefficient of thermal expression

$$\alpha_P = \frac{1}{V}\left(\frac{\partial V}{\partial T}\right)_{P,N} = \frac{\left(\frac{\partial}{\partial T}\left(\frac{\partial G}{\partial P}\right)_{T,N}\right)_{P,N}}{\left(\frac{\partial G}{\partial P}\right)_{T,N}} \tag{A.12}$$

where in the numerator of (A.12) we first take the derivative of $G(T, P, N)$ with respect to P, then a second derivative with respect to T.

[†] Sometimes the notation $\mu_i = (\partial G/\partial N_i)_{T,p,N_j \neq N_i}$ is used. This is ambiguous. First, because one does not know how many of the N_j's are constants. Second, what if the number of say N_i happens to be equal to say N_j? For this reason, it is better to use the notation as in equation (A.8).

Partial molar quantities are defined for any extensive quantity E, as the partial derivative of E with respect to N_i, at constant T, P, N_i' i.e.,

$$\overline{E}_i = \left(\frac{\partial E}{\partial N_i}\right)_{T,P,N_i'}. \tag{A.13}$$

The chemical potential defined in (A.8) is the partial molar Gibbs energy.

From the Euler theorem for homogenous functions of order one, one gets for any extensive quantity, the identity

$$E = \sum_{i=1}^{c} N_i \overline{E}_i \tag{A.14}$$

and of special importance is the expression for the Gibbs energy, in terms of the chemical potentials μ_i:

$$G = \sum_{i=1}^{c} N_i \mu_i. \tag{A.15}$$

Taking the total differential of G, we get

$$dG = \sum N_i \, d\mu_i + \sum \mu_i \, dN_i. \tag{A.16}$$

Comparing with (A.4), we obtain the Gibbs–Duhem equation

$$-S\,dT + V\,dP + \sum_{i=1}^{c} N_i \, d\mu_i = 0 \tag{A.17}$$

which is essentially a statement on the dependence of the variables T, P, μ, i.e., one cannot vary all of these variables independently.

Very often, we want to switch from a derivative of a quantity in one set of variables into a derivative with respect to another set of variables. For example, in a two-component system, we can choose the independent variables, say T, ρ_A, ρ_B where ρ_i is the number density of i (in moles per liter or number of particles per unit of volume).

In this set of independent variables, the total differential of say, μ_A, is

$$d\mu_A = \left(\frac{\partial \mu_A}{\partial T}\right)_{\rho_A,\rho_B} dT + \left(\frac{\partial \mu_A}{\partial \rho_A}\right)_{T,\rho_B} d\rho_A + \left(\frac{\partial \mu_A}{\partial \rho_B}\right)_{T,\rho_A} d\rho_B. \tag{A.18}$$

Now, suppose we choose T, μ_A, ρ_A as our independent variables and need, say the derivative of ρ_B with respect to ρ_A at constant T, μ_A. From (A.18), we have

$$0 = \left(\frac{\partial \mu_A}{\partial \rho_A}\right)_{T,\mu_A} = \left(\frac{\partial \mu_A}{\partial \rho_A}\right)_{T,\rho_B} + \left(\frac{\partial \mu_A}{\partial \rho_B}\right)_{T,\rho_A} \left(\frac{\partial \rho_B}{\partial \rho_A}\right)_{T,\mu_A} \tag{A.19}$$

or rearranging (A.19) to obtain the identity

$$\left(\frac{\partial \mu_A}{\partial \rho_A}\right)_{T,\rho_B}\left(\frac{\partial \rho_B}{\partial \mu_A}\right)_{T,\rho_A}\left(\frac{\partial \rho_A}{\partial \rho_B}\right)_{T,\mu_A} = -1. \tag{A.20}$$

Note that T is constant in all three derivatives in (A.20). This is a very general relation. If a system is described by the variables $(X, Y, Z, \ldots)$, we can always write the relation

$$\left(\frac{\partial X}{\partial Y}\right)_{Z\ldots}\left(\frac{\partial Z}{\partial X}\right)_{Y\ldots}\left(\frac{\partial Y}{\partial Z}\right)_{X\ldots} = -1 \tag{A.21}$$

where " . . . " stands for all the variables that are kept constants in all derivatives.

Another useful relation may be obtained from (A.18). Suppose we want to choose T, μ_B, ρ_A as independent variables, and we need the derivative of μ_A with respect to ρ_A at T, μ_B constants.

From (A.18) and (A.20), we have

$$\begin{aligned}\left(\frac{\partial \mu_A}{\partial \rho_A}\right)_{T,\mu_B} &= \left(\frac{\partial \mu_A}{\partial \rho_A}\right)_{T,\rho_B} + \left(\frac{\partial \mu_A}{\partial \rho_B}\right)_{T,\rho_A}\left(\frac{\partial \rho_B}{\partial \rho_A}\right)_{T,\mu_B}\\ &= \left(\frac{\partial \mu_A}{\partial \rho_A}\right)_{T,\rho_B} - \left(\frac{\partial \mu_A}{\partial \rho_B}\right)_{T,\rho_A}\frac{(\partial \mu_B/\partial \rho_A)_{T,\rho_B}}{(\partial \mu_B/\partial \rho_B)_{T,\rho_A}}\end{aligned} \tag{A.22}$$

where we now have on the rhs of (A.22) only derivatives with respect to the set of independent variables T, ρ_A, ρ_B. Relation (A.22) can be generalized to any variable X instead of μ_B.

Finally, we mention the chain rule for differentiation; for instance, if we have the derivative of μ_A with respect to the density ρ_A at some set of constant variables C, we now want the derivative of μ_A with respect to, say, the mole fraction x_A, at the same set of constant variables C. We have

$$\left(\frac{\partial \mu_A}{\partial \rho_A}\right)_C = \left(\frac{\partial \mu_A}{\partial x_A}\right)_C\left(\frac{\partial x_A}{\partial \rho_A}\right)_C. \tag{A.23}$$

We could choose any other variable instead of x_A, say ρ_B. Then we have

$$\left(\frac{\partial \mu_A}{\partial \rho_A}\right)_C = \left(\frac{\partial \mu_A}{\partial \rho_B}\right)_C\left(\frac{\partial \rho_B}{\partial \rho_A}\right)_C \tag{A.24}$$

which can be rewritten as

$$\left(\frac{\partial \mu_A}{\partial \rho_A}\right)_C\left(\frac{\partial \rho_B}{\partial \mu_A}\right)_C\left(\frac{\partial \rho_A}{\partial \rho_B}\right)_C = 1. \tag{A.25}$$

Note that the pattern of the derivatives is the same as in (A.20). But here, all the derivatives are taken holding the same set of constants C.

APPENDIX B
Functional derivative and functional Taylor expansion

We present here a simplified definition of the operations of functional derivative and functional Taylor expansion. It is based on a formal generalization of the corresponding operations applied to functions of a finite number of independent variables.

Consider first a function of one variable

$$y = f(x). \tag{B.1}$$

The derivative of this function with respect to x is defined as[†]

$$\frac{dy}{dx} = \lim_{\varepsilon \to 0} \frac{f(x+\varepsilon) - f(x)}{\varepsilon} \tag{B.2}$$

The derivative itself is a function of x and may be evaluated at any point x_0 in the region where there it is defined. For example

$$\begin{aligned} y &= f(x) = ax \\ \frac{dy}{dx} &= \lim_{\varepsilon \to 0} \frac{a(x+\varepsilon) - a(x)}{\varepsilon} = a. \end{aligned} \tag{B.3}$$

Here, the derivative is a constant and has the value a at any point $x = x_0$.

A second example is

$$y = f(x) = x^2 \tag{B.4}$$

The derivative of this function is

$$\frac{dy}{dx} = 2x \tag{B.5}$$

and its value at $x = x_0$ is $2x_0$.

[†] In this rather qualitative presentation, we shall not discuss the conditions such as differentiability and integrability under which the operations of integration and differentiation are valid.

Next, consider a function of n independent variables $x_1, \ldots, x_n$. A simple example of such a function is

$$f(x_1, \ldots, x_n) = \sum_{i=1}^{n} a_i x_i. \tag{B.6}$$

The partial derivative of f with respect to, say x_j, is

$$\frac{\partial f}{\partial x_j} = \sum_{i=1}^{n} a_i \frac{\partial x_i}{\partial x_j} = \sum_{i=1}^{n} a_i \delta_{ij} = a_j \tag{B.7}$$

where δ_{ij} is the Kronecker delta function.

A further generalization of (B.6) is the case of the vector $\boldsymbol{y} = (y_1, \ldots, y_n)$ which is a function of the vector $\boldsymbol{x} = (x_1, \ldots, x_n)$. This connection can be written symbolically as

$$\boldsymbol{y} = \boldsymbol{F}\boldsymbol{x} \tag{B.8}$$

where $\boldsymbol{F}$ is a matrix operating on a vector $\boldsymbol{x}$ to produce a new vector $\boldsymbol{y}$.

We now generalize (B.8) as follows. We first rewrite the vector $\boldsymbol{x}$ in a new notation

$$\boldsymbol{x} = (x_1, \ldots, x_n) = [x(1), \ldots, x(n)] \tag{B.9}$$

i.e., we rewrite the *component* x_i as $x(i)$, where i is a discrete variable, $i = 1, 2, \ldots, n$. We now let i take any value in a continuous range of real numbers $a \leq i \leq b$. This procedure gives us a new vector $\boldsymbol{x}$, with an *infinite* number of components $x(i)$. The relation (B.8) is now reinterpreted as a relation between a function $\boldsymbol{x}$ and a function $\boldsymbol{y}$.

In equation (B.8), $\boldsymbol{x}$ and $\boldsymbol{y}$ are vectors and $\boldsymbol{F}$ is a matrix operating in a vector space (of finite dimensions). When $\boldsymbol{x}$ and $\boldsymbol{y}$ are functions of a continuous variable, say t, they are viewed as vectors with infinite number of components. In this case, $\boldsymbol{F}$ is an operator acting in a functional space rather than on a finite-dimensional vector space. A simple relation between two such functions is

$$y(t) = \int_a^b K(s, t) x(s)\, ds \tag{B.10}$$

That is for each *function* $\boldsymbol{x}$ whose components are $x(s)$, we get a new *function* $\boldsymbol{y}$, whose components are $y(t)$. The function $K(s, t)$ is presumed to be known. Relation (B.10) can also be written symbolically in the form (B.8) which reads: $\boldsymbol{F}$ operates on $\boldsymbol{x}$ to give the result $\boldsymbol{y}$.

In the discrete case, for any two components x_i and x_j of the vector $\boldsymbol{x}$, we have the relation

$$\frac{\partial x_i}{\partial x_j} = \delta_{ij}. \tag{B.11}$$

Similarly, viewing $x(t)$ and $x(s)$ as two "components" of the functions $\boldsymbol{x}$, we have

$$\frac{\delta x(t)}{\delta x(s)} = \delta(t - s) \tag{B.12}$$

where the Dirac delta function replaces the Kronecker delta function in (B.11).

In (B.7), the quantity $\partial x_i/\partial x_j$ is referred to as the *partial derivative* of f with respect to the component x_j. Similarly, the functional derivative of $\boldsymbol{y}$ in (B.10) with respect to the "component" $x(s')$ is

$$\frac{\delta y(t)}{\delta x(s')} = \int K(s,t)\frac{\delta x(s)}{\delta x(s')}\,ds = \int K(s,t)\delta(s-s')\,ds = K(s',t). \tag{B.13}$$

In section 3.6, we encountered the following example of a functional. The average volume of a system in the T, P, $\mathbf{N}$ ensemble is written as

$$V(N) = \int_0^\infty \phi N(\phi)\,d\phi. \tag{B.14}$$

(Here, we use N instead of the more cumbersome notation $\mathbf{N}_\psi^{(1)}$ of section 3.6.)

The functional derivative of V with respect to the component $N(\phi')$ is thus (Ben-Naim 1974)

$$\frac{\delta V(N)}{\delta N(\phi')} = \int_0^\infty \phi\frac{\delta N(\phi)}{\delta N(\phi')}\,d\phi = \int_0^\infty \phi\delta(\phi-\phi')\,d\phi = \phi'. \tag{B.15}$$

As a second example of the application of the functional derivatives, we show that the pair distribution function can be obtained as a functional derivative of the configurational partition function. For a system of N spherical particles, with pairwise additive potential, we write

$$Z(\boldsymbol{U}) = \int\cdots\int d\boldsymbol{R}^N \exp\left[-\beta\sum_{i<j} U(\boldsymbol{R}_i,\boldsymbol{R}_j)\right]. \tag{B.16}$$

In (B.16), we view Z as a functional of the function $\boldsymbol{U}$ (i.e., the pair potential which is considered here as a continuous function of six variables).

The functional derivative of Z with respect to the "component" $U(\boldsymbol{R}',\boldsymbol{R}'')$ is

$$\begin{aligned}
&\frac{\delta Z(\boldsymbol{U})}{\delta U(\boldsymbol{R}',\boldsymbol{R}'')}\\
&= \int\cdots\int d\boldsymbol{R}^N\left[-\beta\sum_{i<j}\frac{\delta U(\boldsymbol{R}_i,\boldsymbol{R}_j)}{\delta U(\boldsymbol{R}',\boldsymbol{R}'')}\right]\exp\left[-\beta\sum_{i<j}U(\boldsymbol{R}_i,\boldsymbol{R}_j)\right]\\
&= \int\cdots\int d\boldsymbol{R}^N\left[-\beta\sum_{i<j}\delta(\boldsymbol{R}_i-\boldsymbol{R}')\delta(\boldsymbol{R}_j-\boldsymbol{R}'')\right]\exp\left[-\beta\sum_{i<j}U(\boldsymbol{R}_i,\boldsymbol{R}_j)\right]\\
&= -\beta\rho^{(2)}(\boldsymbol{R}',\boldsymbol{R}'')Z(\boldsymbol{U})/2
\end{aligned} \tag{B.17}$$

where we have used the definition of the pair distribution function (see section 2.2). Relation (B.17) can be written as

$$\rho^{(2)}(\mathbf{R}', \mathbf{R}'') = -2kT\delta[\ln Z(\mathbf{U})]/\delta U(\mathbf{R}', \mathbf{R}''). \tag{B.18}$$

Before turning to functional Taylor expansion, we note that many operations with ordinary derivatives can be extended to functional derivatives. We note, in particular, the chain rule of differentiation.

For functions of one variable $y = f(x)$ and $x = g(t)$, we have

$$\frac{dy}{dx}\frac{dx}{dt} = \frac{dy}{dt} \tag{B.19}$$

and, in particular,

$$\frac{dy}{dx}\frac{dx}{dy} = 1. \tag{B.20}$$

In the case of functions of n variables, say

$$y_k = f_k(x_1, \ldots, x_n), \quad k = 1, \ldots, n \tag{B.21}$$

we have[†]

$$\frac{dy_k}{dt} = \sum_{i=1}^{n} \frac{\partial y_k}{\partial x_i}\frac{dx_i}{dt} \tag{B.22}$$

and, in particular

$$\delta_{kj} = \frac{dy_k}{dy_j} = \sum_{i=1}^{n} \frac{\partial y_k}{\partial x_i}\frac{\partial x_i}{\partial y_j}. \tag{B.23}$$

The generalization of (B.23), in the case of functional space, is straightforward. We view $\mathbf{y} = \mathbf{F}(\mathbf{x})$ as a connection between the two functions whose components are $y(t)$ and $x(t)$, respectively, and write by analogy with (B.23)

$$\frac{\delta y(t)}{\delta y(v)} = \int \frac{\delta y(t)}{\delta x(s)}\frac{\delta x(s)}{\delta y(v)}\, ds = \delta(t - v) \tag{B.24}$$

where integration replaces the summation in (B.23).

We now consider the functional Taylor expansion. We start with a simple function of one variable $f(x)$ for which the Taylor expansion about $x = 0$ is

$$f(x) = f(0) + \left.\frac{\partial f}{\partial x}\right|_{x=0} x + \frac{1}{2}\left.\frac{\partial^2 f}{\partial x^2}\right|_{x=0} x^2 + \cdots \tag{B.25}$$

As an example, $f(x) = a + bx$. Then, we have

$$f(0) = a, \quad \frac{\partial f}{\partial x} = b. \tag{B.26}$$

[†] We assume here that the functions and their inverse are differentiable.

Hence, from (B.25) we get

$$f(x) = a + bx \tag{B.27}$$

i.e., the expansion to first order in x is, in this case, *exact* for any x. In general, if we take the first-order expansion

$$f(x) = f(0) + \left.\frac{\partial f}{\partial x}\right|_{x=0} x \tag{B.28}$$

we get an approximate value for $f(x)$. The quality of the approximation depends on x and on the function $\boldsymbol{f}$.

For a function of n variables $f(x_1, \ldots, x_n)$, the Taylor expansion about the point $\boldsymbol{x}=0$ is

$$f(x_1, \ldots, x_n) = f(0, \ldots, 0) + \sum_{i=1}^{n} \left.\frac{\partial f}{\partial x_i}\right|_{x=0} x_i + \cdots \tag{B.29}$$

where all the derivatives are evaluated at the point $\boldsymbol{x}=0$. Again, the quality of the approximation (B.29) depends both on $\boldsymbol{f}$ and all the x_i's.

The generalization to the continuous case is, by analogy to (B.29),

$$f(\boldsymbol{x}) = f(0) + \int \left.\frac{\delta f(\boldsymbol{x})}{\delta x(t)}\right|_{\boldsymbol{x}=0} x(t)\, dt + \cdots \tag{B.30}$$

where the partial derivative has become the functional derivative and the summation over i has become the integration over t. We note again that the first-order expansion in (B.30) can be viewed as an approximation to $f(\boldsymbol{x})$. The quality of the approximation depends on both $\boldsymbol{x}$ and on the function $\boldsymbol{f}$. In Appendix D, we present an application of such a first-order Taylor expansion.

As for the nomenclature, the quantity $\partial f/\partial x_i|_{\boldsymbol{x}=0}$ in (B.29) is referred to as the *partial* derivative of $\boldsymbol{f}$ with respect to the *component* x_i, evaluated at the point $x=0$. Similarly, in (B.30), we have functional derivative of $\boldsymbol{f}$ with respect to the "component" $x(t)$, at the point $\boldsymbol{x}=0$.

In the theory of variations, one starts with a functional $f(\boldsymbol{x})$ where $\boldsymbol{x}$ is a function of say t, and asks for the effect of variation of the function $\boldsymbol{x}$ on $\boldsymbol{f}$. To first order the variation in $\boldsymbol{f}$ is

$$\delta f(\boldsymbol{x}) = \int \left.\frac{\delta f(\boldsymbol{x})}{\delta x(t)}\right|_{\boldsymbol{x}=0} \delta x\, dt. \tag{B.31}$$

This is a generalization of the *total differential* of a function $f(x_1, \ldots, x_n)$

$$df = \sum_{i=1}^{n} \frac{\partial f}{\partial x_i} dx_i. \tag{B.32}$$

Thus, the partial derivative in (B.32) is replaced by the functional derivative in (B.31). The sum in (B.32) over the discrete index i is replaced by an integral in (B.31). $\delta f(\boldsymbol{x})$ is referred to either as the differential or the first-order variation of the functional $f(\boldsymbol{x})$.

Thus, both (B.32) and (B.31) give the first-order effect of the variation of the whole vector $\boldsymbol{x}$ on the function $\boldsymbol{f}$. The vector in (B.32) is a finite-dimensional vector $(x_1, \ldots, x_n)$, whereas in (B.31), it is an infinite dimensional vector $\boldsymbol{x}$ whose "components" are $x(t)$.

Note again that as the partial derivatives in (B.32) depend on both the index i and on the point of evaluation $\boldsymbol{x}$, the functional derivative depends on the index of the component $x(t)$, as well as on the entire vector $\boldsymbol{x}$.

One easy way of performing the functional derivative is to add to the function with respect to which we take the derivative, the "variation" function $\varepsilon\delta(x - x')$, and take the partial derivative with respect to ε.

For instance

$$f[N(\phi)] = \int \phi N(\phi)\, d\phi \tag{B.33}$$

$$\frac{\delta f}{\delta N(\phi)} = \frac{\partial}{\delta \varepsilon}[\int \phi[N(\phi) + \varepsilon\delta(\phi - \phi')]\, d\phi$$
$$= \int \phi\delta(\phi - \phi')]d\phi = \phi'. \tag{B.34}$$

APPENDIX C

The Ornstein–Zernike relation

The original derivation of the Ornstein–Zernike relation (Ornstein and Zernike 1914) employs arguments on local density fluctuations in the fluid. We present here a different derivation based on the method of functional derivatives (Appendix B). A very thorough discussion of this topic is given by Münster (1969), and by Gray and Gubbins (1984).

Consider the grand partition function of a system of spherical particles exposed to an external potential $\boldsymbol{\psi}$:

$$\Xi(\boldsymbol{\psi}) = \sum_{N=0}^{\infty} (z^N/N!) \int \cdots \int d\boldsymbol{R}^N \exp[-\beta U(\boldsymbol{R}^N, \boldsymbol{\psi})] \tag{C.1}$$

where

$$U(\boldsymbol{R}^N, \boldsymbol{\psi}) = U_N(\boldsymbol{R}^N) + \sum_{i=1}^{N} \psi(\boldsymbol{R}_i) \tag{C.2}$$

and

$$z = q[\exp(\beta\mu)]/\Lambda^3. \tag{C.3}$$

Here, $U_N(\boldsymbol{R}^N)$ is the total potential energy due to interactions among the particles N, and the second term on the rhs of (C.2) is due to interaction of the system at configuration $\boldsymbol{R}^N$ with the external potential. As in Appendix B, we use the symbol $\boldsymbol{\psi}$ to designate the whole function whose "components" are $\psi(\boldsymbol{R}_i)$.

The functional derivative of In Ξ with respect to the component $\psi(\boldsymbol{R}')$ is

$$\begin{aligned}
&\frac{\delta \ln \Xi(\boldsymbol{\psi})}{\delta\psi(\boldsymbol{R}')} \\
&= \Xi(\boldsymbol{\psi}) \sum_N \left(\frac{Z}{N!}\right) \int \cdots \int d\boldsymbol{R}^N \left[-\beta \sum_{i=1}^{N} \frac{\delta\psi(\boldsymbol{R}_i)}{\delta\psi(\boldsymbol{R}')}\right] \exp[-\beta U(\boldsymbol{R}^N, \boldsymbol{\psi})] \\
&= -\beta\Xi^{-1}(\boldsymbol{\psi}) \sum_N \left(\frac{z^N}{N!}\right) \int \cdots \int d\boldsymbol{R}^N \left[\sum_{i=1}^{N} \delta(\boldsymbol{R}_i - \boldsymbol{R}')\right] \exp[-\beta U(\boldsymbol{R}^N, \boldsymbol{\psi})] \\
&= -\beta\rho^{(1)}(\boldsymbol{R}'|\boldsymbol{\psi}).
\end{aligned} \tag{C.4}$$

In the last step on the rhs, we used the definition of the singlet molecular distribution function of a system in an external potential ψ

Next, consider the second functional derivative of ln Ξ with respect to ψ ($\mathbf{R}'$), which can be obtained from (C.4):

$$\begin{aligned}\frac{\delta^2 \ln \Xi(\psi)}{\delta\psi(\mathbf{R}')\delta\psi(\mathbf{R}'')} &= -\beta\frac{\delta\rho^{(1)}(\mathbf{R}'|\psi)}{\delta\psi(\mathbf{R}'')}\\ &= \beta^2\Xi^{-1}(\psi)\left\{\sum_N\left(\frac{z^N}{N!}\right)\int\cdots\int d\mathbf{R}^N\left[\sum_{i,j=1}^{N}\delta(\mathbf{R}_i-\mathbf{R}')\delta(\mathbf{R}_j-\mathbf{R}'')\right]\right.\\ &\quad\left.\times\exp[-\beta U(\mathbf{R}^N,\psi)]-\rho^{(1)}(\mathbf{R}'|\psi)\rho^{(1)}(\mathbf{R}''|\psi)\right\}\\ &= \beta^2\{\rho^{(2)}(\mathbf{R}',\mathbf{R}''|\psi)+\rho^{(1)}(\mathbf{R}'|\psi)\delta(\mathbf{R}'-\mathbf{R}'')\\ &\quad -\rho^{(1)}(\mathbf{R}'|\psi)\rho^{(1)}(\mathbf{R}''|\psi)\}\end{aligned} \tag{C.5}$$

where $\rho^{(2)}$ ($\mathbf{R}',\mathbf{R}' \mid \psi$) is the pair distribution function in the presence of the external potential ψ. In the second step on the rhs of (C.5), we have separated the double sum over i and j into two terms, the first containing all terms for which $i \neq j$, and the second, all terms with $i = j$.

We next define the *total* correlation function by

$$h(\mathbf{R}', \mathbf{R}'') = g(\mathbf{R}', \mathbf{R}'') - 1 \tag{C.6}$$

and rewrite (C.5) when evaluated at $\psi = 0$ as

$$-kT\frac{\delta\rho^{(1)}(\mathbf{R}'|\psi)}{\delta\psi(\mathbf{R}'')}\bigg|_{\psi=0} = \rho^2 h(\mathbf{R}', \mathbf{R}'') + \rho^{(1)}(\mathbf{R}')\delta(\mathbf{R}' - \mathbf{R}'') \tag{C.7}$$

i.e., the functional derivative of the singlet molecular distribution function, evaluated at $\psi = \mathbf{0}$ is "almost" equal to the total correlation function. The singular case arises when $\mathbf{R}' = \mathbf{R}''$.

We now introduce the so-called *direct* correlation function, which is defined in terms of the inverse functional derivative in (C.7). To do this, we view ψ as a functional of the density, which we write symbolically as $\psi(\mathbf{R}' \mid \boldsymbol{\rho}^{(1)})$. Now, the "external" potential is produced by preparing a system with an arbitrary local density $\boldsymbol{\rho}^{(1)}$. It is for this reason that it is necessary to work in the grand ensemble where an arbitrary density change may be envisaged; see Percus (1964)†.

The direct correlation function is defined by

$$c(\mathbf{R}', \mathbf{R}'') = \beta\frac{\delta\psi(\mathbf{R}'|\boldsymbol{\rho}^{(1)})}{\delta\rho^{(1)}(\mathbf{R}'')} + \frac{\delta(\mathbf{R}' - \mathbf{R}'')}{\rho^{(1)}(\mathbf{R}')}. \tag{C.8}$$

† Recently the corresponding Ornstein–Zernike equation in a closed system has been derived by White and Velasco (2001)

Next, we apply the chain rule of functional derivatives (see Appendix B), which, for the present case, takes the form

$$\int \frac{\delta\psi(\mathbf{R}'|\boldsymbol{\rho}^{(1)})}{\delta\boldsymbol{\rho}^{(1)}(\mathbf{R}''')} \frac{\delta\boldsymbol{\rho}^{(1)}(\mathbf{R}'''|\boldsymbol{\psi})}{\delta\psi(\mathbf{R}'')}\bigg|_{\boldsymbol{\psi}=\mathbf{0}} d\mathbf{R}''' = \delta(\mathbf{R}' - \mathbf{R}''). \tag{C.9}$$

Substituting (C.7) and (C.8) in (C.9), we obtain

$$-\int [c(\mathbf{R}', \mathbf{R}''') - \rho^{-1}\delta(\mathbf{R}' - \mathbf{R}''')][\rho^2 h(\mathbf{R}'', \mathbf{R}''') + \rho\delta(\mathbf{R}'' - \mathbf{R}''')]\, d\mathbf{R}''' \\ = \delta(\mathbf{R}' - \mathbf{R}'') \tag{C.10}$$

which yields, upon rearrangement, and using the basic property of the Dirac delta function, the result

$$h(\mathbf{R}', \mathbf{R}'') = c(\mathbf{R}', \mathbf{R}'') + \rho \int c(\mathbf{R}', \mathbf{R}''')h(\mathbf{R}'', \mathbf{R}''')\, d\mathbf{R}''' \tag{C.11}$$

which is the Orsntein–Zernike relation for a system of spherical particles. One straightforward interpretation of this integral may be obtained by substituting the whole the rhs of (C.11) into the integrand on the rhs of (C.11), and continuing this process, we get

$$h(\mathbf{R}', \mathbf{R}'') = c(\mathbf{R}', \mathbf{R}'') + \rho \int c(\mathbf{R}', \mathbf{R}''')c(\mathbf{R}'', \mathbf{R}''')d\mathbf{R}''' \\ + \rho^2 \int c(\mathbf{R}', \mathbf{R}''')c(\mathbf{R}''', \mathbf{R}'''')c(\mathbf{R}'''', \mathbf{R}'')d\mathbf{R}'''d\mathbf{R}'''' + \cdots \tag{C.12}$$

In this form, the total correlation function is viewed as a sum of "chains" of direct correlation functions between the two points $\mathbf{R}'$ and $\mathbf{R}''$.

Relation (C.11) may be viewed as an implicit definition of the direct correlation $c(\mathbf{R})$ in terms of the total correlation $h(\mathbf{R})$. It can be made explicit by taking the Fourier transform of both sides of equation (C.11), and noting that the integral on the rhs of (C.11) is a convolution (for spherical particles); hence, applying the convolution theorem, we obtain

$$\hat{\boldsymbol{h}} = \hat{\boldsymbol{c}} + \rho\hat{\boldsymbol{c}} \cdot \hat{\boldsymbol{h}} \tag{C.13}$$

where $\hat{\boldsymbol{f}}$ is the Fourier transform of $\boldsymbol{f}$. Hence, we can *define* the direct correlation function as

$$\hat{\boldsymbol{c}} = \frac{\hat{\boldsymbol{h}}}{1 + \rho\hat{\boldsymbol{h}}}. \tag{C.14}$$

Taking the inverse transform of (C.14), we get back the function **c**. It should be noted that the physical meaning of $c(\mathbf{R})$ is not clear†, i.e., to what extent is

† Rushbrook (1968) wrote: "The Introduction of $c(R)$ merely enriches the language we can use in discussing the structure of fluids, without necessarily adding to our understanding."

the function $c(\boldsymbol{R})$ a *correlation function*, in the probabilistic sense – correlation between which quantities?

Relations (C.11), or equivalently (C.13) or (C.14), are relations between the *functions* $c(\boldsymbol{R})$ and $h(\boldsymbol{R})$. A simpler relation between the integrals of these function may be obtained by integrating both sides of (C.11) over $\boldsymbol{R}''$, noting that both $\boldsymbol{c}$ and $\boldsymbol{h}$ are functions of the scalar distance R. We obtain

$$[1+\rho\int h(R)4\pi R^2 dR]\times[(1-\rho\int c(R)4\pi R^2\, dR)]=1. \tag{C.15}$$

Using the compressibility equation and equation (C.15), we obtain

$$1-\rho\int c(R)4\pi R^2\, dR=(kT\rho\kappa_T)^{-1}. \tag{C.16}$$

Thus, even when κ_T diverges to infinity (at the critical point), the integral over $c(R)$ does not diverge. In fact, since $\rho=\rho_c$ and $T=T_c$ are finite at the critical point, it follows that at this point

$$\rho_c\int c(R)4\pi R^2\, dR=1. \tag{C.17}$$

For mixtures of spherical particles, the generalization of the Ornstein–Zernike equation is

$$h_{ij}(\boldsymbol{R}',\boldsymbol{R}'')=c_{ij}(\boldsymbol{R}',\boldsymbol{R}'')+\sum_k\rho_k\int c_{ik}(\boldsymbol{R}',\boldsymbol{R}''')h_{kj}(\boldsymbol{R}''',\boldsymbol{R}'')\, d\boldsymbol{R}''' \tag{C.18}$$

which is the generalization of equation (C.11). Again, integrating over $\boldsymbol{R}'$, and noting that the functions c_{ij} and h_{ij} are functions of the distance only, one can obtain the relations between the integrals

$$H_{ij}=G_{ij}=\int h_{ij}(R)4\pi R^2\, dR \tag{C.19}$$

$$C_{ij}=\int c(R)4\pi R^2\, dR \tag{C.20}$$

$$G_{ij}=C_{ij}+\sum_k\rho_k C_{ik}G_{kj} \tag{C.21}$$

which can be written as

$$\mathbf{G}=\mathbf{C}+\mathbf{C}\boldsymbol{\rho}\mathbf{G} \tag{C.22}$$

where $\boldsymbol{\rho}$ is a diagonal matrix

$$\boldsymbol{\rho}=\begin{pmatrix}\rho_1 & & & \\ & \rho_2 & & \\ & & \rho_3 & \\ & & & \ddots\end{pmatrix}. \tag{C.23}$$

This can be transformed into a form analogous to (C.15) as follows. Multiply (C.22) by $\boldsymbol{\rho}$ and rearrange:

$$\boldsymbol{\rho G}(\boldsymbol{I} - \boldsymbol{\rho C}) = \boldsymbol{\rho C}. \tag{C.24}$$

where $\boldsymbol{I}$ is a unit matrix of the same dimensions as the number of components. Hence, (C.24) can be rewritten as

$$\boldsymbol{\rho G} = \boldsymbol{\rho C}(\boldsymbol{I} - \boldsymbol{\rho C})^{-1} \tag{C.25}$$

or equivalently

$$\boldsymbol{I} + \boldsymbol{\rho G} = (\boldsymbol{I} - \boldsymbol{\rho C})^{-1}. \tag{C.26}$$

As we have expressed the compressibility equation in terms of the integral over the direct correlation function in (C.16), one can write the KB theory in terms of C_{ij} instead of G_{ij} the two are equivalent formulations. O'Connel (1971) has expressed the view that the formulation in terms of C_{ij} might be more useful for numerical work since the direct correlation function is considered to be "shorter range" than the pair correlation function[†]. For further applications of this approach, see O'Connel (1971), Perry and O'Connel (1984), and Hamad et al. (1987, 1989, 1990a, b, 1993, 1997, 1998).

[†] One can also argue that the *direct correlation* functions might be of "shorter range", but their meaning as *correlations* is not clear.

APPENDIX D

The Percus–Yevick integral equation

The history of the search for an integral equation for the pair correlation function is quite long. It probably started with Kirkwood (1935), followed by Yvon (1935, 1958), Born and Green (1946), and many others. For a summary of these efforts, see Hill (1956), Fisher (1964), Rushbrooke (1968), Münster (1969), and Hansen and McDonald (1976). Most of the earlier works used the superposition approximation to obtain an integral equation for the pair correlation function. It was in 1958 that Percus and Yevick developed an integral equation that did not include explicitly the assumption of superposition, i.e., pairwise additivity of the higher order potentials of mean force. The Percus–Yevick (PY) equation was found most useful in the study of both pure liquids as well as mixtures of liquids.

We present here a derivation of the Percus–Yevick equation based on the material of Appendices B and C. As in Appendix C, we consider a system in an "external" potential ψ. In the present case, the external potential is produced by a *particle* (identical to the other particles of the system) fixed at $\boldsymbol{R}_0$:

$$\psi(\boldsymbol{R}_1) = U(\boldsymbol{R}_0, \boldsymbol{R}_1) \tag{D.1}$$

i.e., the "external" potential at $\boldsymbol{R}_1$ is equal to the potential produced by placing a particle at $\boldsymbol{R}_0$. When $\psi = 0$, the particle at $\boldsymbol{R}_0$ is "switched off."

Consider the singlet density at $\boldsymbol{R}_1$ in the presence and in the absence of ψ. Clearly, we have (using the notation of Appendix C)

$$\rho^{(1)}(\boldsymbol{R}_1|\psi) = \rho(\boldsymbol{R}_1/\boldsymbol{R}_0) \tag{D.2}$$

$$\rho^{(1)}(\mathrm{R}_1|\psi = \mathbf{0}) = \rho^{(1)}(\boldsymbol{R}_1). \tag{D.3}$$

Both the vertical and the slashed lines can be read as "given that." On the lhs of (D.2) and (D.3), we have the singlet density given and external potential ψ, whereas on the rhs, we have the conditional density given a particle at $\boldsymbol{R}_0$.

Viewing $\rho^{(1)}(\boldsymbol{R}_1 \mid \boldsymbol{\psi})$ as a functional of $\boldsymbol{\psi}$, we write the functional Taylor expansion (see Appendix B).

$$\rho^{(1)}(\boldsymbol{R}_1|\boldsymbol{\psi})=\rho^{(1)}(\boldsymbol{R}_1|\boldsymbol{\psi}=0)+\int \left.\frac{\delta\rho^{(1)}(\boldsymbol{R}_1|\boldsymbol{\psi})}{\delta\psi(\boldsymbol{R}_2)}\right|_{\boldsymbol{\psi}=0}\psi(\boldsymbol{R}_2)d\boldsymbol{R}_2+\cdots \tag{D.4}$$

This particular expansion does not prove to be useful. The reason, as explained in Appendix B, is that a first-order Taylor expansion is expected to be useful when the "increment" here, $\boldsymbol{\psi}$, is "small." For instance, in equation (B.28) of Appendix B, if x is very large, we cannot expect that a first-order Taylor expansion will lead to a good approximation. In (D.4), $\boldsymbol{\psi}$ replaces x (of the one-dimensional example). Since $\psi(\boldsymbol{R}') \to \infty$ as $\boldsymbol{R}' \to \boldsymbol{R}_0$, the increment $\boldsymbol{\psi}$ cannot be considered to be "small."

Instead, Percus (1962) suggested a different expansion of a functional of a function which is everywhere finite, hence expecting a better approximation.

Consider the following two functionals of $\boldsymbol{\psi}$:

$$\xi(\boldsymbol{R}_1|\boldsymbol{\psi}) = \rho^{(1)}(\boldsymbol{R}_1|\boldsymbol{\psi}) \exp[\beta\psi(\boldsymbol{R}_1)] \tag{D.5}$$

$$\eta(\boldsymbol{R}_1|\boldsymbol{\psi}) = \rho^{(1)}(\boldsymbol{R}_1|\boldsymbol{\psi}) \tag{D.6}$$

where we have

$$\xi(\boldsymbol{R}_1|\boldsymbol{\psi} = \boldsymbol{0}) = \rho^{(1)}(\boldsymbol{R}_1) \qquad \eta(\boldsymbol{R}_1|\boldsymbol{\psi} = 0) = \rho^{(1)}(\boldsymbol{R}_1). \tag{D.7}$$

We now view $\boldsymbol{\xi}$ as a functional of $\boldsymbol{\eta}$ which itself is a functional of $\boldsymbol{\psi}$. In this way, we avoid the possibility of an infinite increment as in (D.4). Thus, the first-order functional Taylor expansion is

$$\xi(\boldsymbol{R}_1|\boldsymbol{\psi})=\xi(\boldsymbol{R}_1|\boldsymbol{\psi}=\boldsymbol{0})+\int\frac{\delta\xi(\boldsymbol{R}_1)}{\delta\eta(\boldsymbol{R}_2)}[\eta(\boldsymbol{R}_2|\boldsymbol{\psi})-\eta(\boldsymbol{R}_2|\boldsymbol{\psi}=\boldsymbol{0})]\,d\boldsymbol{R}_2. \tag{D.8}$$

The functional derivative[†] in (D.8) is of $\boldsymbol{\xi}$, viewed as a functional of $\boldsymbol{\eta}$, taken at the point $\boldsymbol{\eta}=\boldsymbol{\rho}^{(1)}$, i.e., at $\boldsymbol{\psi}=0$; see (D.7). Using (D.1), (D.2)–(D.6), we rewrite (D.8) as

$$\begin{aligned}&\rho(\boldsymbol{R}_1/\boldsymbol{R}_0)\exp[\beta\psi(\boldsymbol{R}_1)]\\&=\rho^{(1)}(\boldsymbol{R}_1)+\int[\delta\xi(\boldsymbol{R}_1)/\delta\eta(\boldsymbol{R}_2)][\rho(\boldsymbol{R}_2/\boldsymbol{R}_0)-\rho^{(1)}(\boldsymbol{R}_2)]\,d\boldsymbol{R}_2.\end{aligned} \tag{D.9}$$

Using the chain rule for the functional derivative (Appendix B), we find

$$\frac{\delta\xi(\boldsymbol{R}_1)}{\delta\eta(\boldsymbol{R}_2)}=\int\frac{\delta\xi(\boldsymbol{R}_1)}{\delta\psi(\boldsymbol{R}_3)}\frac{\delta\psi(\boldsymbol{R}_3)}{\delta\eta(\boldsymbol{R}_2)}\,d\boldsymbol{R}_3. \tag{D.10}$$

[†] In (D.8) and the following equation, we use a shorthand notation whenever possible. For instance, the notation $\delta\xi(\boldsymbol{R}_1)/\delta\eta(\boldsymbol{R}_2)$ means that we view $\boldsymbol{\xi}$ as a functional of $\boldsymbol{\eta}$ and take the derivative with respect to $\eta(\boldsymbol{R}_2)$, the derivative being evaluated at the point $\boldsymbol{\eta}=\boldsymbol{\rho}^{(1)}$, corresponding to $\boldsymbol{\psi}=\boldsymbol{0}$.

The two functional derivatives on the rhs can be obtained from (D.5) and (D.6):

$$\frac{\delta\xi(\mathbf{R}_1)}{\delta\psi(\mathbf{R}_3)} = \frac{\delta\rho^{(1)}(\mathbf{R}_1|\boldsymbol{\psi})}{\delta\psi(\mathbf{R}_3)} \exp\left[\beta\psi(\mathbf{R}_1)\right] + \beta\rho^{(1)}(\mathbf{R}_1|\boldsymbol{\psi}) \exp[\beta\boldsymbol{\psi}(\mathbf{R}_1)]\,\delta(\mathbf{R}_1 - \mathbf{R}_3) \tag{D.11}$$

which, at the point $\boldsymbol{\psi} = \mathbf{0}$, reduces to (see also equation (C.7) of Appendix C)

$$\left.\frac{\delta\xi(\mathbf{R}_1)}{\delta\psi(\mathbf{R}_3)}\right|_{\psi=0} = -\beta\left[\rho^{(2)}h(\mathbf{R}_1,\mathbf{R}_3) + \rho^{(1)}(\mathbf{R}_1)\,\delta(\mathbf{R}_1 - \mathbf{R}_3)\right] + \beta\rho^{(1)}(\mathbf{R}_1)\,\delta(\mathbf{R}_1 - \mathbf{R}_3). \tag{D.12}$$

Similarly, from (D.6) and from equation (C.8) of Appendix C, we get

$$\left.\frac{\delta\psi(\mathbf{R}_3)}{\delta\eta(\mathbf{R}_2)}\right|_{\psi=0} = \left.\frac{\delta\psi(\mathbf{R}_3)}{\delta\rho^{(1)}(\mathbf{R}_2)}\right|_{\psi=0} = \beta^{-1}\left[c(\mathbf{R}_2,\mathbf{R}_3) - \frac{\delta(\mathbf{R}_2 - \mathbf{R}_3)}{\rho^{(1)}(\mathbf{R}_2)}\right]. \tag{D.13}$$

Substituting (D.12) and (D.13) into (D.10), we get, after rearrangement,

$$\left.\frac{\delta\xi(\mathbf{R}_1)}{\delta\eta(\mathbf{R}_2)}\right|_{\psi=0} = \rho h(\mathbf{R}_1,\mathbf{R}_2) - \rho^2\int h(\mathbf{R}_1,\mathbf{R}_3)\,c(\mathbf{R}_2,\mathbf{R}_3)\,d\mathbf{R}_3 = \rho c(\mathbf{R}_1,\mathbf{R}_2) \tag{D.14}$$

where in the second step on the rhs we have used the Ornstein–Zernike relation (Appendix C). Substituting (D.14) into (D.9), we get

$$\rho g(\mathbf{R}_1,\mathbf{R}_0)\exp[\beta\psi(\mathbf{R}_1)] = \rho + \rho^2\int c(\mathbf{R}_1,\mathbf{R}_2)h(\mathbf{R}_2,\mathbf{R}_0)\,d\mathbf{R}_2. \tag{D.15}$$

Using the notation

$$f(\mathbf{R}_1,\mathbf{R}_2) = \exp[-\beta U(\mathbf{R}_1,\mathbf{R}_2)] - 1 \tag{D.16}$$

$$y(\mathbf{R}_1,\mathbf{R}_2) = g(\mathbf{R}_1,\mathbf{R}_2)\exp[\beta U(\mathbf{R}_1,\mathbf{R}_2)] \tag{D.17}$$

and the Ornstein–Zernike relation, we can rewrite (D.15) as

$$c(\mathbf{R}_1,\mathbf{R}_2) = y(\mathbf{R}_1,\mathbf{R}_2)\,f(\mathbf{R}_1,\mathbf{R}_2). \tag{D.18}$$

Here, we obtained $c(R)$ as a *result* of the first-order expansion. When (D.18) is used in the Ornstein–Zernike relation, we get the PY equation. Alternatively, one can *assume* relation (D.18) rewritten as*

$$c(R) = y(R)\exp[-\beta U(R)] - y(R) = g(R) - y(R). \tag{D.19}$$

$g(R)$ is referred to as the “total” correlation and $y(R)$ is referred to as the “indirect” correlation. Then the difference should have the meaning of the “direct” correlation. It is also clear that with this definition of $c(R)$, it has the same range as $U(R)$, i.e., $c(R)$ vanishes when $U(R)$ vanishes.

* Here $R = |\mathbf{R}_2 - \mathbf{R}_1|$ is a scalar

Alternatively, one can expand $y(R)$ to first order in $\beta[W(R) - U(R)]$ to obtain

$$y(R) = \exp\{-\beta[W(R) - U(R)]\} \approx 1 - \beta[W(R) - U(R)]. \qquad (D.20)$$

Then, redefine $c(R)$ as

$$\begin{aligned} c(R) &= g(R) - 1 + \beta[W(R) - U(R)] \\ &= g(R) - 1 - \ln y(R). \end{aligned} \qquad (D.21)$$

This, when used in the Ornstein–Zernike equation, gives the so-called hyper-netted- chain equation for $g(R)$.
Relation (D.18) is often referred to as the Percus–Yevick approximation. If we use (D.18) in the Ornstein–Zernike relation, we get an integral equation for $\boldsymbol{y}$:

$$\begin{aligned} y(\boldsymbol{R}_1, \boldsymbol{R}_2) = 1 + \rho \int & y(\boldsymbol{R}_1, \boldsymbol{R}_2) f(\boldsymbol{R}_1, \boldsymbol{R}_3) \\ & \times [y(\boldsymbol{R}_2, \boldsymbol{R}_3) f(\boldsymbol{R}_2, \boldsymbol{R}_3) + y(\boldsymbol{R}_2, \boldsymbol{R}_3) - 1] \, d\boldsymbol{R}_3. \end{aligned} \qquad (D.22)$$

This is the Percus–Yevick integral equation for $\boldsymbol{y}$. Once a solution for y is obtained, one can calculate $\boldsymbol{g}$ from (D.17).

Another simpler and useful form of this equation is obtained by transforming to bipolar coordinates

$$u = |\boldsymbol{R}_1 - \boldsymbol{R}_3|, \qquad v = |\boldsymbol{R}_2 - \boldsymbol{R}_3| \qquad R = |\boldsymbol{R}_1 - \boldsymbol{R}_2|. \qquad (D.23)$$

The element of volume is

$$d\boldsymbol{R}_3 = 2\pi uv \, du \, dv / R \qquad (D.24)$$

and for spherical particles, (D.22) is transformed into

$$y(R) = 1 + 2\pi\rho R^{-1} \int_0^\infty y(u) f(u) u \, du \int_{|R-u|}^{R+u} [y(v) f(v) + y(v) - 1] v \, dv. \qquad (D.25)$$

Another equivalent equation that was useful in the numerical solution of the *PY* equation is for the function

$$z(R) = y(R)R. \qquad (D.26)$$

With (D.26) and (D.25), we get an integral equation for $z(R)$, which reads

$$z(R) = R + 2\pi\rho \int_0^\infty z(u) f(u) \, du \int_{|R-u|}^{R+u} [z(v) f(v) + z(v) - v] \, dv. \qquad (D.27)$$

This equation was found to be a convenient form for a numerical solution. This is further discussed in Appendix E.

APPENDIX E

Numerical solution of the Percus–Yevick equation

The exact solution of the Percus–Yevick (PY) equation is known for a one-component system of hard spheres (Wertheim 1963; Thiele 1963) and for mixtures of hard spheres (Lebowitz 1964). Numerical solutions of the PY equation (for Lennard-Jones particles) have been carried out by many authors, e.g., Broyles (1960, 1961), Broyles et al. (1962), Throop and Bearman (1966), Baxter (1967), Watts (1968), Mandel et al. (1970), Grundke and Henderson (1972a, b)

We present here a brief account of the numerical procedure employed for the computations of $g(R)$ which we have used for our illustrations in this book. We start with the integral equation for the function $z(R)$ (see Appendix D)

$$z(R) = R + 2\pi\rho \int_0^\infty z(u)f(u)\,du \int_{|R-u|}^{R+u} [z(v)f(v) + z(v) - v]\,dv. \quad \text{(E.1)}$$

We begin the iterative procedure by substituting the initial function

$$z_0(R) = R \quad \text{(E.2)}$$

in the rhs of (E.1), to obtain

$$z_1(R) = R + 2\pi\rho \int_0^\infty uf(u)\,du \int_{|R-u|}^{R+u} vf(v)\,dv \quad \text{(E.3)}$$

Next, $z_1(R)$ from (E.3) is substituted in the rhs of (1) to obtain $z_2(R)$, and so forth. It turns out that for high densities ρ, such a procedure does not lead to a convergent solution. Instead, one uses a "mixing" parameter λ, $0 \le \lambda \le 1$ (Broyles 1960, 1962; Throop and Bearman 1966; Ben-Naim 1972a, b, 1974) so that the $(k+1)$th input function is constructed from the kth input and the kth output, as follows:

$$z_{k+1}^{\text{in}}(R) = \lambda z_k^{\text{out}}(R) + (1-\lambda)z_k^{\text{in}}(R). \quad \text{(E.4)}$$

In practice, it is found that as ρ increases, one is compelled to use smaller values of λ in (E.4), and large numbers of iterations to get a convergent result.

As a criterion of convergence, we can choose the quantity

$$\zeta = N_R^{-1} \sum_{i=1}^{N_R} |z_k(i) - z_{k-1}(i)| \tag{E.5}$$

where N_R is the number of division points at which the function z is evaluated. $z_k(i)$ denotes the value of the function z_k at the point R_i. The iterative procedure is terminated when ζ falls below a certain small value, say 10^{-5} or 10^{-6}, depending on the required accuracy.

For mixtures, of say, two components (equation E.1) is generalized to

$$\begin{aligned} z_{\alpha\beta}(R) &= R + \sum_{\gamma=\alpha,\beta} 2\pi\rho_\gamma \int_0^\infty z_{\alpha\gamma}(u) f_{\alpha\gamma}(u)\, du \\ &\times \int_{|R-u|}^{R+u} [z_{\gamma\beta}(v) f_{\gamma\beta}(v) + z_{\gamma\beta}(v) - v]\, dv \end{aligned} \tag{E.6}$$

where the sum includes two terms $\gamma = \alpha, \beta$. The numerical procedure is similar to the case for one component. One starts with

$$z_{\alpha\beta}(R) = R \tag{E.7}$$

for all the four functions $z_{\alpha\beta}(R)$ and proceeds to solve the four integral equations (E.6) by iteration.

In most of the illustrations for this book, we calculated the pair correlation functions at volume densities of 0.4 and 0.45. For hard spheres Yau et al. (1999) reported calculations of PY up to $\eta = 0.52$. Even at these, relatively far from the close-packing densities, the convergence of the solutions is very slow and requires up to 1000 iterations.

APPENDIX F

Local density fluctuations

In chapters 1 and 4, we have discussed fluctuations in the number of particles in the entire system. Here, we shall be interested in the fluctuations of the density in a given region S within the system.

Consider, for example, a system in the T, V, N ensemble. We select a region S within the system and inquire as to the number of particles that fall within S for a given configuration[†] $\mathbf{R}^N$:

$$N(\mathbf{R}^N, S) = \sum_{i=1}^{N} \int_S \delta(\mathbf{R}_i - \mathbf{R}')\, d\mathbf{R}'. \tag{F.1}$$

Each term in the sum over i is unity whenever $\mathbf{R}_i$ is in S, and is zero otherwise. Therefore, the sum over i *counts* all the particles that are within S at a given configuration $\mathbf{R}^N$. The average number of particles in S is

$$\begin{aligned}\langle N(S)\rangle &= \sum_{i=1}^{N} \int \underset{V}{\cdots} \int d\mathbf{R}^N\, P(\mathbf{R}^N) \int_S \delta(\mathbf{R}_i - \mathbf{R}')\, d\mathbf{R}' \\ &= N \int \underset{V}{\cdots} \int d\mathbf{R}^N\, P(\mathbf{R}^N) \int_S \delta(\mathbf{R}_1 - \mathbf{R}')\, d\mathbf{R}' \\ &= \int_S d\mathbf{R}' N \int \underset{V}{\cdots} \int d\mathbf{R}^N\, P(\mathbf{R}^N)\delta(\mathbf{R}_1 - \mathbf{R}') \\ &= \int_S d\mathbf{R}'\, \rho^{(1)}(\mathbf{R}') = \rho V(S).\end{aligned} \tag{F.2}$$

In (F.2), we have used the definition of the singlet molecular distribution function. The last relation holds for a homogenous fluid, where $V(S)$ is the volume of the region S, and ρ is the the bulk density $\rho = N/V$.

Next, consider the average of the square of $N(\mathbf{R}^N, S)$:

$$\langle N(S)^2\rangle = \int \underset{V}{\cdots} \int d\mathbf{R}^N P(\mathbf{R}^N) \int_S \sum_{i=1}^{N} \delta(\mathbf{R}_i - \mathbf{R}')\, d\mathbf{R}' \int_S \sum_{j=1}^{N} \delta(\mathbf{R}_j - \mathbf{R}'')\, d\mathbf{R}''. \tag{F.3}$$

[†] The results of this section apply to a system of particles which are not necessarily spherical. Here, we use $\mathbf{R}^N$ for describing the locations of all the *centers* of the particles. The orientations are of no importance for the present considerations.

Rearranging (F.3), we obtain

$$\begin{aligned}
\langle N(S)^2\rangle &= \sum_{i=1}^{N}\sum_{j=1}^{N}\int_S d\mathbf{R}'\int_S d\mathbf{R}''\int \underset{V}{\cdots}\int d\mathbf{R}^N\, P(\mathbf{R}^N)\,\delta(\mathbf{R}_i-\mathbf{R}')\,\delta(\mathbf{R}_j-\mathbf{R}'') \\
&= \sum_{i=1}^{N}\int_S d\mathbf{R}'\int_S d\mathbf{R}''\,\delta(\mathbf{R}'-\mathbf{R}'')\int \underset{V}{\cdots}\int d\mathbf{R}^N\, P(\mathbf{R}^N)\,\delta(\mathbf{R}_i-\mathbf{R}') \\
&\quad + \sum_{i\neq j}\int_S d\mathbf{R}'\int_S d\mathbf{R}''\int \underset{V}{\cdots}\int d\mathbf{R}^N\, P(\mathbf{R}^N)\,\delta(\mathbf{R}_i-\mathbf{R}')\,\delta(\mathbf{R}_j-\mathbf{R}'') \\
&= \int_S d\mathbf{R}'\int_S d\mathbf{R}''\,\delta(\mathbf{R}'-\mathbf{R}'')\rho^{(1)}(\mathbf{R}') + \int_S d\mathbf{R}'\int_S d\mathbf{R}''\rho^{(2)}(\mathbf{R}',\mathbf{R}'') \\
&= \langle N(S)\rangle + \int_S d\mathbf{R}'\int_S d\mathbf{R}''\rho^{(2)}(\mathbf{R}',\mathbf{R}''). \qquad \text{(F.4)}
\end{aligned}$$

In the second step on the rhs of (F.4), we have split the double sum over i and j into two sums; the first over all the terms with $i=j$, and the second over the terms with $i\neq j$. We have also used the identity of the product of two Dirac functions[†]

$$\delta(\mathbf{R}_i-\mathbf{R}')\delta(\mathbf{R}_i-\mathbf{R}'') = \delta(\mathbf{R}_i-\mathbf{R}')\delta(\mathbf{R}'-\mathbf{R}''). \qquad \text{(F.5)}$$

In the third step of (F.4), we have used the definition of the pair distribution function.

The fluctuations in the number of particles within S are given by

$$\begin{aligned}
\langle \Delta N(S)^2\rangle &= \langle (N(S)-\langle N(S)\rangle)^2\rangle \\
&= \langle N(S)^2\rangle - \langle N(S)\rangle^2 \\
&= \langle N(S)\rangle + \int_S d\mathbf{R}'\int_S d\mathbf{R}''\rho^{(2)}(\mathbf{R}',\mathbf{R}'') \\
&\quad - \int_S d\mathbf{R}'\,\rho^{(1)}(\mathbf{R}')\int_S d\mathbf{R}'\rho^{(1)}(\mathbf{R}'') \\
&= \langle N(S)\rangle + \rho^2\int_S d\mathbf{R}'\int_S d\mathbf{R}''\,[g(R)-1] \qquad \text{(F.6)}
\end{aligned}$$

where $R=|\mathbf{R}'-\mathbf{R}''|$.

Relation (F.6) holds for any region S within the system. Care must be exercised in applying (F.6) for the whole volume V. If we are in the T, V, μ ensemble, then, taking S as the whole volume of the system, we get the compressibility relation. However, if we are in the T, V, N ensemble, then N is fixed

[†] The analog of (F.5) in the discrete case is $\delta_{ik}\delta_{ij}=\delta_{ik}\delta_{kj}$.

and we have

$$\langle N(V)\rangle = N(V) = N \tag{F.7}$$

and (F.6) reduces to

$$N + \rho^2 \int_V d\mathbf{R}' \int_V d\mathbf{R}'' \, [g(R) - 1] = 0 \tag{F.8}$$

which is the correct normalization condition for $g(R)$ in the T, V, N ensemble.

Another limiting case is that in which S is an infinitesimally small region such that *at most* one particle can occupy S at any given time. For instance, if the maximum diameter of S is smaller than σ (the effective diameter of the particles), then $g(R)$ is zero in the integrand of (F.6), and (F.6) reduces to

$$\langle \Delta N(S)^2\rangle = \langle N(S)\rangle - \rho^2 V(S)^2 = \rho V(S)\,[1 - \rho V(S)]. \tag{F.9}$$

Thus, if $V(S)$ is infinitesimally small, then fluctuations in the number of particles are dominated by $\rho^2 V(S)$.

Next, consider two regions S_1 and S_2 within the system, Figure F.1. The cross fluctuations in the number of particles in the two regions are given by

$$\begin{aligned}
\langle \Delta N(S_1)\Delta N(S_2)\rangle &= \langle [N(S_1) - \langle N(S_1)\rangle][N(S_2) - \langle N(S_2)\rangle]\rangle \\
&= \langle N(S_1)N(S_2)\rangle - \langle N(S_1)\rangle\langle N(S_2)\rangle \\
&= \int \underset{V}{\cdots} \int d\mathbf{R}^N P(\mathbf{R}^N) \int_{S_1} \sum_{i=1}^{N} \delta(\mathbf{R}_i - \mathbf{R}')\, d\mathbf{R}' \int_{S_2} \sum_{j=1}^{N} \delta(\mathbf{R}_j - \mathbf{R}'')\, d\mathbf{R}'' \\
&\quad - \int_{S_1} d\mathbf{R}' \rho^{(1)}(\mathbf{R}') \int_{S_2} d\mathbf{R}'' \rho^{(1)}(\mathbf{R}'') \\
&= \sum_{i=1}^{N} \int_{S_1} d\mathbf{R}' \int_{S_2} d\mathbf{R}'' \delta(\mathbf{R}' - \mathbf{R}'') \int \underset{V}{\cdots} \int d\mathbf{R}^N P(\mathbf{R}^N) \delta(\mathbf{R}_i - \mathbf{R}') \\
&\quad + \sum_{i \neq 1} \int_{S_1} d\mathbf{R}' \int_{S_2} d\mathbf{R}'' \int \underset{V}{\cdots} \int d\mathbf{R}^N P(\mathbf{R}^N) \delta(\mathbf{R}_i - \mathbf{R}') \delta(\mathbf{R}_j - \mathbf{R}'') \\
&\quad - \int_{S_1} d\mathbf{R}' \rho^{(1)}(\mathbf{R}') \int_{S_2} d\mathbf{R}'' \rho^{(1)}(\mathbf{R}'') \\
&= \int_{S_1} d\mathbf{R}' \int_{S_2} d\mathbf{R}'' \rho^{(1)}(\mathbf{R}') \delta(\mathbf{R}' - \mathbf{R}'') \\
&\quad + \int_{S_1} d\mathbf{R}' \int_{S_2} d\mathbf{R}'' [\rho^{(2)}(\mathbf{R}', \mathbf{R}'') - \rho^{(1)}(\mathbf{R}') \rho^{(1)}(\mathbf{R}'')]
\end{aligned} \tag{F.10}$$

where we have used again the definitions of the singlet and the pair distribution functions.

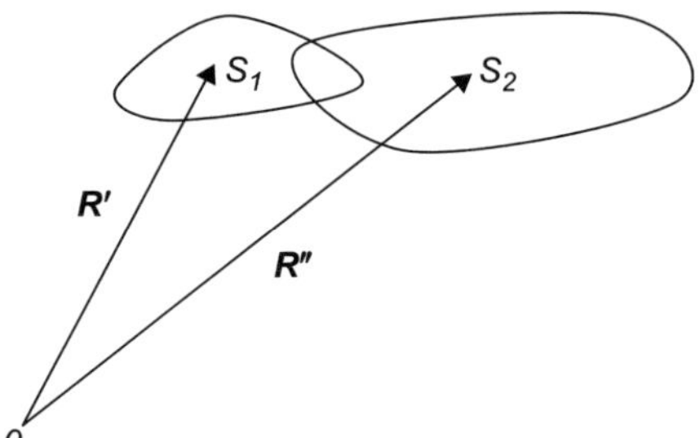

Figure F.1 Two regions S_1 and S_2 within the system. R' and R'' are two points within S_1 and S_2, respectively.

We now distinguish between two cases:

(1) If the two regions S_1 and S_2 do not overlap, the first term on the rhs of (F.10) is zero. This follows from the property of the Dirac delta function:

$$\int_{S_2} d\mathbf{R}''\delta(\mathbf{R}' - \mathbf{R}'') = 0 \qquad \text{if } \mathbf{R}' \notin S_2. \tag{F.11}$$

Since $\mathbf{R}'$ is always within S_1, relation (F.11) will hold whenever S_1 and S_2 do not intersect.

(2) If S_1 and S_2 do intersect, the first term on the rhs of (F.10) is

$$\int_{S_2} d\mathbf{R}''\delta(\mathbf{R}' - \mathbf{R}'') = \begin{cases} 0 & \text{if } \mathbf{R}' \notin S_2 \\ 1 & \text{if } \mathbf{R}' \in S_2. \end{cases} \tag{F.12}$$

But $\mathbf{R}'$ is always within S_1, hence

$$\int_{S_1} d\mathbf{R}'\rho^{(1)}(\mathbf{R}') \int_{S_2} d\mathbf{R}''\delta(\mathbf{R}' - \mathbf{R}'') = \int_{S_1 \cap S_2} d\mathbf{R}'\rho^{(1)}(\mathbf{R}') = \rho V(S_1 \cap S_2) \tag{F.13}$$

where $S_1 \cap S_2$ is the intersection region between S_1 and S_2. The last equality on the rhs of (F.13) holds for homogeneous fluids. For this case, we write the final form of (F.10) as

$$\langle \Delta N(S_1)\Delta N(S_2)\rangle = \rho V(S_1 \cap S_2) + \rho^2 \int_{S_1} d\mathbf{R}' \int_{S_2} d\mathbf{R}''[g(R) - 1]. \tag{F.14}$$

Two special cases of (F.14) are the following:

(a) If S_1 and S_2 are identical, i.e., $S_1 = S_2 = S$, then (F.14) reduces to the previous result (equation F.6).

(b) If S_1 and S_2 are *infinitesimal, nonoverlapping* regions of volumes, say $d\mathbf{R}'$ and dR'', then

$$\langle \Delta N(d\mathbf{R}')\Delta N(d\mathbf{R}'')\rangle = \rho^2[g(R) - 1]d\mathbf{R}'\, d\mathbf{R}''. \tag{F.15}$$

On the other hand, dividing (F.15) by $d\mathbf{R}'\,d\mathbf{R}''$ and using the second equality on the rhs of (F.10), we get

$$\frac{\langle \Delta N(d\mathbf{R}')\Delta N(d\mathbf{R}'')\rangle}{d\mathbf{R}'d\mathbf{R}''} = \frac{\langle N(d\mathbf{R}')N(d\mathbf{R}'')\rangle - \langle N(d\mathbf{R}')\rangle\langle N(d\mathbf{R}'')\rangle}{d\mathbf{R}'d\mathbf{R}''}$$
$$= \langle \rho(\mathbf{R}')\rho(\mathbf{R}'')\rangle - \langle \rho(\mathbf{R}')\rangle\langle \rho(\mathbf{R}'')\rangle \tag{F.16}$$

where $\rho(\mathbf{R})$ is for the local density at $\mathbf{R}$ for a *given* configuration $\mathbf{R}^N$ of the system. Note the difference between $\rho(\mathbf{R})$ and $\rho^{(1)}(\mathbf{R})$. Combining (F.16) and (F.15), we get

$$\langle \rho(\mathbf{R}')\rho(\mathbf{R}'')\rangle - \langle \rho(\mathbf{R}')\rangle\langle \rho(\mathbf{R}'')\rangle = \rho^2[g(R) - 1] \tag{F.17}$$

or

$$\langle \rho(\mathbf{R}')\rho(\mathbf{R}'')\rangle = \rho^2 g(R), \quad R = |\mathbf{R}'' - \mathbf{R}'| \ \text{ and } \ \mathbf{R}' \neq \mathbf{R}''. \tag{F.18}$$

The last result can also be viewed as a definition of $g(R)$, i.e., this function conveys the *correlation* in the *local densities* at two points $\mathbf{R}'$ and $\mathbf{R}''$ in the fluid. Note that if we also allow the case $\mathbf{R}'' = \mathbf{R}'$, then (F.18) should be modified to read

$$\langle \rho(\mathbf{R}')\rho(\mathbf{R}'')\rangle = \rho\delta(\mathbf{R}' - \mathbf{R}'') + \rho^2 g(R). \tag{F.19}$$

APPENDIX G

The long-range behavior of the pair correlation function

In sections 3.5 and 4.2 we inferred from the normalization condition on the pair correlation function that the pair correlation function must have different behaviors in an open (O) and in a closed (C) system. In this appendix, we further elaborate on this aspect of the pair correlation function.

We start with a (theoretical) ideal gas, i.e., when there are *no* intermolecular interactions between the particles. For such a system, the pair correlation function may be calculated exactly from the corresponding partition functions.

The pair distribution function in a closed (C) system of N particles is

$$\rho_C^{(2)} = \frac{N(N-1)V^{N-1}}{V^N} = \frac{N(N-1)}{V^2}. \tag{G.1}$$

This can be obtained directly from the definition of the pair distribution function (section 2.2) by putting $U_N \equiv 0$ for all configurations of the N particles.

The singlet distribution function is

$$\rho_C^{(1)} = \frac{NV^{N-1}}{V^N} = \frac{N}{V} = \rho. \tag{G.2}$$

Hence, the pair correlation, in a closed system is

$$g_C^{(2)}(R) = \frac{\rho_C^{(2)}}{(\rho_C^{(1)})^2} = 1 - \frac{1}{N}. \tag{G.3}$$

In an open system (O), the singlet and the pair distribution functions are obtained from the grand partition function

$$\begin{aligned}\rho_O^{(2)} &= \frac{1}{\Xi}\sum_{N\geq 2}\frac{z^N}{(n-2)!}\int d\mathbf{R}_3\ldots d\mathbf{R}_N \exp[-\beta U_N]\\ &= z^2\end{aligned} \tag{G.4}$$

and

$$\rho_O^{(1)} = z \tag{G.5}$$

where $z = \exp[\beta\mu]/\Lambda^3$ is the activity.

Hence, from (G.4) and (G.5), we obtain for an ideal gas

$$g_O^{(2)}(R) = 1. \tag{G.6}$$

For most practical purposes, when we are interested in the behavior of the pair correlation function itself, the difference between (G.3) and (G.6) is negligible for macroscopic system, where $N \approx 10^{23}$. However, this small difference becomes important when we integrate over *macroscopic volumes*. This is clear from the following two exact normalization conditions (see section 3.5)

$$\rho G_C = \rho \int [g_C^{(2)}(\mathbf{R}) - 1] d\mathbf{R} = -1 \tag{G.7}$$

$$\rho G_O = \rho \int [g_O^{(2)}(R) - 1] d\mathbf{R} = -1 + kT\rho\kappa_T. \tag{G.8}$$

Here, the integrations extend over the entire volume of the system. The interpretation of (G.7) and (G.8) is straightforward. The quantity ρG is the change in the number of particles in the entire system caused by placing one particle at some fixed point, say $\mathbf{R}_0$. When N is constant, this change is exactly -1; the particle we have placed at. No such *closure* condition is imposed in an open system. Equation (G.8) is just the compressibility equation.

The probabilistic interpretation of (G.3) and (G.6) is also straightforward. Placing a particle at in an ideal gas system of exactly N particles, changes the conditional probability (or density) of finding a particle at any location in the system from N/V into $(N-1)/V$; hence, the correlation function has the form (G.3). In an open system, placing a particle at a fixed position does not affect the density at any other point in the system. This is so since the chemical potential μ, rather than N is fixed, hence the density at any point in the system is constant, $\langle N \rangle / V$, figure G.1.

The aforementioned arguments are exact for a *theoretical* ideal gas at any concentration. Of course, such a system does not exist. We now turn to a real gas, which in the limit of $\rho \to 0$ behaves as an ideal gas.

For a *real* gas, as $\rho \to 0$, we also have the following well-known results

$$g_C^{(2)}(R) = \left(1 - \frac{1}{N}\right) \exp[-\beta U(R)] \tag{G.9}$$

and

$$g_O^{(2)}(R) = \exp[-\beta U(R)]. \tag{G.10}$$

Here, we can say that the *short*-range behavior is dominated by $\exp[-\beta U(R)]$ in both the open and the closed system. However, the difference between the

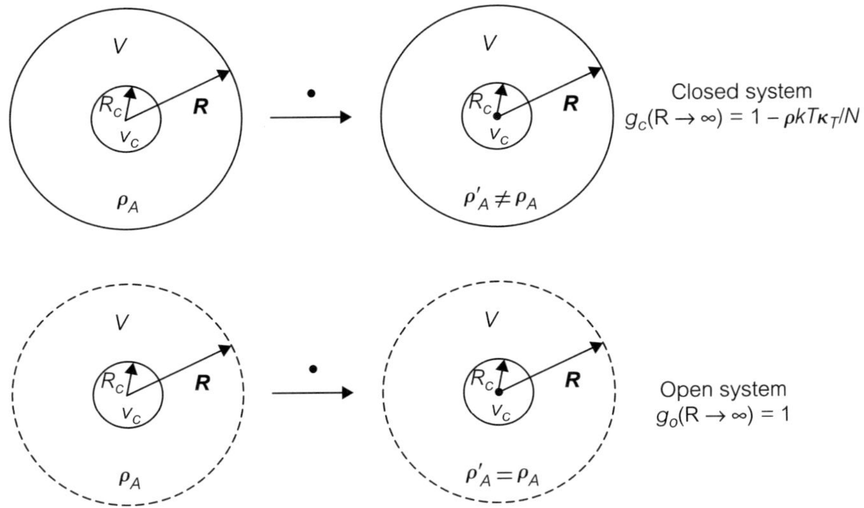

Figure G.1 The difference in the response of the density of an *A* particle at large distances to placing a particle (dot) at the center of the correlation sphere, in a closed and in an open system. R_C is the correlation distance; R is the macroscopic length of the system.

two systems is in the "long range" behavior, i.e., for R large enough so that $U(R)$ is practically zero, we have for an open system $g_O^{(2)}(R) = 1$, while for a closed system $g_C^{(2)}(R) = 1 - N^{-1}$.

Note, however, that while the first contribution to the pair correlation is justifiably referred to as the *short-range* behavior, the second contribution applies for any R. It is true that the second contribution is negligible at short distances when $U(R)$ is finite, and it has an effect only when integration extends over long range. However, strictly speaking, the behavior $1 - 1/N$ is valid everywhere, as in the case of a theoretical gas. For this reason, we prefer to refer to the first contribution as the effect of the *direct* interaction, and since the direct interactions, even the long-range ones, are always *local*, we shall refer to this part as the local correlation (LC). The second contribution due to the closure condition will be referred to as closure correlation (CC). This is nonlocal correlation. The terms *local* and *nonlocal* correlations might better be applied than *short range* and *long range* (the latter is also referred to as asymptotic behavior).

We now turn to a system of interacting particles. The "long range" or the "asymptotic" behavior of the pair correlation functions are well known for closed and open systems:

$$g_C^{(2)}(R) \to 1 - \frac{kT\rho\kappa_T}{N} \tag{G.11}$$

$$g_O^{(2)}(R) \to 1. \tag{G.12}$$

Relations (G.11) and (G.12) are referred to as either the asymptotic behavior† or the long-range behavior of the correlation function.

Relations (G.11) and (G.12) strictly hold (for systems of interacting particles) at large distances. Therefore, the terms long-range or asymptotic behavior for this case are appropriate. However, since for ideal gases equations (G.3) and (G.6) hold true for any R, not only in the limit $R \to \infty$, a more descriptive term should apply for both the ideal gas case as well as for system of interacting particles. This would have the advantage of *indicating the source of the correlation.*

We shall now derive relation (G.11) using a more intuitive, albeit less rigorous argument, based on the two exact normalization conditions.

$$\rho G_C = \rho \int_0^\infty [g_C(R) - 1)]4\pi R^2 \, dR = -1 \tag{G.13}$$

$$\rho G_O = \rho \int_0^\infty [(g_O(R) - 1)]4\pi R^2 \, dR = -1 + kT\rho\kappa_T. \tag{G.14}$$

We have dropped the superscript (2) from $g(R)$ for convenience. We have also changed from integration over the volume V to integration over the entire range of distances $(0, \infty)$. We shall discuss the *pair* correlation function only. Since both (G.13) and (G.14) are exact, it follows that g_C and g_O must be different functions of R.

We now recognize that correlation between densities at different locations can arise from two sources. One is due to *direct* intermolecular interactions, and the other is due to the *closure* condition. We assume that the first is operative only at short distances‡ say $0 \le R \le R_{COR}$, where R_{COR} is the *correlation distance* beyond which a particle placed at fixed position does not have any influence on the density at any other position. This is the distance R_{COR}, beyond which $g(R)$ is nearly equal to 1. The second, is normally referred to as long-range correlation. We shall refer to it as the *closure correlation* (CC). Since this part has the N^{-1} dependence on N, it has a negligible effect on $g(R)$ at $R \le R_{COR}$. It becomes important when integration extends to $R \to \infty$.

† Hill (1956) has derived this *asymptotic* behavior based on cluster expansion argument due to DeBoer (1940). Lebowitz and Percus (1961, 1963), who also generalized these equations, for the correlation between two groups of n and m particles, referred to this behavior as the *long-range* correlations, and specifically refer to equation (G.11) as the Ornstein–Zernike relation (Ornstein and Zernike 1914).

‡ For charged particles, the interactions are also long-ranged, but still there exists some R_{COR} beyond which no direct influence of the interaction is noticeable. There are several studies of the manner that the pair correlation function decays to unity [see for example Fisher and Widom (1969), Perry and Throup (1972)]. In any case, even when $g(R)$ is of relatively long-range, we can still find a radius R_C beyond which the pair correlation is practically unity.

Denoting by g(LC) the *local* correlation due to direct interactions, and by g(CC) the *closure* correlation, we rewrite (G.13) and (G.14) as

$$\rho G_C = \rho \int_0^{R_{COR}} [(g_C(LC) - 1)] 4\pi R^2 \, dR + \rho \int_{R_{COR}}^{\infty} [g_C(CC) - 1] 4\pi R^2 \, dR = -1 \tag{G.15}$$

$$\rho G_O = \rho \int_0^{R_{COR}} [(g_O(LC) - 1)] 4\pi R^2 \, dR = -1 + kT\rho\kappa_T. \tag{G.16}$$

The existence of such a correlation distance is intuitively clear and is confirmed by experimental data. There exists no formal proof of this contention, however. Lebowitz and Percus (1961) argued that such a proof would be extremely difficult to obtain. Since the existence of such a correlation length would be violated for the solid state, a proof of the existence of such a correlation length is tantamount to a criterion for distinction between a fluid and a crystal.

Note that in an *open* system, all the correlation is due to the direct interactions, no closure condition is in effect; beyond $R > R_{COR}$, $g_O(LC) = 1$. Next, we assume that within the correlation distance, both g_C and g_O are the same (or nearly the same, but the difference is negligible) function of R. Subtracting (G.16) from (G.15) we obtain

$$\rho \int_{R_{COR}}^{\infty} [(g_C(CC) - 1)] 4\pi R^2 \, dR = -1 - (-1 + kT\rho\kappa_T) = -kT\rho\kappa_T. \tag{G.17}$$

Finally, we assume that g_C(CC) in the second integral on the rhs of (G.15) is independent of the distance R (this is the reason why we prefer to refer to this part as the nonlocal correlation). Hence,

$$\rho[g_C(CC) - 1][V - V_{COR}] = -kT\rho\kappa_T. \tag{G.18}$$

Since $V \gg V_{COR} = (4\pi R_{COR}^3)/3$ we get from (G.18) the final form

$$g_C(CC) = 1 - \frac{kT\rho\kappa_T}{N} \tag{G.19}$$

which is the required result.

We next turn to the probabilistic interpretation of (G.19). We have seen that for an ideal gas there is a very simple probabilistic interpretation of behavior (G.3), and we have seen that the N^{-1} term is a result of the closure correlation.

The probabilistic interpretation of g_C(CC) in (G.19) is not as obvious, and it is a little more tricky than in the case of an ideal gas. We provide here the appropriate probabilistic interpretation of the closure correlation in (G.19).

We recall the interpretation of ρG_O. This is the change in the average number of particles in V_{COR} due to placing a particle at the center of V_{COR}. We write this as

$$\rho G_O = \Delta N(V_{COR}). \tag{G.20}$$

From the compressibility equation we have

$$\rho G_O = -1 + kT\rho\kappa_T. \tag{G.21}$$

Hence, from (G.19), (G.20) and (G.21) we find

$$\begin{aligned} g_C(CC) &= 1 - \frac{1+\rho G_O}{N} \\ &= \frac{N-1-\rho G_O}{N} \\ &= \frac{N-1-\Delta N(V_{COR})}{N} \\ &= \left(1-\frac{1}{N}\right) - \frac{\Delta N(V_{COR})}{N} \end{aligned} \tag{G.22}$$

We have seen that in the ideal gas, the closure correlation is due to the change of the density at any point $\boldsymbol{R}$, $\boldsymbol{R} \neq \boldsymbol{R}_0$ from N/V into $(N-1)/N$. This is also true for system of interacting particles, i.e., placing one particle at a fixed point $\boldsymbol{R}_0$ changes the density at any other point from N/V into $(N-1)/V$. This produces the first term in brackets on the rhs of (G.22) which is the same as in equation (G.3). In systems of interacting particles, we have an *additional* contribution to the *closure* correlation. Placing a particle at some fixed position changes the average number of particles in V_{COR} by the amount $\Delta N(V_{COR})$ (equation G.20). Because of the closure condition, this change in the number of particles must change the average number of particles in the volume outside V_{COR} i.e., in $V - V_{COR}$ by exactly the amount $-\Delta N(V_{COR})$. This causes a change in the average density in $V - V_{COR}$ of $-\Delta N(V_{COR})/V$.

Thus, the change in the conditional density due to the closure condition has two contributions: one is due to the missing particle that has been placed at the center of V_{COR}; the second is due to the change of the density caused by the direct interaction of the particle at the center with its surroundings.

We can now rewrite (G.22) as

$$g_C(CC) = \frac{[N-1-\Delta N(V_{COR})]/V}{N/V} = \frac{\rho^{(1)}(\boldsymbol{R}/\boldsymbol{R}_0)}{\rho} \tag{G.23}$$

where ρ is the bulk density, and $\rho^{(1)}$ $(\boldsymbol{R}/\boldsymbol{R}_0)$ is the conditional density at R (beyond V_{COR}), given a particle placed at the center of V_{COR}, say at R_0.

Clearly, when $\rho \rightarrow 0$, we recover the ideal-gas behavior for $g_C(\mathrm{CC})$. Since in this case, $\Delta N(V_{\mathrm{COR}}) = 0$, we have

$$g_C(\mathrm{CC}) = 1 - \frac{1}{N}. \tag{G.24}$$

For a theoretical ideal gas, $U(R) = 0$, equation (G.24) is also the *total* correlation as in (G.3). But for *real* gas as $\rho \rightarrow 0$, we have an additional contribution due to the intermolecular interaction as in (G.9). The latter, is in general of short range, hence the former is referred to as the long-range correlation. As we have noted before, this part of the correlation for an ideal gas holds true at any distance, not necessarily for $R \rightarrow \infty$. Therefore, we feel that the term closure correlation is more appropriate for $g_C(\mathrm{CC})$.

In writing equation (G.22), we have used both g_C on the lhs, and g_O (within G_O) on the rhs. There exists no inconsistency however. We recall that G_O has only a contribution from the *local* correlations (no long-range correlation). The *local* correlation in G_O is assumed to be the same as the *local* correlation in G_C (see equations G.15 and G.16). Hence, in equation (G.22) we can actually replace ρG_O by the first integral on the rhs of equation (G.15), using $g_C(\mathrm{LC})$ within V_{COR}.

Hence, equation (G.22) can also be rewritten as

$$g_C(\mathrm{CC}) = \left(1 - \frac{1}{N}\right) - \frac{1}{N}\int_0^{R_{\mathrm{COR}}} [g_C(\mathrm{LC}) - 1] 4\pi R^2\, dR. \tag{G.25}$$

Here, we use only g_C (in the closed system). It clearly shows how the *local* correlation, due to direct interaction, $g_C(\mathrm{LC})$ affects the closure correlation, $g_C(\mathrm{CC})$.

Another interesting way of interpreting (G.22) is to use the relation for the partial molar volume of a particle placed at a fixed position in a system characterized by *T, P, N*. The relevant relation is (see section 7.5 and appendix O)

$$\rho V^* = -\rho G_O = -\Delta N(V_{\mathrm{COR}}). \tag{G.26}$$

Thus, in a constant-pressure system, placing a particle at a fixed position causes a change in the volume of the system V^*. Relation (G.26) shows that the average change in the number of particles in V_{COR} in a *T, V, N* system is the same as the average number of particles occupying V^* (i.e., ρV^*) in a *T, P, N* system.

We now extend the result obtained above for two-component system of *A* and *B*. We first write the general result from the inversion of the Kirkwood–Buff theory (section 4.4).

$$G_{O,\alpha\beta} = kT\kappa_T - \frac{\delta_{\alpha\beta}}{\rho_\alpha} + \frac{kT}{V}\frac{(1 - \rho_\alpha \overline{V_\alpha})(1 - \rho_\beta \overline{V_\beta})}{\rho_\alpha \rho_\beta \mu_{\alpha\beta}} \tag{G.27}$$

where κ_T is the isothermal compressibility of the system, V is the total volume of the system, and $\mu_{\alpha\beta} = (\partial\mu_\alpha/\partial N_\beta)_{T,\,P,\,N'_\beta}$.

For closed system with respect to both A and B, we have

$$G_{C,\alpha\beta} = \frac{-\delta_{\alpha\beta}}{\rho_\alpha}. \tag{G.28}$$

We now write the analogs of equations (G.15) and (G.16) for each of the pairs $\alpha\beta$. We make the same assumptions as before. First, that the local correlation (LC) due to the direct interactions is *local*, i.e. it extends to a distance R_{COR} which is on the order of a few molecular diameters.[†] Second, that the integral over $[g(R) - 1]$ in the correlation volume V_{COR} is the same for the open and the closed systems, and finally that the correlation due to the closure (CC) is independent of R for $R > R_{COR}$.

Thus, we have the analogs of (G.15) and (G.16):

$$\begin{aligned} G_{C,\alpha\beta} &= \int_0^{R_{COR}} [g_{C,\alpha\beta}(\mathrm{LC}) - 1]4\pi R^2\,dR + \int_{R_{COR}}^{\infty} [g_{C,\alpha\beta}(\mathrm{CC}) - 1]4\pi R^2\,dR \\ &= \frac{-\delta_{\alpha\beta}}{\rho_\alpha} \end{aligned} \tag{G.29}$$

$$\begin{aligned} G_{O,\alpha\beta} &= \int_0^{R_{COR}} [g_{O,\alpha\beta}(\mathrm{LC}) - 1]4\pi R^2\,dR \\ &= kT\kappa_T - \frac{-\delta_{\alpha\beta}}{\rho_\alpha} - \frac{kT(1-\rho_\alpha\overline{V}_\alpha)(1-\rho_\beta\overline{V}_\beta)}{V\rho_\alpha\rho_\beta\mu_{\alpha\beta}}. \end{aligned} \tag{G.30}$$

From (G.29) and (G.30) we obtain

$$\int_{R_{COR}}^{\infty} [g_{C,\alpha\beta}(\mathrm{CC}) - 1]4\pi R^2\,dR = -kT\kappa_T - \frac{kT(1-\rho_\alpha\overline{V}_\alpha)(1-\rho_\beta\overline{V}_\beta)}{V\rho_\alpha\rho_\beta\mu_{\alpha\beta}}. \tag{G.31}$$

Again, assuming that the correlation due to closure is independent of R outside V_{COR}, and assuming that $V \gg V_{COR}$, we get the final result

$$g_{C,\alpha\beta}(\mathrm{CC}) = 1 - \frac{kT\kappa_T}{V} - \frac{kT(1-\rho_\alpha\overline{V}_\alpha)(1-\rho_\beta\overline{V}_\beta)}{V^2\rho_\alpha\rho_\beta\mu_{\alpha\beta}} \tag{G.32}$$

which is the generalization of equation (G.19) for the closure correlation in a two-component system.

Since the inversion of the KB theory can be done to any multicomponent system to obtain all the $G_{\alpha\beta}$'s in an open system, one can repeat the same procedure to obtain the closure correlation for any multicomponent system.

[†] Note that for the mixtures, R_{COR} might be different for each pair of species and might depend on the composition. It is assumed that there exists a large enough correlation distance R_{COR} which serves for all the pairs $\alpha\beta$, and is still on the order of a few molecular diameters. In other words, is $g_{\alpha\beta}(R)$ is practically unity for $R > R_{COR}$ for each pair of species.

In the case of the one-component system, we have noted that the probabilistic interpretation of (G.19) is not so obvious as in the case of an ideal gas. It is less obvious for the case of mixtures. In order to interpret (G.30) probabilistically, we proceed to do a similar transformation of equation (G.30). From (G.28) and (G.30), we obtain the general result

$$g_{C,\alpha\beta}(\mathrm{CC}) = 1 - \frac{\delta_{\alpha\beta}}{N_\alpha} - \frac{G_{O,\alpha\beta}}{V} \tag{G.33}$$

which is the analog of (G.22).

The probabilistic interpretation is now straightforward. For the case $\alpha = \beta = A$, we have

$$\begin{aligned} g_{C,AA}(\mathrm{CC}) &= 1 - \frac{1}{N_A} - \frac{\rho_A G_{O,AA}}{N_A} \\ &= \frac{(N_A - 1 - \rho_A G_{O,AA})/V}{N_A/V} \\ &= \frac{\rho_A^{(1)}(\boldsymbol{R}'_A/\boldsymbol{R}_A)}{\rho_A^{(1)}(\boldsymbol{R}'_A)} \end{aligned} \tag{G.34}$$

where $g_{C,AA}$ is the ratio of the conditional density of A at $\boldsymbol{R}'_A$ given A at $\boldsymbol{R}_A$, and the bulk density of A. Similarly, for $\alpha \neq \beta$, we have two possibilities. Suppose we place A at $\boldsymbol{R}_A$, and ask for the change in the density of B at $\boldsymbol{R}'_B$ given A at $\boldsymbol{R}_A$, the result is

$$g_{C,AB}(\mathrm{CC}) = 1 - \frac{\rho_B G_{O,AB}}{N_B} = \frac{(N_B - \rho_B G_{O,AB})/V}{N_B/V} = \frac{\rho_B^{(1)}(\boldsymbol{R}'_B/\boldsymbol{R}_A)}{\rho_B^{(1)}(\boldsymbol{R}'_{B\cdot})}. \tag{G.35}$$

Note that in the case $\alpha \neq \beta$, placing A at $\boldsymbol{R}_A$ does *not* reduce by one the number of B's in the system. In (G.33), also the change in the average number of B's in V_{COR} around A contributes to the closure correlation. A similar result is obtained for placing B at $\boldsymbol{R}_B$ and asking for the conditional density of A at some point $\boldsymbol{R}'_A$.

It is interesting to note that if the mixture is symmetrical ideal, then

$$\mu_{AA} = \frac{kTx_B}{x_A N_T} \tag{G.36}$$

$$\mu_{AB} = \frac{-kT}{N_T} \tag{G.37}$$

where $N_T = N_A + N_B$ Hence, the closure correlations in this case are

$$g_{C,AA}(\mathrm{CC}) = 1 - \frac{kT\kappa_T}{V} - \frac{\rho_T \overline{V}_B^2}{V} \tag{G.38}$$

$$g_{C,AB}(\mathrm{CC}) = 1 - \frac{kT\kappa_T}{V} - \frac{\rho_T \overline{V_A}\,\overline{V_B}}{V} \tag{G.39}$$

$$g_{C,BB}(\mathrm{CC}) = 1 - \frac{kT\kappa_T}{V} - \frac{\rho_T \overline{V}_A^2}{V}. \tag{G.40}$$

If in addition, the system has no change of volume upon mixing, then the partial molar volumes $\overline{V}_A$ and $\overline{V}_B$, turn into the molar volumes of the pure A and B, respectively. A further simplification arises when the molar volumes are equal, i.e., $V_A = V_B = V_m$ in which case $V = N_T V_m$ and we have

$$g_{C,\alpha\beta}(\mathrm{CC}) = 1 - \frac{kT\kappa_T}{V} + \frac{1 - 2\delta_{\alpha\beta}}{N_T}. \tag{G.41}$$

To summarize, we emphasize again that whenever using the compressibility equation or the KB theory, one must take the pair correlation function as defined in an *open* system. Alternatively, one can use the pair correlation defined in the closed system, but first one must take the thermodynamic limit

$$g_C^{\infty} = \lim_{\substack{N\to\infty \\ V\to\infty \\ N/V=\text{cons tan t}}} g_C \tag{G.42}$$

before carrying the integration over the distance R.

APPENDIX H

Thermodynamics of mixing and assimilation in ideal-gas systems

In this appendix, we discuss briefly the concept of "free energy of mixing" and "entropy of mixing," and the relatively newly defined concept of assimilation. A more detailed discussion of this topic may be found elsewhere (Ben-Naim 1987a,b).

Consider a general mixture of c components at some given temperature T and pressure P, and composition $N_1, \ldots, N_c$. The Gibbs energy of this system is

$$G^l = \sum_i N_i \mu_i$$
$$= \sum_i N_i W(i|l) + \sum_i N_i kT \ln \rho_i^l \Lambda_i^3 \quad \text{(H.1)}$$

where $W(i \mid l)$ is the coupling work of i to the liquid mixture l, and for simplicity we assume that the molecules do not possess any internal degrees of freedom. The sum over i is over all the species that are present in the system.

Let μ_i^p be the chemical potential of the pure i th species at the same P and T. The total Gibbs energy of the combined pure species before the mixing, is then

$$G^p = \sum N_i \mu_i^p$$
$$= \sum N_i W(i|i) + \sum N_i kT \ln \rho_i^p \Lambda_i^3 \quad \text{(H.2)}$$

where $W(i \mid i)$ is the coupling work of i at the same P and T, and ρ_i^p is the density of i in this pure state. Thus, the Gibbs energy change for the process formation of the mixture from the pure liquids is given by

$$\Delta G^M = G^l - G^p = \sum N_i [W(i|l) - W(i|i) + \sum N_i kT \ln(\rho_i^l / \rho_i^p). \quad \text{(H.3)}$$

A particular case of equation (H.3), which is discussed in almost all textbooks of thermodynamics, is the case of mixing ideal gases. In this case, the first sum on the rhs of equation (H.3) is zero (no interactions). Also, since the total pressure of the mixture is the same as the pressure of each pure component i before the process, we have

$$P = kT \sum \rho_i^l = kT\rho_i^p. \tag{H.4}$$

Hence,

$$\rho_i^l/\rho_i^p = \rho_i^l / \sum \rho_i^l = x_i \tag{H.5}$$

where x_i is the mole fraction of the i th species in the ideal-gas mixture. In this case, equation (H.3) reduces to

$$\Delta G^M = \sum_i N_i kT \ln x_i \tag{H.6}$$

This quantity is commonly referred to as the "free energy of mixing" (or Gibbs energy of mixing). Similarly, the quantity

$$\Delta S^M = -\sum_i N_i k \ln x_i \tag{H.7}$$

is referred to as the "entropy of mixing." These terms are applied to quantities of the form (H.6) and (H.7) even for system of interacting molecules. Thus, in the general equation (H.3), the first term is referred to as the interaction Gibbs energy while the second term, or a modified form of it, is referred to as the mixing Gibbs energy.

In this appendix, we shall discuss a system of ideal gases. We shall examine the suitability of the terms "mixing entropy" and "mixing Gibbs energy," and the validity of the statement that "the mixing process is essentially reversible", see, for example, Denbigh (1966).

In the next appendix (I), we shall discuss systems with interacting particles. These are more relevant to solution chemistry.

H.1 Simple case mixing of two components

We begin by considering a classical process of mixing as described by most textbooks of thermodynamics and statistical thermodynamics. For simplicity, we discuss only a two-component system and also take the simplest case as depicted in process I (figure H.1). In this process, we have two compartments, one of which contains N_A molecules A in volume V_A, and the second,

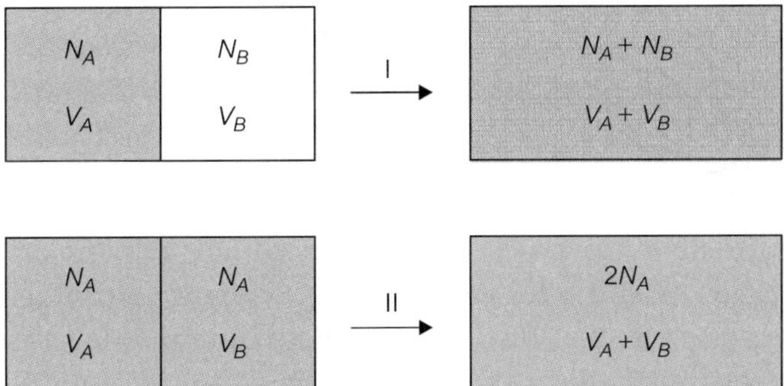

Figure H.1 Process I: process of mixing two ideal gases *A* and *B*. Initially, the two gases are separated by a partition. The mixing occurs after removing the partition. It is assumed that $N_A = N_B$ and $V_A = V_B$. Process II: the same as process I, but now the two compartments contain the same gas, say, *A*.

N_B molecules *B* in volume V_B. Furthermore, for simplicity we assume that

$$N = N_A = N_B \qquad V = V_A = V_B. \tag{H.8}$$

An elementary calculation leads to the result that the Gibbs energy change in process I (figure H.1) is

$$\Delta G_I = N_A kT \ln x_A + N_B kT \ln x_B \tag{H.9}$$

and the corresponding entropy change is

$$\Delta S_I = -N_A k \ln x_A - N_B k \ln x_B \tag{H.10}$$

where x_A and x_B are the mole fractions of *A* and *B*, respectively. In our particular example, equations (H.9) and (H.10) reduce to

$$\Delta G_I = 2NkT \ln 1/2 < 0 \tag{H.11}$$

and

$$\Delta S_I = -2Nk \ln 1/2 > 0. \tag{H.12}$$

The results (H.9) and (H.11) are referred to as "free energy of mixing," and (H.10) and (H.12) as "entropy of mixing." Since the first is always negative, and the second is always positive, most authors conclude that the *mixing process is inherently an irreversible process*, i.e., the *mixing* is the cause for the increase of the entropy and the decrease in free energy.

We shall now show that mixing is *not* an irreversible process, and the *entropy change*, in the process depicted in figure H.1, is not due to the *mixing process*, but to expansion. Therefore, the reference to the quantity (H.10) or (H.12) as entropy of mixing is inappropriate and should be avoided.

The simplest and the most straightforward argument is to calculate the entropy change for the process depicted in figure H.1. Taking the partition function of the system before and after the process, we arrive at the result

$$\Delta S_I = 2Nk \ln \frac{2V}{V} = 2Nk \ln 2. \tag{H.13}$$

Clearly, the change in entropy in process I is due to *expansion*, not to mixing. Each particle, which initially could access the volume V, is accessing the volume $2V$ at the final state of the process. One can also show that the *mixing* process could be carried out *reversibly* and the *demixing* process can be irreversible; therefore, it is clear that the "mixing" in itself is not reversible or irreversible. It simply does not contribute anything to the change of the entropy or the free energy of the process.

This conclusion, though clear and irrefutable, leaves us uneasy when comparing processes I and II in figure H.1. One can easily compute the change in the entropy in process II and find

$$\Delta S_{II} = 0. \tag{H.14}$$

In other words, one *observes mixing* in process I with accompanying increase in entropy, whilst in process II *nothing happens*, and no change in entropy. The natural conclusion is that *mixing* in I must be responsible for the increment in entropy. This conclusion is erroneous, however. First, because we have already seen that (H.10) and (H.12) are due to expansion, not to the mixing. Second, it is not true that nothing happens in process II. In fact, we can easily calculate the change in entropy for process II. The result is

$$\Delta S_{II} = 2Nk \ln \frac{2V}{V} - k \ln \frac{(2N)!}{(N!)^2} > 0. \tag{H.15}$$

This entropy change is always positive. This follows from the identity

$$2^{2N} = (1+1)^{2N} = \sum_{i=0}^{2N} \binom{2N}{i}. \tag{H.16}$$

If we now replace the sum of positive binomial coefficients on the rhs of (H.16) by the maximal term, we must have the inequality

$$2^{2N} > \binom{2N}{N} = \frac{(2N)!}{(N!)^2} \tag{H.17}$$

or

$$2N \ln 2 > \ln \frac{(2N)!}{(N!)^2}. \tag{H.18}$$

It is only in the limit of macroscopic system that the two terms on the rhs of (H.15) cancel out, and we have

$$\Delta S_{II} = 2Nk \ln \frac{2V}{V} - 2Nk \ln 2 = 0. \tag{H.19}$$

We can now conclude that in process II, there are two contributions to the entropy change; these two contributions are, in general, not equal in magnitude. However, for a large system they become equal in magnitude and cancel each other. Conceptually, they are of different origins; one is due to expansion from V to $2V$. The second is due to assimilation, i.e., initially, we have two compartments with N indistinguishable particles; in the final state we have $2N$ indistinguishable particles in one compartment. The latter contributes negatively to the entropy change in process II.

Thus, in comparing the two processes in I and II, we have a positive change in entropy in I due to *expansion*, but in II, we have both *expansion* and *assimilation* which contributes equal quantities, but of opposite signs to produce a net zero change in entropy, in process II. In no case does the *mixing*, in itself, contribute anything to the change in entropy.

H.2 Process involving assimilation and deassimilation

We have seen that the assimilation in process II contributes negatively to the entropy change. Therefore, we expect that the reverse of the assimilation process, i.e., the deassimilation process, will *increase* the entropy. Figure H.2 shows a "pure" process of deassimilation where the entropy change is positive.

Consider an ideal gas of $2N$ molecules, each containing one chiralic center. Initially, we prepare the system in such a way that all the molecules are in one of the enantiomeric forms, say the d enantiomer. We then introduce a catalyst which induces a racemization process in adiabatic conditions. At equilibrium, we obtain N molecules of the d enantiomer and N molecules of the l enantiomer.

The entropy change in this spontaneous process is well known:

$$\Delta S = 2kN \ln 2 > 0. \tag{H.20}$$

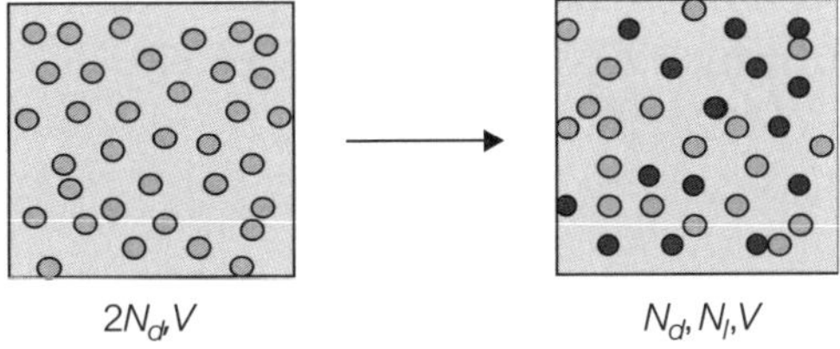

Figure H.2 A pure process of deassimilation. We start with a system of $2N$ molecules of type d, and let the system evolve into a mixture of l and d. If d and l are two enantiomers of the same molecule, the final equilibrium number of d and of l molecules will be the same.

If we analyze carefully the various factors involved in the expression of the chemical potential of the d and l molecules, we find that the momentum partition function, the internal partition function of each molecule, and the volume accessible to each molecule are unchanged during the entire process. The only change that does take place is the deassimilation process; hence, the entropy change is due to the deassimilation of $2N$ identical molecules into two subgroups of distinguishable molecules; N of one kind, and N of a second kind. This process, along with an alternative route for its realization, is described in detail in Ben-Naim (1987a, b).

The aforementioned example clearly involves *neither* proper mixing *nor* inter-diffusion. The spontaneous process which occurs is the transformation of *one* species into *two distinguishable* species, and this process evolves in a homogeneous phase with no apparent mutual diffusion of one species into the other. Therefore, the term "mixing" may not be used here either causatively or even descriptively. In spite of this, and curiously enough, the entropy change in this process, e.g., equation (H.20), is traditionally referred to as the "mixing entropy." If the process of racemization is referred to as "mixing," should we refer to the reverse as "demixing?" The only reason which leads to such misinterpretation is probably the *lack* of recognition of the role of the deassimilation phenomenon.

The deassimilation process may also occur in more complex reactions. For example, in a *cis–trans* reaction, we have the equilibrium condition

$$N_{cis}/N_{trans} = \exp(-\Delta G^0/kT). \qquad \text{(H.21)}$$

Suppose we start with $2N$ molecules in the *cis* form of dichloroethylene and allow the system to reach equilibrium, where we have N_{cis} and N_{trans} molecules, and $N_{cis} + N_{trans} = 2N$. In this case, the entropy change will have one contribution due to the *reaction*, and another contribution due to the deassimilation of $2N$ molecules into two subgroups, i.e.,

$$\Delta S = \Delta S_{\text{reaction}} + \Delta S_{\text{deassimilation}} \qquad \text{(H.22)}$$

The process of racemization described in figure H.2 is a special case of a chemical reaction for which $\Delta G^0 = 0$; hence $N_d = N_l = N$. More details can be found in Ben-Naim (1987a, b).

APPENDIX I

Mixing and assimilation in systems with interacting particles

In Appendix H, we have examined processes involving mixing and assimilation in ideal-gas systems. We have seen that mixing in itself does not contribute anything to the thermodynamics of the process, whereas assimilation and deassimilation do. We now examine similar processes in nonideal systems where intermolecular interactions exist. We shall examine the change in the Gibbs energy, rather than the entropy. But the conclusions are the same.

We first consider the following processes, figure I.1: two compartments of the same volume V and temperature T; one contains N_1 and the second contains N_2 particles of the *same* species, say A. Upon removing the partition between the two compartments, the change in Gibbs energy between the initial state i to the final state f is:

$$\begin{aligned}\Delta G = G^f - G^i &= N_1\mu_1^f + N_2\mu_2^f - N_1\mu_1^i - N_2\mu_2^i \\ &= [N_1 W(A|f) - N_1 W(A|i_1) + N_2 W(A|f) - N_2 W(A|i_2)] \\ &\quad + \left[N_1 kT \ln\frac{V}{2V} + N_2 kT \ln\frac{V}{2V}\right] \\ &\quad + \left[N_1 kT \ln\frac{N_1+N_2}{N_1} + N_2 kT \ln\frac{N_1+N_2}{N_2}\right]. \end{aligned} \tag{I.1}$$

We have included in brackets three different types of contributions to the change in the free energy. The first arises from the changes in the *interactions*

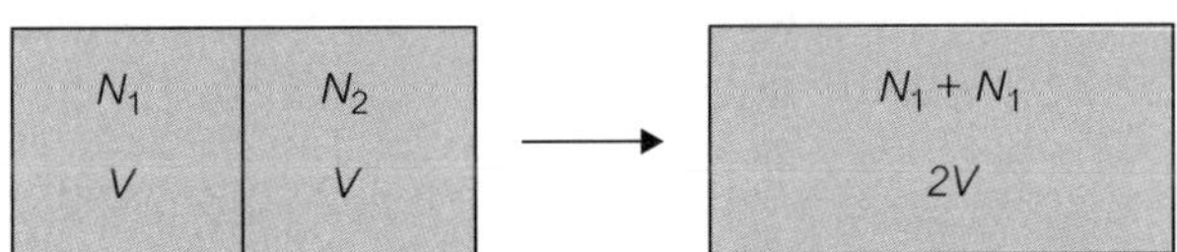

Figure I.1 Two compartments of equal volumes, V, containing N_1 and N_2 molecules ($N_1 \neq N_2$) of the same kind. The system is not an ideal gas system.

among the molecules from their initial states (i_1 and i_2) to their common final state (f). The second is the expansion term from V to $2V$ (for simplicity we choose $V_1 = V_2 = V$). The third term is the contribution due to assimilation. This classification of the various terms is independent of any assumption of ideality of the system. To emphasize the advantage of this point of view, consider the same processes carried out with ideal gases. Here, relation (I.1) reduces to

$$\Delta G^{ig} = N_1 kT \ln \frac{V}{2V} + N_2 kT \ln \frac{V}{2V} + N_1 kT \ln \frac{N_1 + N_2}{N_1} + N_2 kT \ln \frac{N_1 + N_2}{N_2}$$
$$= N_1 kT \ln \frac{P_f}{P_{i_1}} + N_2 kT \ln \frac{P_f}{P_{i_2}}. \tag{I.2}$$

On the rhs of (I.2), we expressed ΔG^{ig} as it might appear in a typical textbook of thermodynamics, i.e., N_1 molecules are transferred from an initial pressure P_{i_1} into the final pressure P_f. Likewise, N_2 molecules are transferred from P_{i_2} into P_f. This way of computing ΔG is *only* valid for ideal gases. Once interactions among the particles are present, the transfer from P_{i_1} to P_f does not tell us anything about the Gibbs energy change.

In fact, thermodynamics cannot tell us anything about the significance of the various terms on the rhs of relation (I.1), whether or not the system is an ideal gas. On the other hand, by introducing the concept of assimilation, we have a clear-cut classification of the three terms in relation (I.1). This classification is valid, independent of any assumption of ideality.

Next, consider the process depicted in figure I.2. Here again, thermodynamics would have taught us that the system is compressed with a Gibbs energy change equal to

$$N_1 kT \ln\left(\frac{P_f}{P_{i_1}}\right) + N_2 kT \ln\left(\frac{P_f}{P_{i_2}}\right). \tag{I.3}$$

This is again valid only for ideal gases. Once we have interacting particles, the validity of expression (I.3) as a measure of the Gibbs energy change is lost, and we cannot claim that the compression of the system is the only cause of the Gibbs energy change.

Considering the same process from the molecular point of view, we see that each particle is allowed to wander in the *same* volume V, before and after the process. Therefore, here we only have two contributions to ΔG, one due to a change in interactions and the second due to assimilation. If the gas is ideal, the only contribution is due to assimilation and amounts to

$$N_1 kT \ln \frac{N_1 + N_2}{N_1} + N_2 kT \ln \frac{N_1 + N_2}{N_2}. \tag{I.4}$$

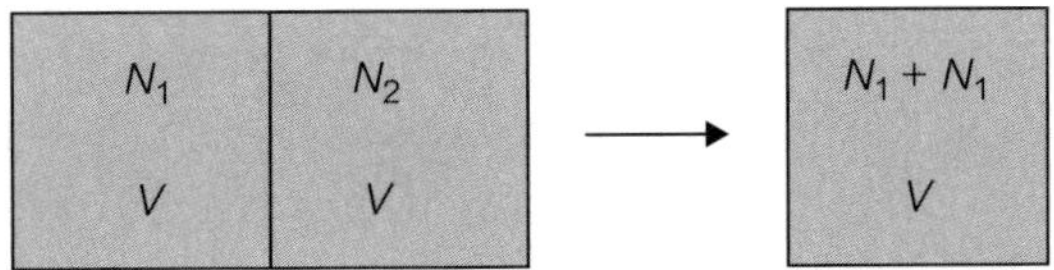

Figure I.2 The initial state is the same as in figure I.1. In the final state, the gas is contained in a volume V.

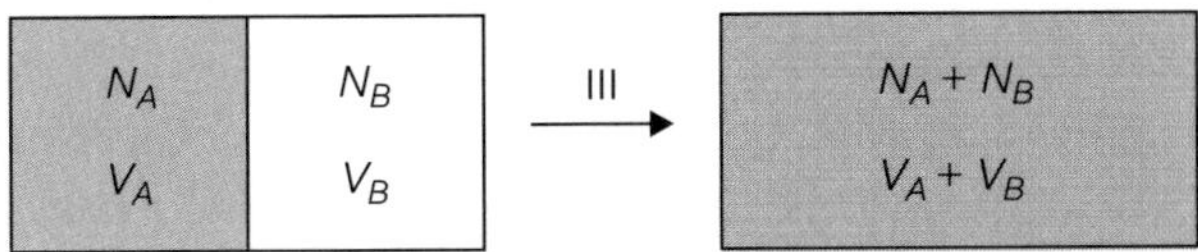

Figure I.3 The same as in figure I.1, but now the two compartments contain N_A A particles and N_B B particles, respectively. Process III is the same as Process I of figure H.1 but for nonideal gases.

This very term would have been referred to as "compression" in the traditional thermodynamic language.

Next, we perform the same experiment as in figure I.1, but when we have N_A molecules of type A, and N_B molecules of type B, figure I.3, process III. The Gibbs energy change for this process is

$$\begin{aligned}\Delta G_{III} &= G^f - G^i \\ &= N_A[W(A|A+B) - W(A|A)] \\ &\quad + N_B[W(B|A+B) - W(B|B)] - [N_A kT \ln\frac{2V}{V} + N_B kT \ln\frac{2V}{V}]\end{aligned} \tag{I.5}$$

Here, the first two terms are due to the change in the coupling work in the process. Initially, each A particle interacted with A particles only; its coupling work has changed from $W(A\,|\,A)$ into $W(A\,|\,A+B)$. Similarly, the coupling work of each B particle has changed from $W(B\,|\,B)$ into $W(B\,|\,A+B)$. The last term on the rhs of (I.5) is due to the change of the accessible volume for each particle from V to $2V$. Note, however, that there *is no* contribution to assimilation as we had in process I of figure I.1. Again, we stress that the last term on the rhs of (I.5) is traditionally referred to as the mixing free energy. If we have started from two different volumes V_A and V_B, the last term on the rhs of (I.5) can be written as

$$-N_A kT \ln\frac{V_A + V_B}{V_A} - N_B kT \ln\frac{V_A + V_B}{V_A} = N_A kT \ln y_A + N_B kT \ln y_B \tag{I.6}$$

where y_A and y_B are the *volume fractions*, as defined in (I.6). This term certainly cannot be interpreted as the mixing free energy. We have had the same expansion term in process I where we have no mixing at all. Perhaps, the *first* two terms in (I.5) could be more appropriately referred to as the mixing free energy, in the sense that they arise from the change of the *environments* of A

and B upon mixing. But clearly, this term would be zero when no interactions exist, i.e., an ideal gas this "mixing free energy" would be zero.

Next, we consider the analog of process II, but when we start with N_A in V, and N_B molecules in V, the temperature is kept constant T. The change in the Gibbs energy for this case is obtained from (I.6) by letting the volume fractions $y_A = y_B = 1$. Clearly, since there is no change in the accessible volume for each particle, the last term on the rhs of (I.5) is zero, and we are only left with the change due to the changes in the coupling work of A and B. Again, we stress that in this case, although we observe mixing of A and B, there exists no mixing free energy in the conventional form of (I.6). Also in this case, there is no assimilation term as we had in process II.

We shall briefly discuss now two special cases of ideal solutions.

(1) *Symmetrical ideal solution.* Let A and B be similar particles, in the sense discussed in chapter 5. We perform the same process III as in figure I.3. The corresponding change in Gibbs energy is (I.5). If we perform the same process, but under the same pressure before and after the mixing, we have for the chemical potentials of A and B, in the final states

$$\mu_A^f = \mu_A^p(P,T) + kT \ln x_A \tag{I.7}$$

$$\mu_B^f = \mu_B^p(P,T) + kT \ln x_B \tag{I.8}$$

where $\mu_A^p(P,T)$ and $\mu_B^p(P,T)$ are the chemical potential of pure A and pure B, under the same T and P. In this case, the change of free energy in the process is

$$\begin{aligned}\Delta G^{\text{SI}} &= N_A[\mu_A^f - \mu_A^p] + N_B[\mu_B^f - \mu_B^p] \\ &= N_A kT \ln x_A + N_B kT \ln x_B\end{aligned} \tag{I.9}$$

Note that the coupling terms in equation (I.5) does not appear in this case. The change in free energy is due only to the expansion from V_A to $V_A + V_B$ for A, and from V_B to $V_A + V_B$ for B. It should be stressed that the origin of the mole fractions x_A and x_B is

$$x_A = \frac{\rho_A^l}{\rho_A^p} = \frac{N_A V_A}{(V_A + V_B)N_A} = \frac{V_A}{V_A + V_B} \tag{I.10}$$

where ρ_A^l is the density of A in the mixture, and ρ_A^p is the density of A in pure A at the same P, T. Since we can form the SI solution by replacing all B's by A's, i.e., $\rho_T = \rho_A + \rho_B = \rho_A^P$, we can also interpret x_A as the mole fraction of A, i.e.,

$$x_A = \frac{\rho_A^l}{\rho_A^l + \rho_B^l} = \frac{N_A}{N_A + N_B}. \tag{I.11}$$

We conclude that whereas in process III, the driving force for the mixing consists of both the change in the coupling work and the expansion, in the SI case, though interactions exist, there is no contribution due to the changes in the coupling work. The only contribution comes from the expansion and not to the mixing.

(2) *The dilute-ideal* (DI) *solutions.* This is the case which has attracted the most attention in solution chemistry. Unfortunately, this is also the case where most of the misconceptions have been involved. Here, we only focus on one aspect of the problem, namely the identification of the so-called mixing terms. Consider the case of a very dilute solution of A in B, such that $N_A \ll N_B$. In this case, when we mix A and B, we may assume that

$$W(A \mid A+B) \approx W(A \mid B), \quad W(B \mid A+B) \approx W(B \mid B). \tag{I.12}$$

Also, one can approximate $\rho_B = \rho_B^p$, i.e., the addition of a small amount of A to pure B causes a negligible effect on ρ_B^p and on the coupling work of B. In this case, the general expression for ΔG reduces to

$$\Delta G^{\mathrm{DI}} = N_A[W(A \mid B) - W(A \mid A)] + N_A kT \ln(\rho_A / \rho_A^p). \tag{I.13}$$

In this process, a particle A which "sees" only A particles in the pure A, is "seeing" now only pure B in its environment. Here, there are two contributions to the Gibbs energy change: one is due to the (extreme) change in the environment of A (from pure A into pure B), and the second, to the change of the accessible volume available to A.

Neither of these terms is an adequate candidate to be referred to as mixing term. However, if one insists on using the term "mixing Gibbs energy," it is perhaps more appropriate to refer to the first term on the rhs of equation (I.13), but certainly not to the second term. The traditional view is, quite astonishingly, the opposite. It is very common to refer to the *second* term as the "mixing Gibbs energy." A more conventional form of equation (I.13) is obtained after transforming into mole fractions. For this case, the transformation of variables is

$$x_A = \rho_A/(\rho_A + \rho_B) \approx \rho_A/\rho_B^p. \tag{I.14}$$

We can rewrite ΔG^{DI} in the conventional thermodynamic way as

$$\Delta G^{\mathrm{DI}} = N_A[\mu_A^{0x} - \mu_B^p] + N_A kT \ln x_A \tag{I.15}$$

where μ_A^{0x} is the standard chemical potential of A in the mole fraction scale, and μ_A^p is the chemical potential of pure A.

The reference to $N_A kT \ln x_A$ as a "mixing Gibbs energy" term is so ubiquitous that it is really difficult to find exceptions. This erroneous interpretation of the term $kT \ln x_A$ also leads to an erroneous interpretation of the term $\mu_A^{0x} - \mu_A^p$, which is very common in solution chemistry.

When investigating dilute solutions of, say, argon in water or xenon in organic solvents, the stated purpose of the investigators is to study the Gibbs energy of interaction (or solvation) of the solute in the solvent (chapter 7). Usually, the primary quantity that is obtained in the experiment is a partition coefficient, i.e., a ratio of the number of densities of the solute between two phases. This ratio is directly related to the quantity $W(A \mid A + B)$ which measures the solvation Gibbs energy of the solute A in the solvent (chapter 7).

However, because of the prevalent erroneous interpretation of the term $kT \ln x_A$ as a "mixing term", sometimes referred to as "cratic" term† (Gurney 1953; Tanford 1973), one transforms the concentration units from number densities into mole fractions and then compute a so-called "standard Gibbs energy of solution." This is also referred to as the "unitary" term (Gurney 1953). Unfortunately, the whole procedure involves superfluous work and, in fact, misses the principal aim of the investigation. As we have seen, $kT \ln x_A$ is *not* a mixing Gibbs energy, and the quantity left after subtracting this term from ΔG^{DI} does not necessarily have the meaning so often assigned to it.

† The terms "cratic" and "unitary" were introduced by Gurney (1953). In my opinion, these terms were misused. A detailed discussion of these terms can be found in Ben-Naim (1978).

APPENDIX J

Delocalization process, communal entropy and assimilation

The concept of communal entropy has featured within the lattice models of liquids and mixtures. We show in this appendix that this entropy change is due to a combination of assimilation and expansion.

In figure J.1, we depict a process of delocalization. Initially, we have N particles each confined to a cell of size v. We remove all the partitions and the particles are allowed to occupy the entire volume V. The entropy change upon the removal of all the partitions is

$$\begin{aligned}\Delta S &= S^{ig} - S^{loc} \\ &= kN\ln(V/\Lambda^3) - k\ln N! + \tfrac{3}{2}kN - kN\ln(v/\Lambda^3) - \tfrac{3}{2}kN \end{aligned} \tag{J.1}$$

where $v = V/N$ is the volume of each cell. Application of the Stirling approximation yields the well-known result

$$\Delta S = kN. \tag{J.2}$$

This is known as the communal entropy. Closer examination reveals that ΔS in equation (J.2) is a net result of two effects; the increase of the accessible volume

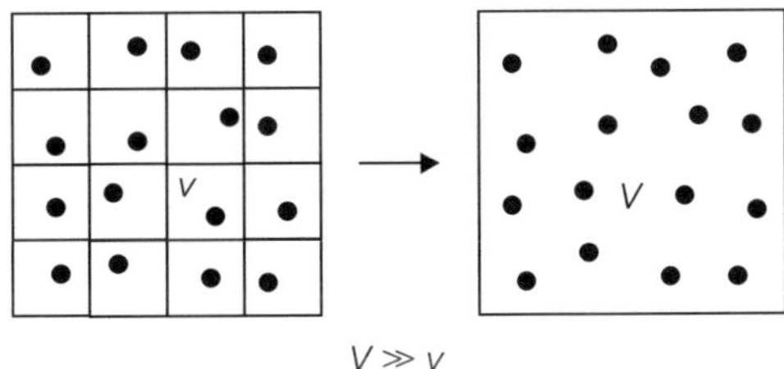

Figure J.1 A delocalization experiment. Initially, there are N particles of the same kind, each in a cell of volume v. Upon removal of the partitions between the cells, each particle can access the entire volume $V = Nv$. The entropy change in this process, for an ideal gas, is the so-called communal entropy.

for each particle from v to V, *and* the assimilation of N particles. The two contributions are explicitly

$$\Delta S = kN \ln(V/v) - k \ln N!. \tag{J.3}$$

Note that since we chose $v = V/N$, the first term on the rhs is the dominating one, i.e., the contribution of the volume change accessible to each particle is larger than the assimilation contribution (the latter is always negative).

Within the Stirling approximation, we have partial cancellation leading to

$$\Delta S = kN \ln N - [kN \ln N - kN] = kN. \tag{J.4}$$

Originally, Hirschfelder et al. (1937) introduced the concept of communal entropy to explain the entropy of melting of solids. They specifically stated that "this communal sharing of volume gives rise to an entropy of fusion." This idea was later criticized by Rice (1938) and by Kirkwood (1950) and now the whole concept of communal entropy in the context of the theory of liquids is considered to be obsolete.

Here we have cited this example to stress the point that the communal entropy in equation (J.2) is not a result of volume change only but a combination of volume change and assimilation.

APPENDIX K
A simplified expression for the derivative of the chemical potential

In section 4.2, we have derived the expression for the derivative of the chemical potential with respect to the number of particles

$$\mu_{ij} = \left(\frac{\partial \mu_i}{\partial N_j}\right)_{T,\,P,\,N';} \tag{K.1}$$

For a *c*-component system at a given *T* and *P*, the general expression for μ_{ij} within the Kirkwood–Buff theory is (see section 4.2)

$$\mu_{ij} = \frac{kT}{V|\boldsymbol{B}|} \frac{B^{ij} \sum_{\alpha,\beta} \rho_\alpha \rho_\beta B^{\alpha\beta} - \sum_{\alpha,\beta} \rho_\alpha \rho_\beta B^{\alpha j} B^{i\beta}}{\sum_{\alpha\beta} \rho_\alpha \rho_\beta B^{\alpha\beta}} \tag{K.2}$$

where $\rho_\alpha = N_\alpha / V$ is the average density of α in the open system. The summations extend over all species in the system $\alpha, \beta = 1, 2, 3, \ldots, c$.

The matrix $\boldsymbol{B}$ is constructed from the elements

$$B_{ij} = \rho_i \rho_j G_{ij} + \rho_i \delta_{ij} \tag{K.3}$$

where G_{ij} are the Kirkwood–Buff integrals

$$G_{ij} = \int_0^\infty [g_{ij}(R) - 1] 4\pi R^2 dR \tag{K.4}$$

and $g_{ij}(R)$ is the pair correlation function for the pair of species *i* and *j*.

The quantity B^{ij} denotes the cofactor of the element B_{ij} in the determinant $|\boldsymbol{B}|$. Namely, B^{ij} is a determinant obtained from $|\boldsymbol{B}|$ by deleting the *i*th row and the *j*th column, and the result is multiplied by the sign $(-1)^{i+j}$.

Relation (K.2) is quite simple for $c = 2$ and becomes somewhat cumbersome for $c = 3$. For $c \geq 3$, the application of (K.2) is impractical since it requires handling of a large number of terms.

Fortunately, it was noticed that when one fully expands the expression (K.2) in terms of G_{ij}, many terms cancelled out. This led to a search of a more compact, and easier to use expression for μ_{ij}. Here, we shall present a schematic proof of the procedure of simplification of expression (K.2). More details are available elsewhere (Ben-Naim 1975a).

First, define the matrix $\mathbf{G}$ by

$$\mathbf{G} = \begin{pmatrix} G_{11} + \rho_1^{-1} & G_{12} & G_{13} \cdots \\ G_{21} & G_{22} + \rho_2^{-1} & G_{23} \cdots \\ \vdots & \vdots & \vdots \end{pmatrix}. \tag{K.5}$$

Note that the elements of the matrix are

$$(\mathbf{G})_{ij} = G_{ij} + \delta_{ij}\rho_i^{-1} \tag{K.6}$$

where G_{ij} are given by (K.4). Thus, care must be exercised to distinguish between G_{ij} and $(\mathbf{G})_{ij}$. Note also that $\mathbf{G}$ is a symmetric matrix.

By extracting ρ_i from the ith row, and ρ_j from the jth column, one can easily get the following relations:

$$|\boldsymbol{B}| = \rho^2|\mathbf{G}| \tag{K.7}$$

$$B^{ij} = \frac{\rho^2 G^{ij}}{\rho_i\rho_j} \tag{K.8}$$

$$\sum_{\alpha,\beta=1}^{c} \rho_\alpha\rho_\beta B^{\alpha j}B^{i\beta} = \rho^4 \sum_{\alpha,\beta} \frac{G^{\alpha j}G^{i\beta}}{\rho_i\rho_j} \tag{K.9}$$

$$\sum_{\alpha,\beta=1}^{c} \rho_\alpha\rho_\beta B^{\alpha\beta} = \rho^2 \sum_{\alpha,\beta} G^{\alpha\beta}. \tag{K.10}$$

Note that we have denoted the product of all the densities by

$$\rho = \prod_{i=1}^{c} \rho_i. \tag{K.11}$$

Substituting (K.7) and (K.8) into (K.2), we get[†]

$$\mu_{hk} = \frac{kT}{\rho_h\rho_k V|\mathbf{G}|} \frac{\sum_{\alpha,\beta}[G^{hk}G^{\alpha\beta} - G^{h\beta}G^{\alpha k}]}{\sum_{\alpha,\beta} G^{\alpha\beta}}. \tag{K.12}$$

[†] Note that we use k for the Boltzmann constant as well as a subscript indicating a species.

This is a more convenient form than (K.2) since we have eliminated all the ρ_i's under the summation signs. (Note, however that all the G_{ij}'s are still dependent on the densities.) Note also that some of the terms in the numerator of (K.12) vanish. More specifically,

$$G^{hk}G^{\alpha\beta} - G^{h\beta}G^{\alpha k} = 0 \quad \text{for } \alpha = h (\text{and any } \beta) \text{ or for } \beta = k (\text{and any } \alpha). \tag{K.13}$$

The nonvanishing terms can be viewed as minors of the adjoint determinant of $|\mathbf{G}|$, i.e.,

$$\begin{vmatrix} G^{hk} & G^{h\beta} \\ G^{\alpha k} & G^{\alpha\beta} \end{vmatrix} \tag{K.14}$$

is a minor of order (2) in the adjoint determinant of $|G|$, provided that the indices h, k, α and $\boldsymbol{B}$ fulfill certain conditions. The details of the algebraic steps are presented elsewhere (Ben-Naim 1975a).

The final result for the quantity μ_{hk} in (K.2) is

$$\mu_{hk} = \frac{kT}{\rho_h \rho_k V} \frac{|\boldsymbol{E}(h,k)|}{|\boldsymbol{D}|}. \tag{K.15}$$

Here, $\boldsymbol{D}$ is a matrix of order $c+1$ constructed from $\mathbf{G}$ by appending a row and a column of unities, except for the element D_{11}, which is zero, i.e.,

$$\boldsymbol{D} = \begin{pmatrix} 0 & 1 & 1 & 1 \cdots \\ 1 & & & \\ 1 & & \mathbf{G} & \\ \vdots & & & \end{pmatrix}. \tag{K.16}$$

The general element of $\boldsymbol{D}$ is

$$D_{nm} = \delta_{n1} + \delta_{m1} - 2\delta_{n1}\delta_{m1} + (\mathbf{G})_{n-1,m-1} \qquad n, m = 1, 2, \ldots, c+1. \tag{K.17}$$

Note that $(\mathbf{G})_{ij}$ is defined in equation (K.6) for $i, j = 1, 2, \ldots, c$. We add the definitions $(\mathbf{G})_{ij} = 0$ for any other indices in (K.17).

The determinant in the numerator of (K.15) is constructed from the matrix $\mathbf{G}$ by replacing the hth row and the kth column of $|\mathbf{G}|$ by unities, except for the element hk, which is replaced by zero. The general element of $\boldsymbol{E}(h, k)$ is

$$[\boldsymbol{E}(h,k)]_{\alpha\beta} = \delta_{h\alpha} + \delta_{k\beta} - 2\delta_{h\alpha}\delta_{k\beta} + (\mathbf{G})_{\alpha\beta}(1 - \delta_{h\alpha})(1 - \delta_{k\beta}). \tag{K.18}$$

Relation (K.15) is far easier to apply for a multicomponent system than the original relations (K.2) or (K.12). One application of this simplified expression to examine the solute and solvent effects on chemical equilibrium has been published by Ben-Naim (1975a). In this article, it was also shown that if

$G_{ij} = \frac{1}{2}(G_{ii} + G_{jj})$ for each pair of species, one can always reduce the determinants $|\boldsymbol{E}(h, k)|$ and $|\boldsymbol{D}|$ to a simple form[†], i.e.,

$$|\boldsymbol{E}(h, k)| = \rho_h \rho_k \Big/ \prod_i \rho_i \tag{K.19}$$

$$|\boldsymbol{D}| = -\sum_i \rho_i \Big/ \prod_i \rho_i. \tag{K.20}$$

With this simplification it is easy to prove that the condition $G_{ij} = \frac{1}{2}(G_{ii} + G_{jj})$ (for all i, j) is necessary and sufficient for SI solution for any mixture of c components.

Another useful application is the limiting form of $|\boldsymbol{D}|$ and $|\boldsymbol{E}(A, s)|$ as $\rho_s \to 0$. It is easy to see that in the limit of dilute solution $\rho_s \to 0$, we have

$$\begin{aligned} |\boldsymbol{E}(A, s)| &= \frac{1 + \rho_B(G_{BB} - G_{AB})}{\rho_B} + (G_{As} - G_{Bs}) \\ &= \eta \overline{V}_A / \rho_B + (G_{As} - G_{Bs}) \end{aligned} \tag{K.21}$$

$$|\boldsymbol{E}(B, s)| = \eta \overline{V}_B / \rho_A + (G_{Bs} - G_{As}) \tag{K.22}$$

$$|\boldsymbol{D}| = -\frac{\eta}{\rho_s \rho_A \rho_B}. \tag{K.23}$$

Hence, in this limit

$$\begin{aligned} \left(\frac{\partial \mu_s}{\partial N_A}\right)_{T,P,N_B,N_s} &= \frac{kT}{\rho_s \rho_A V} \frac{|\boldsymbol{E}(A, s)|}{|\boldsymbol{D}|} \\ &= \frac{-kT}{V\eta}[1 + \rho_B(G_{BB} - G_{BA}) + \rho_B(G_{As} - G_{Bs})] \\ &= \frac{-kT}{V\eta}[\eta \overline{V}_A + \rho_B(G_{As} - G_{Bs})]. \end{aligned} \tag{K.24}$$

For the derivative of the pseudo-chemical potential, we have

$$\begin{aligned} \left(\frac{\partial \mu_s^*}{\partial N_A}\right)_{T,P,N_B,N_s} &= \frac{-kT}{V\eta}[\eta \overline{V}_A + \rho_B(G_{As} - G_{Bs})] + \frac{kT\overline{V}_A}{V} \\ &= \frac{-kT\rho_B(G_{As} - G_{Bs})}{V\eta}. \end{aligned} \tag{K.25}$$

The latter can be transformed into derivative with respect to x_A, i.e.,

$$\lim_{\rho_s \to 0} \left(\frac{\partial \mu_s^*}{\partial x_A}\right)_{T,P} = \frac{kT\,(\rho_A + \rho_B)^2}{\eta}(G_{Bs} - G_{As.}). \tag{K.26}$$

[†] This can be achieved simply by adding and subtracting rows and columns of the determinants, leaving the value of the determinants unchanged.

Since for the ideal-gas mixture we have

$$\left(\frac{\partial \mu_s^{*ig}}{\partial x_A}\right)_{T,P} = 0 \tag{K.27}$$

we obtain

$$\lim_{\rho_s \to 0}\left(\frac{\partial \Delta\mu_s^*}{\partial x_A}\right)_{T,P} = \frac{kT\,(\rho_A + \rho_B)^2}{\eta}(G_{Bs} - G_{As}) \tag{K.28}$$

where $\Delta\mu_s^*$ is the solvation Gibbs energy of s in dilute solution of A and B.

Another useful expression for the partial molar volume of s in the limit $\rho_s \to 0$ may be obtained from the Gibbs–Duhem relation

$$\rho_A \mu_{As} + \rho_B \mu_{Bs} + \rho_s \mu_{ss} = 0 \tag{K.29}$$

Using the KB results for μ_{ij} (section 4.2), we can eliminate V_s to obtain

$$\overline{V}_s^0 = \lim_{\rho_s \to 0} \overline{V}_s = kT\kappa_T - \rho_A \overline{V}_A G_{As} - \rho_B \overline{V}_B G_{Bs}. \tag{K.30}$$

For the partial molar volume at fixed position, we have

$$V_s^{*0} = -\rho_A V_A G_{As} - \rho_B \overline{V}_B G_{Bs} \tag{K.31}$$

which is a generalization of equation for V_s^* in pure s (Appendix O).

From equations (K.28) and (K.31), one can solve for G_{As} and G_{Bs}. The results are

$$G_{As} = kT\kappa_T - \overline{V}_s^0 - \frac{\eta \rho_B \overline{V}_B}{kT\rho_T^2}\left(\frac{\partial \Delta\mu_s^*}{\partial x_A}\right)_{T,P} \tag{K.32}$$

$$G_{Bs} = kT\kappa_T - \overline{V}_s^0 + \frac{\eta \rho_A \overline{V}_A}{kT\rho_T^2}\left(\frac{\partial \mu_s^*}{\partial x_A}\right)_{T,P}. \tag{K.33}$$

Since all the quantities on the rhs of (K.32) and (K.33) are measurable (including η which can be determined from the inversion of the KB theory in a two-component mixture of A and B), one can calculate both G_{As} and G_{Bs}.

APPENDIX L

On the first order-deviations from SI solutions

We have seen in section 6.2 that the first-order deviations from symmetrical ideal solution have the form

$$\mu_A^{\text{EX,SI}} = a_2 x_B^2 \tag{L.1}$$

or

$$\mu_A = \mu_A^P + kT \ln x_A + a_2 x_B^2 \tag{L.2}$$

where a_2 can be a function of T, P, but not of x_A.

If (L.2) holds for all $0 \le x_A \le 1$, then the Gibbs–Duhem relations impose a similar expression for μ_B, i.e.,

$$\mu_B = \mu_B^P + kT \ln x_B + a_2 x_A^2 \tag{L.3}$$

with the same constant a_2 as in (L.2). We note that this is also the *first-order* correction and must start as $a_2 x_B^2$. We could have attempted to write

$$\mu_A = \mu_A^P + kT \ln x_A + a_0 + a_1 x_B + a_2 x_B^2. \tag{L.4}$$

Since at $x_B = 0$ we must have $\mu_A = \mu_A^P$, hence $a_0 = 0$. Applying the Gibbs–Duhem relation to (L.4) we, obtain

$$\mu_B = \mu_B^P + kT \ln x_B - a_1 \ln x_B + a_1 x_B + a_2 x_A^2. \tag{L.5}$$

But since as $x_B \rightarrow 0$, the chemical potential must diverge as $kT \ln x_B$, we must have $a_1 = 0$, and we are back to equation (L.3)[†]

[†] Note also that the same expressions (L.2) and (L.3) are obtained had we attempted power series of the form $\mu_A = \mu_A^P + kT \ln x_A + a_0 + a_1 x_A + a_2 x_A^2$; we should find that this must have the form $b_0 (1 - x_A)^2 = b_0 x_B^2$. This is true when we are interested in deviations from SI solutions. On the other hand, if we are interested in deviations from DI solutions, then b_0 together with μ^P will form the new constant of integration, or the standard chemical potential, and one *can* obtain an expression of the form $b_1 x_A + b_2 x_A^2 \cdots$ for the excess chemical potential with respect to DI solutions.

The excess Gibbs energy per mole of the mixture with respect to SI is

$$
\begin{aligned}
g^{EX,SI} &= \frac{G^{EX,SI}}{N_A + N_B} = x_A\mu_A^{EX,SI} + x_B\mu_B^{EX,SI} = a_2 x_A x_B^2 + a_2 x_B x_A^2 \\
&= a_2 x_A x_B.
\end{aligned} \tag{L.6}
$$

Historically, this form of the excess Gibbs energy was suggested as the simplest function which obeys the requirements that $g^{EX,SI}$ must be zero when either $x_A \to 0$ or $x_B \to 0$. This is known as Margules equation, see, e.g., Prausnitz et al (1986). From (L.6), one can obtain both equations (L.3) and (L.4). The latter was derived from theoretical arguments based on lattice model for mixtures (Guggenheim 1952).

Higher-order deviations from SI solutions are equally expressed as power series of the form

$$
g^{EX,SI} = x_A x_B[a + b(x_A - x_B) + c(x_A - x_B)^2 + d(x_A - x_B)^3 + \cdots] \tag{L.7}
$$

where the coefficients are determined by fitting the experimental data. The last equation is known as the Redlich–Kister equation (Redlich and Kister 1948). Clearly, for any expansion of the form (L.7), the excess chemical potentials of A and B are determined by the relations

$$
\mu_A^{EX,SI} = g^{EX,SI} + x_B \frac{\partial g^{EX,SI}}{\partial x_A} \tag{L.8}
$$

and

$$
\mu_B^{EX,SI} = g^{EX,SI} + x_A \frac{\partial g^{EX,SI}}{\partial x_B}. \tag{L.9}
$$

APPENDIX M

Lattice model for ideal and regular solutions

In this appendix, we present a very brief outline of the lattice theory of ideal and regular solutions, developed mainly by Guggenheim (1952). The main reason for doing so is to emphasize the first-order character of the deviations from SI, equation (M.12) below.

The system consists of M lattice sites occupied by a mixture of N_A molecules A and N_B molecules B, such that $N_A + N_B = \mathrm{M}$ (figure M.1). It is assumed that A and B have roughly the same size and that they can exchange sites without disturbing the structure of the lattice.

For each configuration of the system, the canonical partition function is written as

$$Q(N_A, N_B, T) = q_A^{N_A} q_B^{N_B} \sum_j g_j \exp[-\beta E_j] \tag{M.1}$$

where q_A and q_B are the internal partition functions of A and B, respectively. The sum is over all energy levels E_j, and g_j is the degeneracy of that energy level. It is assumed that the energy levels are determined only by the nearest-neighbor

B A B A B A A A B A
A B A B A B B B A A
B A B A A B A A B A
A A A A B B B B A A
B A B A B B A B A B
A B A A B B B B A B
A A B B B A A A B B
A A B A A B B A B B
B B A B A A A A A B
A A A A B B A B A A

Figure M.1 A lattice model of mixture of A and B.

interaction energies $E_{\alpha\beta}$ for pair of species α and β occupying adjacent sites. Hence, for any configuration of the system we have

$$E = N_{AA}E_{AA} + N_{BB}E_{BB} + N_{AB}E_{AB} \tag{M.2}$$

where $N_{\alpha\beta}$ are the number of α–β pairs in that specific configuration. The three $N_{\alpha\beta}$ are connected by the two relations (z is the number of nearest neighbors to each site)

$$\begin{aligned} zN_A &= 2N_{AA} + N_{AB} \\ zN_B &= 2N_{BB} + N_{AB}. \end{aligned} \tag{M.3}$$

Therefore, the energy level E is determined by the single parameters N_{AB}. Hence, we can replace the sum over j in (M.1), by a sum over all possible N_{AB}, i.e.,

$$\begin{aligned} Q &= q_A^{N_A} q_B^{N_B} \sum_{N_{AB}} g(N_{AB}) \exp\left\{-\beta\left[\frac{zN_A - N_{AB}}{2}E_{AA} + \frac{zN_B - N_{AB}}{2}E_{BB} + N_{AB}E_{AB}\right]\right\}. \\ &= Q_A(N_A)Q_B(N_B)\sum g(N_{AB})\exp[-\beta N_{AB}W/2] \end{aligned} \tag{M.4}$$

where Q_A and Q_B are the canonical partition function of pure A and B, respectively, and W is defined as

$$W = E_{AA} + E_{BB} - 2E_{AB}. \tag{M.5}$$

W is referred to as the exchange energy. The sum in (M.4) is over all possible values of N_{AB}.

Although $g(N_{AB})$ is a very complicated function, we know that the sum over all $g(N_{AB})$ must be the total number of configurations, i.e.,

$$\sum_{N_{AB}} g(N_{AB}) = \frac{M!}{N_A!N_B!} \tag{M.6}$$

which is simply the number of ways of arranging N_A particles A, and N_B particles B on the total number of lattice sites $M = N_A + N_B$.

When $W = 0$, it follows immediately from (M.4) and (M.5) that the system will be SI, i.e.,

$$\mu_A = -kT\frac{\partial \ln Q}{\partial N_A} = \mu_A^P + kT\ln x_A. \tag{M.7}$$

Thus, the condition $W = 0$ is a sufficient (but not necessary) condition for SI behavior within this particular lattice model.

The next step is to assume that $\beta N_{AB}W/2$ is small for all configurations, i.e., all N_{AB}'s. Hence, we can expand the exponent in (M.4) to obtain

$$\begin{aligned}\langle \exp[-\beta N_{AB}W/2]\rangle &= \sum \Pr(N_{AB})\exp[-\beta N_{AB}W/2]\\ &= \sum \Pr(N_{AB}) - \frac{\beta W}{2}\sum \Pr(N_{AB})N_{AB}\\ &= \sum \Pr(N_{AB})\left[1 - \frac{\beta W}{2}\langle N_{AB}\rangle\right]\end{aligned} \tag{M.8}$$

where the average is taken with the probability distribution

$$\Pr(N_{AB}) = \frac{g(N_{AB})N_A!N_B!}{M!} \tag{M.9}$$

and the partition function is now written as

$$Q = Q_A(N_A)Q_B(N_B)\frac{M!}{N_A!N_B!}\exp\left[-\frac{\beta W}{2}\langle N_{AB}\rangle\right]. \tag{M.10}$$

Now the average $\langle N_{AB}\rangle$ can be estimated as follows. If each site has z nearest neighbors, and if $\Pr(B/A)$ is the conditional probability of finding a B in the neighborhood of A, then the average number of A–B pairs is

$$\langle N_{AB}\rangle = zN_A\Pr(B/A) = z\frac{N_A N_B}{M} \tag{M.11}$$

where in the last equality we have replaced the conditional probability $\Pr(B/A)$ by the *bulk* probability of finding a site occupied by B. This assumption is sometimes referred to as the randomness assumption. It is clear, however, that this assumption is true only when W is small.

With the assumption (M.11) introduced in (M.10), we get for the chemical potential

$$\mu_A = -kT\frac{\partial \ln Q}{\partial N_A} = \mu_A^P + kT\ln x_A - \frac{zWx_B^2}{2}. \tag{M.12}$$

This behavior has been referred to as *strictly* regular solution (Guggenheim 1952). This should be compared with the first-order deviations from SI behavior discussed in section 6.6. There, the last term on the rhs of (M.12) is replaced by the more general term $kT\rho_T\Delta_{AB}x_B^2/2$. The point to be emphasized here is that the expression (M.12) is valid only for a *small* value of W, i.e., this is only a *first-order* term in the expansion of the excess chemical potential, as shown, for the more general case, in section 6.6. Failure to recognize this fact has misled many scientists to reach invalid conclusions regarding the behavior of the mixture for large values of W (or equivalently low temperatures T). Further discussion of this aspect of large deviations from SI behavior is presented in sections 6.6–6.8.

APPENDIX N

Elements of the scaled particle theory

The scaled particle theory (SPT) was developed in the late 1950s and the early 1960s. It started with the quest for the probability of creating a cavity, or a hole, in the liquid (Hill 1958). It was developed as a theory that was initially designed for hard spheres, and then applied for more realistic fluids and mixtures (Reiss et al. 1959, 1960, 1966; Helfand et al. 1960).

Basically, the SPT is an approximate procedure to compute the work required to create a cavity at a fixed position in a liquid. The work required to create a cavity in the liquid is fundamental in the study of the solvation of solutes in any solvent. The simplest solute is a hard-sphere (HS) particle, and the simplest solvent also consists of HS particles. The solvation process can always be decomposed into two parts; creating a suitable cavity and then turning on the other parts of the solute–solvent interaction.

The basic ingredients of the SPT and the nature of the approximation involved are quite simple. We shall present here only a brief outline of the theory, skipping some of the more complicated details.

The SPT starts by consideration of the work of creating a cavity at some *fixed* position in the fluid. In a fluid consisting of HS particles of diameter a, a cavity of radius r at $\boldsymbol{R}_0$ is nothing but a stipulation that no centers of particles may be found in the sphere of radius r centered at $\boldsymbol{R}_0$. In this sense, creation of a cavity of radius r at $\boldsymbol{R}_0$ is equivalent to placing at R_0 a HS solute of diameter b, such that $r = (a + b)/2$ (figure N.1). A cavity of radius zero is equivalent to placing a HS of negative diameter $b = -a$. The work required to create such a cavity is equivalent to the work required to introduce a HS solute at $\boldsymbol{R}_0$. The work is computed by using a continuous process of building up the particle in the solvent. This is the origin of the name "scaled particle theory." The idea is similar to the Kirkwood's charging process involving the parameter ξ as described in section 3.4.

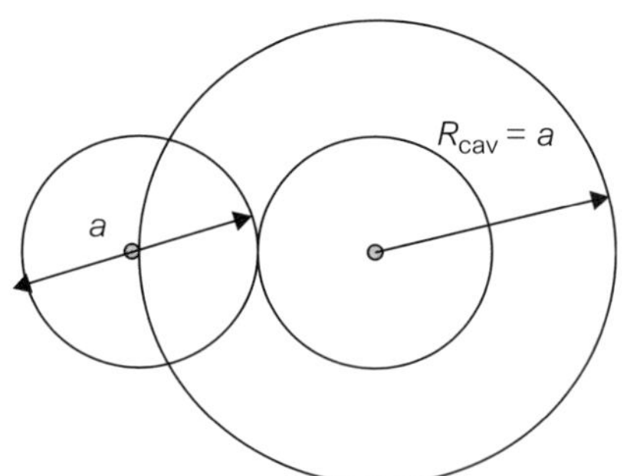

Figure N.1 A particle of diameter *a* produces an excluded volume of radius *a*. This excluded volume is equivalent to a cavity or radius *a*.

In a system of HSs, the molecular parameters that fully describe the particles are the mass m and the diameter a.† It is important to bear this fact in mind when the theory is applied to real fluids, in which case one needs at least three molecular parameters to describe the molecules, and more than three parameters for complex molecules such as water. It is a unique feature of the HS fluid that only two molecular parameters are sufficient for its characterization.

The fundamental distribution function in the SPT is $P_0(r)$, the probability that no molecule has its center within the spherical region of radius r centered at some fixed point $\boldsymbol{R}_0$ in the fluid. Let $P_0(r + dr)$ be the probability that a cavity of radius $(r + dr)$ is empty. (In all the following, a cavity is always assumed to be centered at some fixed point $\boldsymbol{R}_0$, even when this is not mentioned explicitly.) This probability may be written as a product of two factors

$$P_0(r + dr) = P_0(r)P_0(dr/r). \tag{N.1}$$

On the rhs of (N.1), we have introduced the symbol $P_0(dr/r)$ for the *conditional* probability of finding the spherical shell of width dr empty, given that the sphere of radius r is empty. The equality (N.1) is nothing but the well-known definition of a conditional probability in terms of the joint probability.

Next, we define an auxiliary function $G(r)$ by the relation

$$4\pi r^2 \rho G(r) = 1 - P_0(dr/r). \tag{N.2}$$

Since $P_0(dr/r)$ is the conditional probability of finding the spherical shell empty, given that the sphere of radius r is empty, the rhs of (N.2) is the conditional probability of finding the center of at least one particle in this spherical shell, given that the sphere of radius r is empty. This clearly follows from the fact that the spherical shell can be either empty or occupied.

Expanding $P_0(dr/r)$ to first order in dr, we get

$$P_0(r + dr) = P_0(r) + \frac{\partial P_0}{\partial r}\,dr + \cdots \tag{N.3}$$

† The mass is a molecular parameter that enters into the momentum partition function, but this parameter does not enter in the calculation of the work required to create a cavity.

From (N.1), (N.2), and (N.3), we obtain

$$\frac{\partial \ln P_0(r)}{\partial r} = -4\pi r^2 \rho G(r) \tag{N.4}$$

Thus, the function $G(r)$ may be defined either through (N.2) or through (N.4); the latter may also be written in an integral form

$$\ln P_0(r) - \ln P_0(r=0) = -\rho \int_0^r 4\pi\lambda^2 G(\lambda)\, d\lambda. \tag{N.5}$$

Clearly, the probability of finding a cavity of radius zero is unity, hence

$$\ln P_0(r) = -\rho \int_0^r 4\pi\lambda^2 G(\lambda)\, d\lambda. \tag{N.6}$$

There is a general relation between $P_0(r)$, and the work $W(r)$ required to form a cavity of radius r (Ben-Naim 1992), which we write as

$$W(r) = A(r) - A = -kT \ln P_0(r) = kT\rho \int_0^r 4\pi\lambda^2 G(\lambda)\, d\lambda. \tag{N.7}$$

Since the work required to create a cavity of radius r is the same as the work required to insert a hard sphere of diameter $b = 2r - a$ at R_0, we can write the pseudo-chemical potential of the added "solute" in the solvent as

$$\mu_b^* = W(r) = kT\rho \int_0^{(a+b)/2} 4\pi\lambda^2 G(\lambda)\, d\lambda. \tag{N.8}$$

In order to get the chemical potential of the solute having diameter b, we have to add to (N.8) the liberation free energy† namely

$$\mu_b = \mu_b^* + kT \ln \rho_b \Lambda_b^3. \tag{N.9}$$

It should be noted that $\rho_b = 1/V$ is the "solute" density, whereas $\rho = N/V$ is the "solvent" density.

A particular case is obtained when we insert a "solute" having a diameter $b = a$, i.e., a solute which is indistinguishable from other particles in the system. In this case we have

$$\mu_a^* = W(r=a) = kT\rho \int_0^a 4\pi\lambda^2 G(\lambda) d\lambda. \tag{N.10}$$

† In the original publication of the SPT, this quantity has been referred to as the "mixing free energy." As explained in section 3.4, the term liberation free energy is more appropriate for this term (see also Appendix H). Note also that the mass of the particles enter only in Λ_b^3. Note that as long as the particle is not fully coupled, it is distinguishable from all the other particles. It becomes indistinguishable when it is fully coupled to the system.

The integral on the rhs of (N.10) describes the work of coupling a new particle to the system using a continuous "charging," or coupling parameter λ.

At this stage, it is interesting to cite the equation of state for a system of hard spheres of diameter a, namely

$$\frac{P}{kT} = \rho + \tfrac{2}{3}\pi a^3 \rho^2 G(a) \tag{N.11}$$

i.e., the equation of state is determined by the function $G(r)$ at a single point $r = a$. Note that for the chemical potential, one needs the entire function $G(\lambda)$ and not just its value at single point.

The SPT aims at providing an approximate expression for $P_0(r)$ or, equivalently, for $G(\lambda)$. Before presenting this expression, we note that an exact expression is available for $P_0(r)$ at very small r. If the diameter of the HS particles is a, then in a sphere of radius $r < a/2$, there can be at most one center of a particle at any given time. Thus, for such a small r, the probability of finding the sphere occupied is $4\pi r^3\rho/3$. Since this sphere may be occupied by at most one center of an HS, the probability of finding it empty is simply

$$P_0(r) = 1 - \frac{4\pi r^3}{3}\rho \quad \left(\text{for } r \le \frac{a}{2}\right). \tag{N.12}$$

For spheres with a slightly larger radius, namely for $r < a/\sqrt{3}$, there can be at most two centers of HSs in it; the corresponding expression for $P_0(r)$ is

$$P_0(r) = 1 - \frac{4\pi r^3}{3}\rho + \frac{\rho^2}{2}\int\int_{v(r)} g(\mathbf{R}_1, \mathbf{R}_2)\, d\mathbf{R}_1\, d\mathbf{R}_2 \tag{N.13}$$

where $g(R_1, R_2)$ is the pair correlation function, and the integration is carried out over the region defined by the sphere of radius r denoted by $v(r)$. The last equation is valid for a radius smaller than $a/\sqrt{3}$. In a formal fashion, one can write expressions similar to (N.13) for larger cavities, but these involve higher-order molecular distribution functions, and therefore are not useful in practical applications.

Using relation (N.4), we obtain the equivalent of (N.12) in terms of $G(r)$, i.e.,

$$G(r) = \left(1 - \frac{4\pi r^3}{3}\rho\right)^{-1} \quad \text{for } r \le \frac{a}{2} \tag{N.14}$$

and the corresponding work, $W(r)$ is

$$W(r) = -kT\ln\left(1 - \frac{4\pi r^3}{3}\rho\right) \quad \text{for } r \le \frac{a}{2}. \tag{N.15}$$

We now turn to the other extreme case, i.e., to very large cavities. In this case, the cavity becomes macroscopic and the work required to create it is simply

$$W(r) = Pv(r) \quad (\text{for } r \to \infty) \tag{N.16}$$

where P is the macroscopic pressure and $v(r)$ is the volume of the cavity.

Another way of obtaining (N.16) is to use the basic probability in the grand canonical ensemble. For a very large cavity, we can treat the volume $v(r)$ as the volume of a macroscopic system in the T, v, μ ensemble. The probability of finding the system empty is simply

$$P_0(r) = \Xi(T, v(r), \mu)^{-1} = \exp\left[\frac{-Pv(r)}{kT}\right] \tag{N.17}$$

where Ξ is the grand partition function and the last equality holds for the macroscopic systems. Using relation (N.7), we obtain

$$W(r) = -kT \ln P_0(r) = Pv(r) \quad (r \to \infty) \tag{N.18}$$

which is the same as (N.16).

Equation (N.16) is the leading term for a macroscopic volume, i.e., $r \to \infty$. For finite cavities, one may include a term proportional to $v^{2/3}$ to account for the work required to create the surface area, in which case equation (N.16) is modified as

$$W(r) = P\frac{4\pi r^3}{3} + 4\pi r^2 \sigma_0 \tag{N.19}$$

where σ_0 is the surface tension between the fluid and a hard wall. For a still smaller radius, a correction due to the curvature of the cavity may be introduced into (N.19). Also, from relation (N.7) we may obtain the limit of $G(r)$ as $r \to \infty$,

$$G(r \to \infty) = \frac{P}{kT\rho}. \tag{N.20}$$

At this stage, we have two exact results for $G(r)$: one for very small r (N.14), and for very large r (N.20). This information suggests trying to bridge the two ends by a smooth function of r. In fact, this was precisely the procedure taken by Reiss et al. (1959, 1960). The arguments used by the authors to make a particular choice of such a smooth function are quite lengthy and involved. They assumed that $G(r)$ is a monotonic function of r in the entire range of r. They suggested a function of the form

$$G(r) = A + Br^{-1} + Cr^{-2} \tag{N.21}$$

The coefficients A, B, and C were determined by using all the available information on the behavior of the function $G(r)$ for a fluid of hard spheres. The final expression obtained for $G(r)$, after being translated into $W(r)$, i.e., integration of relation (N.7), is the following:

$$W(r) = K_0 + K_1 r + K_2 r^2 + K_3 r^3 \tag{N.22}$$

with the coefficients given by

$$K_0 = kT[-\ln(1-y) + 4.5z^2] - \tfrac{1}{6}\pi P a^3$$

$$K_1 = -\left(\frac{kT}{a}\right)(6z + 18z^2) + \pi P a^2$$

$$K_2 = \left(\frac{kT}{a^2}\right)(12z + 18z^2) - 2\pi P a$$

$$K_3 = \frac{4\pi P}{3} \tag{N.23}$$

where a is the diameter of the hard spheres and y and z are defined by

$$y = \frac{\pi \rho a^3}{6}, \qquad z = \frac{y}{(1-y)}. \tag{N.24}$$

Thus, in essence, what we have obtained is an approximate expression for the work required to create a cavity of radius r in a fluid of HSs characterized by the diameter a.

In figure N.2, we show $W(r^*)/kT$ as a function of the reduced radius of cavity $r^* = r/a$, where a is the diameter of the hard spheres. Equation (N.15) was used

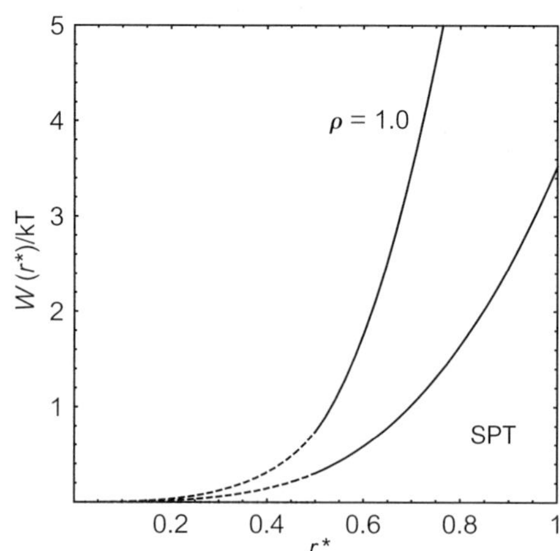

Figure N.2 The work $W(r^*)/kT$ required to create a cavity in a liquid as a function of the reduced distance $r^* = r/a$, for two densities as indicated in the figure.

up to $r^* = 1/2$, and equation (N.22) for $r^* > 1/2$. Note that the monotonic increasing function $W(r^*)$ is a general property of this function, independent of the assumptions of the *SPT*.

Using expression (N.22) for the particular choice of $r = a$, we obtain an expression for the chemical potential:

$$\mu = kT \ln \rho\Lambda^3 + W(r = a). \tag{N.25}$$

Note that at the point $r = a$, the "solute" becomes identical to the solvent molecules and therefore it is assimilated into the system; i.e., $\rho_b = 1/V$ in (N.9) turns into $\rho = N/V$ in (N.25).

Relations (N.22–N.25) is the main result of the SPT. From (N.25), one can get the Gibbs energy $G = N\mu$ as a function of the independent variables T, P, N. In actual application, one uses the SPT, but in addition to specifying T, P, N, one must introduce the density ρ into the theory. In this sense, the SPT is not a purely molecular theory. Normally, if we have $G(T, P, N)$, we should be able to calculate as *output* the the average volume $V = (\partial G/\partial P)_{T,N}$ hence the density. In the SPT, the density is introduced as an additional *input* in the theory.

This comment should be borne in mind when the theory is applied to real fluids. In any real liquid, and certainly for water, we need a few molecular parameters to characterize the molecules, say ε and a in a Lennard-Jones fluid, or in general, a set of molecular parameters $a, b, c, \ldots$. Thus, a proper statistical-mechanical theory of real liquid should provide us with the Gibbs energy as a function of T, P, N and the molecular parameters $a, b, c, \ldots$, i.e., a function of the form $G(T, P, N; a, b, c, \ldots)$. Instead, the SPT makes use of only one molecular parameter, the diameter a. No provision of incorporating other molecular parameters is offered by the theory. This deficiency in the characterization of the molecules is partially compensated for by the use of the measurable density ρ as an input parameter.

The scaled particle theory was extended to mixture of hard spheres by Lebowitz et al. (1965). In a one-component system of hard spheres of diameter a, placing a hard particle of radius R_{HS} produces a cavity of radius $r = R_{HS} + a/2$. When there is a mixture of hard spheres of diameters a_i, the radius of a cavity produced by a hard sphere of radius R_{HS} depends on the species i, i.e., $r_i = R_{HS} + a_i/2$.

The exact result (N.15) for $r \leq a/2$ is rewritten as

$$W(R_{HS}) = -kT \ln \left[1 - \frac{4\pi(R_{HS} + a/2)^3}{3}\right], \quad \text{for } R_{HS} \leq 0 \tag{N.26}$$

and the generalization for a multicomponent system is

$$W(R_{HS}) = -kT\ln\left[1 - \frac{4\pi}{3}\sum_{i=1}^{m}(R_{HS} + a_i/2)^3\right], \quad \text{for } R_{HS} \leq 0. \qquad \text{(N.27)}$$

Thus, in mixture of hard spheres, it is meaningful to place a hard sphere of radius R_{HS} at some fixed position. However, the size of the cavity is different for the different species.

The generalization of equation (N.22), now written for the work required to place a hard sphere of radius R_{HS} at some fixed position, is

$$W(R_{HS}) = K_0 + K_1 R_{HS} + K_2 R_{HS}^2 + K_3 R_{HS}^3 \qquad \text{(N.28)}$$

where

$$K_0 = -kT\ln(1 - \xi_3), \quad K_1 = kT\frac{6\xi_2}{(1 - \xi_3)}$$

$$K_2 = kT\left[\frac{12\xi_1}{1 - \xi_3} + \frac{18\xi_2^2}{(1 - \xi_3)}\right], \quad K_3 = \frac{4\pi P}{3}$$

where

$$\xi_j = \frac{\pi\sum_{i=1}^{m}\rho_i(a_1)^i}{6}. \qquad \text{(N.29)}$$

As can be easily verified, (N.28) reduces to (N.22) for a one-component system, say when $\rho_1 = \rho$ and $\rho_i = 0$ for $i \neq 1$ and $r = R_{HS} + a_1/2$.

APPENDIX O
Solvation volume of pure components

We have noted in section 7.6 that V_s^{*ig} is in general not zero. This is an example of the difference between the theoretical ideal gas and the ideal gas limit of real gas as $P \to 0$. In this appendix we shall first derive a general expression for the partial molar volume at a fixed position then apply the result to the case of ideal gas.

We start from the general expression for the chemical potential for pure A

$$\mu_A^P = \mu_A^{*P} + kT \ln \rho_A \Lambda_A^3. \tag{O.1}$$

The molar volume per particle is

$$\begin{aligned} V_A^P = \left(\frac{\partial \mu_A^P}{\partial P}\right)_T &= V_A^{*P} - kT\left(\frac{\partial \ln V}{\partial P}\right)_T \\ &= V_A^{*P} + kT\kappa_A \end{aligned} \tag{O.2}$$

where κ_A is the isothermal compressibility of pure A. We now use the compressibility equation for the one-component system

$$kT\rho_A\kappa_A = 1 + G_{AA}^P. \tag{O.3}$$

From (O.2) and (O.3), we obtain (note that $\rho_A V_A^P = 1$, for pure A)

$$V_A^{*P} = -G_{AA}^P. \tag{O.4}$$

This is an important result. The change in volume when placing A at a fixed position is equal to the KB integral†. From (O.2) and (O.4), we obtain the result for the molar volume per particle:

$$V_A^P = -G_{AA}^P + kT\kappa_A. \tag{O.5}$$

The value of V_A^* for an ideal gas system depends on which ideal gas system we are referring to.

If we switch off all interactions (i.e., take a theoretical ideal gas), then $g_{AA}(R) \equiv 1$ and $G_{AA} = 0$ (recall that G_{AA} is taken here in the open system, where

† Normally V_s^{*P} is positive, hence G_{AA}^P is negative. However, this does not need to be the case always. e.g. for ionic solute V_s^* might be negative.

all the correlations are due to the intermolecular interactions). Hence,

$$V_A^{*\text{ig}} = 0, \quad \text{(theoretical ideal gas).} \tag{O.6}$$

On the other hand, if we are interested in an ideal gas obtained from a real gas as $\rho_A \to 0$, then we know that in this limit

$$g_{AA}(R) = \exp[-\beta U_{AA}(R)] \tag{O.7}$$

hence

$$\begin{aligned} V_A^{*\text{ig}} &= -\int_0^\infty (\exp[-\beta U_{AA}(R)] - 1) 4\pi R^2 \, dR \\ &= 2B_{AA}, \quad \text{(ideal gas, } \rho_A \to 0) \end{aligned} \tag{O.8}$$

where B_{AA} is the second virial coefficient of pure A. A simple case of (O.8) is when A is a hard sphere of diameter σ_A in which case

$$V_A^{*\text{ig}} = -\int_0^{\sigma_A} (-1) 4\pi R^2 \, dR = \frac{4\pi\sigma_A^3}{3}. \tag{O.9}$$

Thus, placing a single A at a fixed position changes the volume of the system by the exact amount of the excluded volume of A with respect to A particles.

Care must be exercised when calculating $V_A^{*\text{ig}}$ from the derivative of the chemical potential. We have used equation (O.1) to take the derivative with respect to P, then take the limit of $\rho_A \to 0$. If we first take the low-density limit of (O.1) and then differentiate, we get the wrong result. The limiting behavior of the chemical potential is

$$\mu_A \approx kT \ln \frac{P}{kT} \Lambda_A^3 + PB_{AA} \tag{O.10}$$

hence,

$$V_A^{\text{ig}} = \left(\frac{\partial \mu_A}{\partial P}\right)_T = \frac{kT}{P} + B_{AA}. \tag{O.11}$$

If we now take the limiting expression for the compressibility

$$\kappa_A^{\text{ig}} \approx \frac{1}{P} \tag{O.12}$$

and using (O.11) and (O.12) in (O.2) we obtain

$$V_A^{*\text{ig}} = \frac{kT}{P} + B_{AA} - \frac{kT}{P} = B_{AA} \tag{O.13}$$

which is the wrong result (see equation O.8). To obtain the correct result, we should either proceed as we have done in obtaining equation (O.4) and then take the limiting behavior of (O.8) or instead of (O.12) we must also take the first-order expansion of the nondivergent part of κ_A, which can be easily obtained from the virial expansion of $\rho(P)$. The result is

$$kT\kappa_A \approx \frac{kT}{P} - B_{AA}. \tag{O.14}$$

Now using (O.11) and (O.14) in (O.2), we obtain

$$V_A^{*ig} = \frac{kT}{P} + B_{AA} - \left(\frac{kT}{P} - B_{AA}\right) = 2B_{AA} \tag{O.15}$$

which is the correct result.

The solvation volume is thus

$$\Delta V_A^* = V_A^{*l} - V_A^{*ig} = -(G_{AA} - G_{AA}^{ig}) = -(G_{AA} + 2B_{AA}). \tag{O.16}$$

In general, the term $kT\kappa_A$ in (O.2) is small compared with either V_A^P or V_A^{*P}. For instance, for water at 20 °C we have

$$V_W^P = 18.05\,\text{cm}^3\text{mol}^{-1} \quad \text{and} \quad kT\kappa_W = 1.12\,\text{cm}^3\text{mol}^{-1}.$$

For n-heptane at 20 °C

$$V_H^P = 146.6\,\text{cm}^3\text{mol}^{-1} \quad \text{and} \quad kT\kappa_H = 3.39\,\text{cm}^3\text{mol}^{-1}.$$

For benzene at 20 °C

$$V_B^P = 88.86\,\text{cm}^3\text{mol}^{-1} \quad \text{and} \quad kT\kappa_B = 2.28\,\text{cm}^3\text{mol}^{-1}.$$

For argon at 100 K

$$V_A^P = 30.47\,\text{cm}^3\text{mol}^{-1} \quad \text{and} \quad kT\kappa_A = 2.622\,\text{cm}^3\text{mol}^{-1}.$$

We see that in general the term $kT\kappa_A$ is small compared with either V^P or V^*. Using the result (O.4) for argon we can write

$$\begin{aligned} V_A^{*P} &= -G_{AA}^P \\ &= -\int (g_{AA}(R) - 1)4\pi R^2\, dR \\ &= -\int_0^\sigma -4\pi R^2\, dR - \int_\sigma^\infty (g_{AA}(R) - 1)4\pi R^2\, dR \\ &= \frac{4\pi\sigma^3}{3} - \int_\sigma^\infty (g_{AA}(R) - 1)4\pi R^2\, dR \end{aligned} \tag{O.17}$$

Thus, the molar volume (at fixed position) of argon has two components. One is due to the repulsive part of the potential. This is eight times the "actual" molecular volume of argon $4\pi R^3/3$. The second is due to the interaction, usually both repulsive and attractive beyond $R > \sigma$.

APPENDIX P

Deviations from SI solutions expressed in terms of $\rho\Delta_{AB}$ and in terms of P_A/P_A^0

In sections 6.6 and 6.7, we analyzed the conditions of stability using the parameter $\rho\Delta_{AB}$ as a measure of the deviations from SI solutions. When $\Delta_{AB}=0$, we had symmetrical ideal SI solutions. We found that for positive values of Δ_{AB}, the system was always stable. For large *negative* values of Δ_{AB}, we found regions of instability. This conclusion seems to conflict with the experimental results that positive deviations from Raoult's law lead to instability[†]. The classical examples shown in many books are mixtures of water and various normal alcohols. We reproduce the relevant curves in figure P.1. Here, we plot the relative partial pressure of the alcohols in mixtures of water with methanol, ethanol, propanol and n-butanol. Clearly, in all of the four cases, deviations from Raoult's law as measured by either the quantities

$$P_A/P_A^0 = x_A\gamma_A^{\mathrm{SI}}, \quad \mu_A^{\mathrm{EX,SI}} = kT\ln\gamma_A^{\mathrm{SI}} \tag{P.1}$$

are positive in the sense that γ_A^{SI} are larger than unity, or equivalently the partial pressure over the solutions of component A is larger than the partial pressure expected, had the solutions obeyed Raoult's law (i.e., $P_A/P_A^0 = x_A$). We note that the vapor above the solution is assumed to be an ideal-gas mixture. The excess chemical potential with respect to SI is $\mu^{\mathrm{EX,SI}}$.

For the water–methanol, water–ethanol, and water–propanol, systems deviations from Raoult's law are positive. The single phase is stable in the entire range of compositions. In water-n-butanol, we know experimentally that the two components are not miscible in the entire range of compositions. We see from the figure that deviations from Raoult's law are large and positive.

[†] We have also shown that the theoretical results based on the Kirkwood–Buff theory are in conflict with the conclusions based on the *first*-order deviations from SI. This conflict is simply a result of applying first-order expression in $\rho\Delta_{AB}$, for large values of $\rho\Delta_{AB}$.

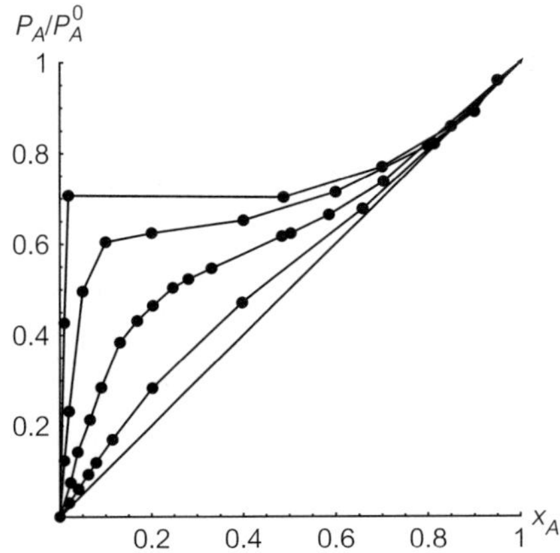

Figure P.1 Values of P_A/P_A^0 as a function of x_A for water–methanol, water–ethanol, water–propanol and water-n-butanol at 25 °C. Based on data from Butler et al (1933). *A* is the alcohol component. Lower curve for methanol, successive higher curves for ethanol, propanol and *n*-butanol.

This fact seems to support the conclusion based on *first-order* deviations from SI solutions (see section 6.6), but in disagreement with the conclusion based on the KB theory (section 6.7). Since at large deviations from SI we must trust the full KB expression (section 6.7) rather than the first-order (in terms of $\rho\Delta_{AB}$) deviation, we are facing a conflict between the conclusions based on the KB theory and the experimental data.

In this appendix, we resolve the conflict by showing that the two measures of the deviations from SI behavior, $\rho\Delta_{AB}$ and γ_A^{SI} (or P_A/P_A^0), are not, in general, equivalent. We start with the first-order expression for the partial pressure, say of water, in the mixture

$$P_A/P_A^0 = x_A\gamma_A^{\text{SI}} = x_A \exp[\rho\Delta_{AB}x_B^2/2]$$

$$\mu_A^{\text{EX,SI}} = kT\ln\gamma_A^{\text{SI}} = kT\rho\Delta_{AB}x_B^2/2. \tag{P.2}$$

This expression, or the equivalent one in the lattice theory of solution, has been traditionally used to analyze deviations from SI solution and stability of the system. It is clear from (P.2) that when $\rho\Delta_{AB}$ is positive (or negative), deviations from Raoult's law will be positive (or negative) and vice versa.

Unfortunately, these simple relationships between $\rho\Delta_{AB}$ and P_A/P_A^0 do not hold for large values of $\rho\Delta_{AB}$. The general relation between the two quantities follows from the general expression for the chemical potential (equation 6.2, chapter 6)

$$\mu_A = \mu_A^P + kT\ln x_A + kT\int_0^{x_B}\frac{\rho x_B\Delta_{AB}}{1+\rho x_A x_B\Delta_{AB}}$$
$$= \mu_A^P + kT\ln x_A + \mu_A^{\text{EX,SI}}. \tag{P.3}$$

Assuming that the vapor above the solution is an ideal gas mixture, we get instead of (P.2) the more general expression

$$P_A/P_A^0 = x_A \exp\left[\int_0^{x_B} \frac{\rho x_B \Delta_{AB}}{1+\rho x_A x_B \Delta_{AB}}\right]. \tag{p.4}$$

It is clear from (P.4) that $\rho\Delta_{AB}$ determines uniquely the value of P_A/P_A^0. In other words, the molecular measure of the deviations from SI, $\rho\Delta_{AB}$, determines not only the stability of the system (see sections 6.6 and 6.7), but also the deviations from Raoult's law as measured by the quantity P_A/P_A^0 (presuming that the vapor is an ideal-gas mixture).

On the other hand, the values of P_A/P_A^0 (whether larger or smaller than x_A) do not determine uniquely either the stability condition or the value of $\rho\Delta_{AB}$.

This can be seen if we expand $\mu_A^{\text{EX,SI}}$ to second order in $\rho\Delta_{AB}$. The result is

$$\mu_A^{\text{EX,SI}} = kT\ln\gamma_A^{\text{SI}} = \frac{kT\rho\Delta_{AB}x_B^2}{2} + kT(\rho\Delta_{AB})^2\left[\frac{x_B^4}{4} - \frac{x_B^3}{3}\right]. \tag{P.5}$$

Clearly, in the first-order expansion (P.2), the sign of $\rho\Delta_{AB}$ is the same as the sign of $\mu_A^{\text{EX,SI}}$. On the other hand, in the second-order expansion in terms of $\rho\Delta_{AB}$, equation (P.5) shows that $\rho\Delta_{AB}$ determines $\mu_A^{\text{EX,SI}}$, but $\mu_A^{\text{EX,SI}}$ (or the value of γ_A^{SI}) does not uniquely determine the value of $\rho\Delta_{AB}$. Figure P.2 shows a plot of $\mu_A^{\text{EX,SI}}/\text{k}T$ as a function of $\rho\Delta_{AB}/\text{k}T$ for one composition $x_A = x_B = \frac{1}{2}$. It is clear that in the first-order expansion, $\mu_A^{\text{EX,SI}}$ and $\rho\Delta_{AB}$ have the same sign. However, for the second-order expansion, the values of $\mu_A^{\text{EX,SI}}$ do not uniquely determine $\rho\Delta_{AB}$. The same conclusion may be drawn by expanding $\mu_A^{\text{EX,SI}}$ to

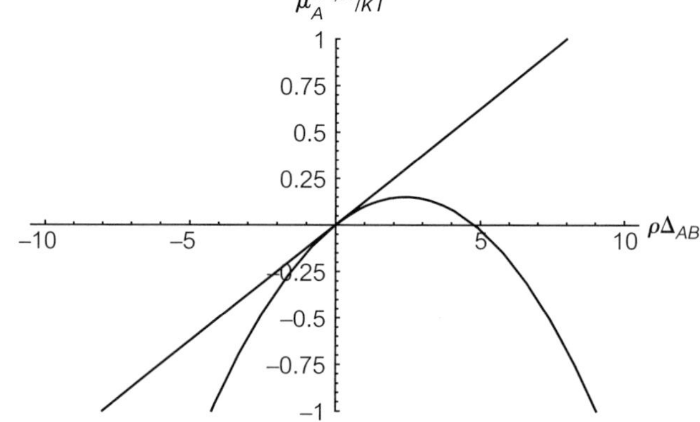

Figure P.2 $\mu_A^{\text{EX,SI}}/kT$ as a function of $\rho\Delta_{AB}$ according to the first- and second-order expansion in $\rho\Delta_{AB}$ at one composition $x_A = x_B = 0.5$.

higher order in $\rho\Delta_{AB}$. Odd and even order show similar behavior to the first- and the second-order expansion in $\rho\Delta_{AB}$.

In the more general case, we have the relation

$$\left(\frac{\partial \mu_A}{\partial x_A}\right)_{P,T} = kT\left(\frac{\partial \ln x_A \gamma_A^{\text{SI}}}{\partial x_A}\right)_{P,T} = \frac{1}{x_A(1 + \rho x_A x_B \Delta_{AB})} \tag{P.6}$$

or equivalently for ideal-gas vapor pressures

$$f(x_A) \equiv \left(\frac{\partial \ln(P_A/P_A^0)}{\partial x_A}\right)_{P,T} = \frac{1}{x_A(1 + \rho x_A x_B \Delta_{AB})} \tag{P.7}$$

where $f(x_A)$ is a measurable quantity. From (P.7) we can eliminate $\rho\Delta_{AB}$ to obtain

$$\rho\Delta_{AB} = \frac{1 - x_A f(x_A)}{x_A^2 x_B f(x_A)}. \tag{P.8}$$

Thus, from (P.6), (P.7), and (P.8), it is clear that the values of either P_A/P_A^0 or of $\mu_A^{\text{EX,SI}}$ do not determine either the stability condition or the value of $\rho\Delta_{AB}$. Instead, the quantity $f(x_A)$, i.e., the derivative of the quantity $\ln[P_A/P_A^0]$ with respect to x_A, determines both the stability condition as well as $\rho\Delta_{AB}$. Thus, if we characterize deviations from SI by the values of P_A/P_A^0 (being larger than x_A), this in itself does not determine the values of the molecular measure of the deviations from SI, $\rho\Delta_{AB}$, nor the stability condition. One needs both P_A/P_A^0 as well as its composition dependence to determine $\rho\Delta_{AB}$ and the stability condition.

References

Alder, B.J. (1964) Phys. Rev. Letter **12**, 317
Almasy, L., Cser, L. and Jancso, G. (2002) *J. Mol. Liquids* **101**, 89
Almasy, L., Jancso, G. and Cser, L. (2002) *Appl. Phys. A* 74S 1376
Arakawa, T. and Timasheff, S.N. (1985) *Methods Enzymol.* **114**, 49
Arnett, E.M. and McKelvey, D.R. (1969) Chapter 6 in Coetzee and Ritchie (1969)
Barker, J.A. (1963) *Lattice Theories of the Liquid State*, Pergamon Press, Oxford
Battino, R. and Clever, H.L. (1966) *Chem Rev.* **66**, 395
Battino, R. (1984) *Fluid Phase Equilibra* **15**, 231
Baxter, R.J. (1970) *J. chem. Phys.* **52**, 4559
Beeby, J.L. (1973) *J. Phys. Chem.* **6**, 2262
Behera, R. (1998) *J. Phys. Chem.* **108**, 3373
Ben-Naim, A. (1972a) *J. Chem. Phys.* **56**, 2864
Ben-Naim, A. (1972b) *J. Chem. Phys.* **57**, 3605
Ben-Naim, A. (1973a) *J. Chem. Phys.* **59**, 6535
Ben-Naim, A. (1973b) *Statistical Phys.* **7**, 3
Ben-Naim, A. (1974) *Water and Aqueous Solutions*, Plenum Press, New York
Ben-Naim, A. (1975a) *J. Chem. Phys.* **63**, 2064
Ben-Naim, A. (1975b) *J. Phys. Chem.* **79**, 1268
Ben-Naim, A. (1977) *J. Chem. Phys.* **67**, 4884
Ben-Naim, A. (1978) *J. Phys. Chem.* **82**, 792
Ben-Naim, A. (1979) *J. Phys. Chem.* **83**, 1803
Ben-Naim, A. (1980) *Hyrdrophobic Interactions*, Plenum Press, New York
Ben-Naim, A. and Marcus, Y. (1984a) *J. Chem. Phys.* **80**, 4438
Ben-Naim, A. and Marcus, Y. (1984b) *J. Chem. Phys.* **81**, 2016
Ben-Naim, A. (1985) *J. Phys. Chem.* **89**, 5738
Ben-Naim, A (1987a), *Am. J. of Phys.* **55**, 725
Ben-Naim, A. (1987b), *Am. J. of Phys.* **55**, 1105
Ben-Naim, A. (1987) *Solvation Thermodynamics*, Plenum Press, New York
Ben-Naim, A. (1989), *J. Phys. Chem.* **93**, 3809
Ben-Naim, A. (1990a) in Matteoli and Mansoori (1990) p. 211
Ben-Naim, A. (1990b), *Pure and Applied Chem.* **62**, 25
Ben-Naim, A. (1992) *Statistical Thermodynamics for Chemists and Biochemists*, Plenum Press, New York
Ben-Naim, A. (2001) *Cooperativity and Regulation in Biochemical Processes*, Kluwer Academic, Plenum Publishers, New York
Benson, G.C. and Kiyohara, O. (1980). *J. sol. Chem.* **10**, 791
Berthelot, D. (1898) *Compt. Rend.* **126**, 1703, 1857
Born, M. and Green, H.S. (1946) *Proc. Roy. Soc.* (London) A **188**, 10
Brey, J.J., Santos, A. and Romero, F. (1982) *J. Chem. Phys.* **77**, 5058
Broyles, A.A. (1960) *J. Chem. Phys.* **33**, 456
Broyles, A.A. (1961) *J. Chem. Phys.* **35**, 493
Broyles, A.A. (1962) *Bull. Am. Phys. Soc.* **7**, 67
Buff, F.P. and Brout, R. (1955), *J. Phys. Chem.* **23**, 458
Buff, F.P. and Schindler, F.M. (1958) *J. Phys. Chem.* **29**, 1075

Butler, J.A.V., Thomson, D.W. and Maclennan, W.H. (1933) *J. Chem. Soc.* 674
Calado, J.C.G. and Staveley, L.A.K. (1971) *Trans Faraday Soc.* **67**, 289
Calado, J.C.G., Jancso, G., Lopes, J.N.C., Marko, L., Nunes da Ponte, M., Rebelo, L.P.N. and Staveley, L.A.K. (1994) *J. Chem. Phys.* **100**, 4582
Calado, J.C.G., Diaz, F.A., Lopes, J.N.C., Nunes de Ponte, M. and Rebelo, L.P. (2000) *Phys. Chem. Chem. Phys.* **2**, 1095
Callen, H.B. (1960) *Thermodynamics*, John Willey, New York
Chandler, D. (1987) *Introduction to Modern Statistical Mechanics*, Oxford University Press, Oxford
Chialvo, A.A. (1990) in Matteoli and Mansoori (1990) p. 131
Chitra, R. and Smith, P.E. (2002) *J. Phys. Chem.* **B106**, 1491
Chui, C.H. and Canfield, F.B. (1971) *Trans Faraday Soc.* **67**, 2933
Cochran, H.D., Lee, L.L. and Pfund, D.M. (1990) in Matteoli and Mansoori (1990) p. 69
Coetzee, J.F. and Ritchie, C.D. (1969) eds. *Solute–Solvent Interactions*, Marcel Dekker, New York
Covington, A.K. and Newman, K.E. (1976) *Adv. Chem. Ser.* **155**, 153
Covington, A.K. and Newman, K.E. (1988) *Faraday Trans.* I. **84**, 1393
Davies, R.H., Duncan, A.G., Saville, G. and Staveley, L.A.K. (1967) *Trans Faraday* Soc. **63**, 855
Debenedetti, P.G. (1987), *J. Phys. Chem.* **87**, 1256
DeBoer, J. (1940) Thesis, Amsterdam
Denbigh, K.G. (1966) *The Principles of Chemical Equilibrium*, Cambridge University Press, London
Denbigh, K.G. (1981) *The Principles of Chemical Equilibrium*, 4th edition Cambridge University Press, Cambridge
Dixit, S., Crain, J., Poon, W.C.K., Finney, J.L. and Soper, A.K. (2002) *Nature*, **416**, 6883
Donkersloot, M.C.A. (1979a) *J. Solution Chem.* **8**, 293
Donkersloot, M.C.A. (1979b) *Chem. Phys. Letter*, **60**, 435
Duh, D.M. and Henderson, D. (1996) J. Chem. Phys. **104**, 6742
Eliel, E.L. (1962) *Stereochemistry of Carbon Compounds*, McGraw-Hill, New York
Eley, D.D. (1939) *Trans Faraday Soc.* **35**, 1281
Eley, D.D. (1944) *Trans Faraday Soc.* **40**, 184
Engberts, J.B.F.N. (1979) in Franks (1979) Chapter 4
Fisher, I.Z. (1964) *Statistical Theory of Liquids*, University of Chicago Press
Fisher, M.A. and Widom, B. (1969) *J. Chem. Phys.* **50**, 3756
Fowler, R. and Guggenheim, E.A. (1956) *Statistical Thermodynamics*, Cambridge University Press, London
Frank, H.S. and Evans, M.W. (1945) *J. Chem. Phys.* **13**, 507
Franks, F. (1973) in *Water: A Comprehensive Treatise* (F. Franks, ed), Vol.2, Plenum Press, New York
Franks, F. (1979) editor, *Water, A Comprehensive Treatise*, Volume 6, Plenum Press, New York
Friedman, L.H. and Ramanathan, P.S. (1970) *J. Chem. Phys.* **74**, 3756
Friedman, L.H. and Krishnan, C.V. (1973) Chapter 1 in Franks (1973)
Friedman, H.L. (1985) *A Course in Statistical Mechanics*, Prentice Hall, New Jersey
Gibbs, J.W. (1906) *Scientific Papers, Vol. I, Thermodynamics*, p. 166 Longmans Green, New York
Gray, C.G. and Gubbins, K.E. (1984) *Theory of Molecular Fluids, Volume 1: Fundamentals*, Clarendon Press, Oxford
Grundke, E.W. and Henderson, D. (1972a) *Molecular Phys.* **24**, 269
Grundke, E.W. and Henderson, D. (1972b) *International J. Quantum Chem.* **6**, 383
Grundke, E.W., Henderson, D. and Murphy, R.D. (1973) *J. Canad de Physique* **51**, 1216
Grunwald, E. (1960) *J. Am. Chem. Soc.* **82**, 5801
Grunwald, E., Baughman, G. and Kohnstam, G. (1960) *J. Am. Chem. Soc.* **82**, 5801

Guggenheim, E.A. (1932) *Proc. Roy. Soc. London A* **135**, 181
Guggenheim, E.A. (1952) *Mixtures*, Clarendon Press, Oxford
Guggenheim, E.A. (1967) *Thermodynamics*, 5th edition, North Holland, Amsterdam
Gurney, R.W. (1953) *Ionic Processes in Solution*, Dover, New York
Hall, D.G. (1972) *Faraday Trans. 2*, **68**, 25
Hamad, E.Z. and Mansoori, G.A. (1987) *Fluid Phase Equilibria* **37**, 255
Hamad, E.Z. and Mansoori, G.A., Matteoli, E. and Lepori, L.Z. (1989) *Phys. Chem. Neue Folge*, Bd. **162**, 27
Hamad, E.Z. and Mansoori, G.A. (1990a) in Matteoli and Mansoori (1990) p. 95
Hamad, E.Z. and Mansoory, G.A. (1990b) *J. Am. Chem. Soc.* **94**, 3148
Hamad, E.Z. and Mansoori, G.A. (1993) *Fluid Phase Equilibria* **85**, 141
Hamad, E.Z. (1997) *J. Chem. Phys.* **106**, 6116
Hamad, E.Z. (1998) *Fluid Phase Equilibria* **142**, 163
Hansen, J.P. and McDonald, I.R. (1976) *Theory of Simple Liquids*, Academic Press, London
Hayashi, H., Nishikawa, K. and Iijima, T. (1990) *J. Phys. Chem.* **94**, 8334
Helfand, E., Reiss, H. and Frish, H.L. (1960) *J. Chem. Phys.* **33**, 1379
Hildebrand, J. (1929) *J. Am. Chem.* Soc. **51**, 66
Hildebrand, J.H., Scott, R.L. (1962) *Regular Solutions*, Prenlice Hall Englewood Cliffs
Hildebrand, J.H. and Scott, R.L. (1964) 3rd edtion, *The Solubility of Non-Electrolytes*, Dover, New York
Hildebrand, J.H., Prausnitz, J.M. and Scott, R.L. (1970) *Regular and Related Solutions*, Van Nostrand, New York
Hill, T.L. (1956) *Statistical Mechanics, Principles and Selected Applications*, McGraw-Hill, New York
Hill, T.L. (1958) *J. Chem. Phys.* **28**, 1179
Hill, T.L. (1960) *Introduction to Statistical Thermodynamics*, Addison Wesley, Reading, Massachusetts
Hill, T.L. (1960) *An Introduction to Statistical Thermodynamics*, Addison-Wesley London
Hirata, F. (2003) editor, *Molecular Theory of Solvation*, Kluwer Academic London Hirschfelder, J.O., Stevenson, D. and Eyring, H. (1937) *J. Chem. Phys.* **5**, 896
Hirschfelder, J.O., Curtiss, C.F. and Bird, R.B. (1963) *Molecular Theory of Gases and Liquids (2nd ed).*, John Wiley, New York
Hoheisel, C. (1993) *Theoretical Treatment of Liquids and Liquid Mixtures*, Elsevier Scientific, Amsterdam
Imai, T. and Hirata, F. (2003) *J. Chem. Phys.* **119**, 5623
Jackson, J.L. and Klein, L.S. (1964), *Phys. Fluids* **7**, 279
Jacques, J., Collet, A. and Wilsen, S.H. (1981) *Enantiomers, Racemates and Resolutions*, John Wiley, New York
Jancso, G. and Jakli, G. (1980) *Aust. J. Chem.* **33**, 2357
Jansco, G. (2004) *Pure and Appl. Chem.* **76**, 11
Jorgensen, W.L. (1982) *J. Chem. Phys.* **77**, 5757
Jorgensen, W.L. and Buckner, J.K. (1987) *J. Phys. Chem.* **91**, 6083
Kato, T., Fujiyama, T. and Nomura, H. (1982) *Bull. Chem. Soc.* Japan **55**, 3368
Kato, T. (1984) *J. Chem. Phys.* **88**, 1248
Kato, T. (1990) in Matteoli, E. and Mansoori, G.A. p. 227
Kirkwood, J.G. (1933) *J. Chem. Phys.* **3**, 300
Kirkwood, J.G. (1935) *J. Chem. Phys.* **3**, 300
Kirkwood, J.G. (1950) *J. Chem. Phys.* **8**, 380
Kirkwood, J.G. and Buff, F.P. (1951) *J. Chem. Phys.* **19**, 774
Kojima, K., Kato, T. and Nomura, H. (1984) *J. Sol.* Chem. **13**, 151

Kusalik, P.G. and Patey, G.N. (1987) *J. Chem. Phys.* **86**, 5110
Lebowitz, J.C (1964) *Phys. Rev.* **133**, A 895
Lebowitz, J.L. and Percus, J.K. (1961) *Phys. Rev.* **122**, 1675
Lebowitz, J.L. and Percus, J.K. (1963) *J. Math. Phys.* **4**, 116
Lebowitz, J.L., Helfand, E. and Praestgaard, E. (1965) *J. Chem. Phys.* **43**, 774
Lee, L.L. and Malijevsky (2001) *J. Chem. Phys.* **114**, 7109
Lepori, L., Matteoli, E., Hamad, E.Z. and Mansoori, G.A. (1990) in Matteoli and Mansoori (1990) p.175
Lewis, G.N. and Randall, M. (1961) *Thermodynamics* (revised by K.S. Pitzer and L. Brewer) McGraw-Hill, New York
Lorentz, H.A. (1881) *Ann. Physik* **12**, 127
Mandel, F., Bearman, R.J. and Bearman, M.Y. (1970) *J. Chem. Phys.* **52**, 3315
Mansoori, G.A. and Ely, J.F. (1985) *Fluid Phase Equilibria* **22**, 253
Mansoori, G.A. and Matteoli, E. (1990) in Matteoli and Mansoori (1990) p. 1
Marcus, Y. (1983) *Aust. J. Chem.* **36**, 1719
Marcus, Y. and Ben-Naim, A. (1985) *J. Chem. Phys.* **84**, 4744
Marcus, Y. (1988) *J.Chem. Soc. Faraday Trans.* **84**, 1465
Marcus, Y. (1989) *J.Chem. Soc. Faraday Trans.* **85**, 381
Marcus, Y. (1999) *Phy. Chem. Chem. Phys.* **1**, 2975
Marcus, Y. (2001) *Monatshefte fur chemie* **132**, 1387
Marcus, Y. (2002) *Solvent Mixtures, Properties and Selective Solvation*, Marcel Dekker, New York
Matteoli, E. (1997) *J. Phys. Chem.* **B101**, 9800
Matteoli, E. and Mansoori, G.A. (1990) *Fluctuation Theory of Mixtures*, Taylor and Francis, New York
Matteoli, E. and Mansoori, G.A. (1995) *J. Chem. Phys.* **103**, 4672
Matteoli, E. and Lepori, L. (1984) *J. Chem. Phys.* **80**, 2856
Matteoli, E. and Lepori, L. (1990a) *J. Molecular Liquids*, **47**, 89
Matteoli, E. and Lepori, L. (1990b) in Matteoli and Mansoori (1990) p. 259
Matteoli, E. and Lepori, L. (1995) *J. Chem. Soc. Faraday Trans.* **91**, 431
Matubayasi, N. and Nakahara, M. (2000) *J. Chem. Phys.* **113**, 6070
Matubayasi, N. and Nakahara, M. (2002) *J. Chem. Phys.* **117**, 3605
Matubayasi, N. and Nakahara, M. (2003) *J. Chem. Phys.* **119**, 9686
Mazo, R.M. (1958) *J. Chem. Phys.* **29**, 1122
McDonald, I.R. (1972) *Mol. Phys.* **24**, 391
McGlashan R.J. and Wingrove M.I. (1956) *Trans. Faraday Soc.* **52**, 470
McMillan, W.G. and Mayer, J.E. (1945) *J. Chem. Phys.* **13**, 276
McQuarrie, D.A. (1976) *Statistical Mechanics*, Harper and Row, New York
Misawa, M. and Yoshida, K. (2000) *J. Phys. Soc.* Japan **69**, 3308
Münster, A. and Sagel, K. (1959) *Z. Phys. Chem. N.F.* **22**, 81
Münster, A. (1969) *Statistical Thermodynamics Volume I.* Springer-Verlag, Berlin
Münster, A. (1974) *Statistical Thermodyanmics*, Vol. 2, Springer-Verlag, Berlin
Nakagawa, T. (2002), *J. Chem. Soc. Japan. Chem. and Ind. Chem. (Nippon Kagaku Kaishi)* **3**, 301
Newman, K.E. (1988) *Faraday Trans.* I. **84**, 1387
Newman, K. (1989a), *J. Chem. Soc. Faraday Trans.* 1, **83**, 485
Newman, K. (1989b) *Faraday Trans.* I. **85**, 485
Newman, K. (1990) in Matteoli and Mansoori (1990) p. 373
Newman, K. (1994) *Chem. Soc. Reviews* p. 31 *Chem. Phys. Letter*, **132**, 50
Nishikawa, K. (1986) *Chem. Phys. Lett.* **132**, 50
Nishikawa, K., Hayashi, It. and Iijama, T. (1989) *J. Phys. Chem.* **93**, 6559
Novák, J.P., Matouš, J. and Pick, J. (1987), *Liquid Liquid Equilibria*, Elsevier, Amsterdam.

O'Connell, J.P. (1971) *Molecular Physics* **20**, 27
O' Connell, J.P. and DeGance, A.E. (1975) *J. Sol. Chem.* **4**, 763
O'Connell, J.P. (1981) *Fluid Phase Equilibria* **6**, 21
O'Connel, J.P. (1990) in Matteoli and Mansoori (1990) p. 45
Ornstein, L.S. and Zernike, F. (1914) *Proc. Amsterdam Acad. Sci.* **17**, 793
Pandey, J.D. and Verna, R. (2001) *Chem. Phys.* **270**, 429
Patil, K.J. (1981) *J. Sol. Chem.* **67**, 4884
Pearson, F.J. and Rushbrooke, G.S. (1957) *Proc. Royal Soc.Edinburgh* **A64**, 305
Percus, J.K. and Yevick G.J. (1958) *Phys. Rev.* **110**, 1
Percus, J.K. (1962) *Phys. Rev. Letters*, **8**, 462
Perera, A. and Sokolic, F. (2004) *J. Chem. Phys.* **121**, 22
Perera, A., Sokolic, F., Almasy, L., Westh, P. and Koga, Y. (2005) *J. Chem. Phys.* **123**, 024503
Perry, P. and Throup, J. (1972) *J. Chem. Phys.* **57**, 1827
Perry, R.L. and O'Connell, J.P. (1984) *Molecular Phys.* **52**, 137
Pollack, G.L. (1981) *J. Chem. Phys.* **75**, 5875
Pollack, G.L., Himm, J.F. and Enyeart, J.J. (1984) *J. Chem. Phys.* **81**, 3239
Prausnitz, J.M., Lichtenthaler, R.N. and Azevedo, E.G. (1986) *Molecular Thermodynamics of Fluid Phase Equilibria*, 2nd ed., Prentice Hall, Englewood Cliffs
Prigogine, I. and Defay, R. (1954) *Chemical Thermodynamics* translated by D.H. Everet, Longmans Green, London
Prigogine I. (1957) *The Molecular Theory of Solution*, North Holland Amsterdam
Randall, L., Perry, L. and O'Connell, J.P. (1984) *Molecular Physics* **52**, 137
Rahman, A. (1964) *Phys. Re. Letters* **12**, 575
Redlich, O. and Kister, A.T. (1948) *Ind. Ing. Chem.* **40**, 345
Reiss, H., Frisch, H.L. and Lebowitz, J.L. (1959) *J. Chem. Phys.* **31**, 369
Reiss, H., Frisch, H.L., Helfand, E. and Lebowitz, J.L. (1960) *J. Chem. Phys.* **32**, 119
Reiss, H. (1966) *Adv. Chem. Phys.* **9**, 1
Reynolds, J.A., Gilbert, D.B. and Tanford, C. (1974) *Proc. Natl. Acad. Sci.*, USA, **71**, 2925
Rice, O.K. (1938) *J. Chem. Phys.* **6**, 476
Rosenberg, R.O., Mikkelineni, R. and Berne, B.J. (1982) *J. Am. Chem. Soc.* **104**, 7647
Rowlinson, J.S. (1959) *Liquids and Liquid Mixtures*, Butterworths Scientific, London
Rowlinson, J.S. and Swinton, F.L. (1982) *Liquids and Liquid Mixtures* 3rd ed. Butterworth, London
Rubio, R.G., Prolongo, M.G., Cabrerizo, U., Pena, M.D. and Renuncio, J.A.R. (1986), *Fluid Phase Equilibria*, **24**, 1
Rubio, R.G., Prolongo, M.G., Pena, M.D. and Renuncio, J.A.R. (1987) *J. Phys. Chem.* **91**, 1177
Rubio, R.G., Pena, M.D. and Patterson, D. (1990), in Matteoli and Mansoori (1990) p. 241
Ruckenstein, E. and Shulgin, I.L. (2001a) *Fluid Phase Equilibria* **180**, 271
Ruckenstein, E. and Shulgin, I.L. (2001b) *Fluid Phase Equilibria* **180**, 281
Ruckenstein, E. and Shulgin, I.L. (2001c) *Fluid Phase Equilibria* **180**, 345
Ruckenstein, E. and Shulgin, I.L. (2002) *Ind. Eng. Chem. Res.* **41**, 4674
Rushbrooke (1968) Chapter 2 in Temperley et al. (1968)
Sandler, S.I. (1994) editor, *Models for Thermodynamic and Phase Equilibria Calculations*, Marcel Dekker, New York
Shimizu, S. (2004) *J. Chem. Phys.* **120**, 4989
Shimizu, S. and Boon, C.L. (2004) *J. Chem. Phys.* **121**, 9147
Shimizu, S. and Smith, D.J. (2004) *J. Chem. Phys.* **121**, 1148
Shulgin, I.L. and Ruckenstein, E. (1999) *J. Phys.* Chem. **103**, 872
Shulgin, I.L. and Ruckenstein, E. (2002) *Ind. Eng. Chem. Res.* **41**, 1689
Shulgin, I.L. and Ruckenstein, E. (2005a) *J. of Chem. Phys.* **35**, 129

Shulgin I.L. and Ruckenstein E. (2005b) *J. Chem. Phys.* **123**, 054909
Shulgin I.L. and Ruckenstein E. (2005c) *Biophys. Chem.* **118**, 128
Singer, J.V.L. and Singer, K. (1972) *Mol. Phys.* **24**, 357
Stell, G. (1963) *Physica* **29**, 517
Tanford, C. (1973) *The Hydrophobic Effect*, John Wiley, New York
Tanford, C. (1979) *J. Phys. Chem.* **83**, 1802
Temperley, H.N.V., Rowlinson, J.S. and Rushbrooke (1968) *Physics of Simple Liquids*, North Holland, Amsterdam
Thiele E. (1963), *J. Chem. Phys.* **38**, 1959
Throop G.J. and Bearman R.J. (1965), *J. Chem. Phys.* **42**, 2408
Throop G.J. and Bearman R.J. (1966), *J. Chem. Phys.* **44**, 1432
Throop G.J. and Bearman R.J. (1967), *J. Chem. Phys.* **47**, 3036
Timasheff, S.N. (1998) *Adv. Protein Chem* **51**, 355
Tobias, D.J. and Brooks III, C.L. (1990) *J. Chem. Phys.* **92**, 2582
Van Ness, H.C. and Abbott, M.M. (1982) *Classical Thermodynamics of Non- Electrolyte Solutions*, McGraw-Hill, New York
Vergara, A., Paduano, L., Capuano, F. and Sartorio, R. (2002) *Phys. Chem. Chem. Phys.* **4**, 4716
Wang H. and Ben-Naim A. (1996), *J. Med. Chem.* **39**, 1531
Wang, H. and Ben-Naim, A. (1997) *J. Phys. Chem.B* **101**, 1077
Watts, R.O. (1969) *J. Chem. Phys.* **50**, 984
Watts, R.O. and McGee, I.J. (1976) *Liquid State Chemical Physics*, John Wiley and Sons, New York
Weerasinghe, S. and Smith, P.E. (2003) *J. Chem. Phys.* **118**, 5801
Wertheim M. (1963) *Phys. Rev. Letters* **8**, 321
White, J.A. and Velasco, S. (2001) *Europhys. Letters*, **54**, 475
Widom, B. (1963) *J. Chem. Phys.* **39**, 2808
Widom, B. (1982) *J. Chem. Phys.* **86**, 869
Wilhelm, E. and Battino, R. (1973) *Chem. Rev.* **73**, 1
Yau, D.H.L., Chank, Y. and Henderson, D. (1999) *Molecular Phys.* **88**, 1237
Yvon, J. (1935) *Actualites Scientifiques et Industrielles*, Herman and Cie, Paris
Yvon, J. (1958) *Nuovo Cimento* **9**, 144
Zaitsev, A.L., Kessler, Y.M. and Petrenko, V.E. (1985) *Russian J. of Phys. Chem.* **59**, 1631
Zaitsev, A.L., Petrenko, V.E. and Kessler, Y.M. (1989) *J. Sol. Chem.* **18**, 115
von Zawidski J. (1900) *Z. Phys. Chem.* **35**, 129
Zichi, D.A. and Rossky, P.J. (1986), *J. Chem. Phys.* **84**, 1712
Zielkiewicz, J. (1995a) *J. Phys. Chem.* **99**, 3357
Zielkiewicz, J. (1995b) *J. Phys. Chem.* **99**, 4787
Zielkiewicz, J. (1998) *J. Chem. Soc. Faraday Trans.* **94**, 1713
Zielkiewicz, J. (2000) *Phys. Chem. Chem. Phys.* **2**, 2925
Zielkiewicz, J. and Mazerski, J. (2002) *J. Phys. Chem.* **B106**, 861
Zielkiewicz, J. (2003a) *Phys. Chem. Chem. Phys.* **5**, 1619
Zielkiewicz, J. (2003b), *Phys. Chem. Chem. Phys.* **5**, 3193

Index